Basic Electronics Engineering

Basic Electronics Engineering

Irshad Ahamad Khilji

Pratap Book Services
Ghaziabad

Edition : 2015

ISBN : 978–93–85084–03–4

Published by :

Pratap Book Services

495, Adarsh Colony

Ghaziabad (UP)

Printed in India.

Preface

Electronics engineering, or electronic engineering, is an engineering discipline where non-linear and active electrical components such as electron tubes, and semiconductor devices, especially transistors, diodes and integrated circuits, are utilized to design electronic circuits, devices and systems, typically also including passive electrical components and based on printed circuit boards. The term denotes a broad engineering field that covers important subfields such as analog electronics, digital electronics, consumer electronics, embedded systems and power electronics. Electronics engineering deals with implementation of applications, principles and algorithms developed within many related fields, for example solid-state physics, radio engineering, telecommunications, control systems, signal processing, systems engineering, computer engineering, instrumentation engineering, electric power control, robotics, and many others. Electronics Engineering is one of the largest and fastest growing fields of engineering. It covers a wide range of applications which make our life easier and enjoyable such as Television, Radio, computers, telecommunication etc. They help us to see, hear and communicate over vast distances and do things faster. Electronics has a major role in improving productivity in industries like oil, energy, agriculture and so many other important sectors of economy. In steel, petroleum and chemical industries it is the electronic devices that direct, control and test production processes. Health care industry depend on electronic instruments to perform chemical tests and to check body functions. The safety in transportation, factories and mines and in homes rely heavily on electronics.

Electronics engineering is a constantly changing and widening branch of the engineering profession. Electronics industry being a labour intensive industry provides many job opportunities for the skilled job seekers. Increased production and demand by government and businesses for communication equipment, computers and military electronics along with consumer demand and increased research and development on robots and other types of automation contributes to the growth of employment opportunities in the field. For most engineers not involved at the cutting edge of system design and development, technical work accounts for only a fraction of the work they do. A lot of time is also spent on tasks such as discussing proposals with clients, preparing budgets and determining project schedules. Many senior engineers manage a team of technicians or other engineers and for this reason project management skills are important. Most engineering projects involve some form of documentation and strong written communication skills are therefore very important. The workplaces of electronics engineers are just as varied as the types of work they do. Electronics engineers may be found in the pristine laboratory environment of a fabrication plant, the offices of a consulting firm or in a research laboratory. During their working life, electronics engineers may find themselves supervising a wide range of individuals including scientists, electricians, computer programmers and other engineers.

This book has been written in a very simple and lucid manner. Every effort has been made to make the treatments simple and comprehensive. The primary aim of the book is to meet the requirements of students pursuing Diploma and degree courses in Electronics and Communication Engineering offered by various Indian polytechnic colleges and Universities.

Book writing-whether original or compilation is a laborious complicated job requiring academic labour.

The present work is purely a research compilation based on authoritative publications and researches authored by eminent electronics engineers and subject experts.

Assistance has also been sought from Websites through Internet.

The author, humbly acknowledges the contributions of all those eminent electronics engineers and subject experts alongwith their respective publishers and Websites in preparation of this book.

—Irshad Ahamad Khilji

Contents

Chapter

1

Diodes and Applications

Diode

In electronics, a diode is a two-terminal electronic component with asymmetric conductance, it has low (ideally zero) resistance to current flow in one direction, and high (ideally infinite) resistance in the other.

A semiconductor diode, the most common type today, is a crystalline piece of semiconductor material with a p–n junction connected to two electrical terminals. A vacuum tube diode has two electrodes, a plate (anode) and a heated cathode.

The most common function of a diode is to allow an electric current to pass in one direction (called the diode's *forward* direction), while blocking current in the opposite direction (the *reverse* direction). Thus, the diode can be viewed as an electronic version of a check valve. This unidirectional behaviour is called rectification, and is used to convert alternating current to direct current, including extraction of modulation from radio signals in radio receivers—these diodes are forms of rectifiers.

However, diodes can have more complicated behaviour than this simple on–off action. Semiconductor diodes begin conducting electricity only if a certain threshold voltage or cut-in voltage is present in the forward direction (a state in which the diode is said to be *forward-biased*). The voltage drop across a forward-biased diode varies only a little with the current, and is a function of temperature; this effect can be used as a temperature sensor or voltage reference.

Semiconductor diodes' nonlinear current–voltage characteristic can be tailored by varying the semiconductor materials and doping, introducing impurities into the materials. These are exploited in special-purpose diodes that perform many different functions. For example, diodes are used to regulate voltage (Zener diodes), to protect circuits from high voltage surges (avalanche diodes), to electronically tune radio and TV receivers (varactor diodes), to generate radio frequency oscillations (tunnel diodes, Gunn diodes, IMPATT diodes), and to produce light (light emitting diodes). Tunnel diodes exhibit negative resistance, which makes them useful in some types of circuits.

Diodes were the first semiconductor electronic devices. The discovery of crystals' rectifying abilities was made by German physicist Ferdinand Braun in 1874. The first semiconductor diodes, called cat's whisker diodes, developed around 1906, were made of mineral crystals such as galena. Today most diodes are made of silicon, but other semiconductors such as selenium or germanium are sometimes used.

History

Thermionic (vacuum tube) diodes and solid state (semiconductor) diodes were developed separately, at approximately the same time, in the early 1900s, as radio receiver detectors. Until the 1950s vacuum tube diodes were more often used in radios because the early point-contact type semiconductor diodes (cat's-whisker detectors) were less stable, and because most receiving sets had vacuum tubes for amplification that could easily have diodes included in the tube (for example the 12SQ7 double-diode triode), and vacuum tube rectifiers and gas-filled rectifiers handled some high voltage/high current rectification tasks beyond the capabilities of semiconductor diodes (such as selenium rectifiers) available at the time.

Discovery of Vacuum Tube Diodes

In 1873, Frederick Guthrie discovered the basic principle of operation of thermionic diodes. Guthrie discovered that a positively charged electroscope could be discharged by bringing a grounded piece of white-hot metal close to it (but not actually touching it). The same did not apply to a negatively charged electroscope, indicating that the current flow was only possible in one direction.

Thomas Edison independently rediscovered the principle on February 13, 1880. At the time, Edison was investigating why the filaments of his carbon-filament light bulbs nearly always burned out at the positive-connected end. He had a special bulb made with a metal plate sealed into the glass envelope. Using this device, he confirmed that an invisible current flowed from the glowing filament through the vacuum to the metal plate, but only when the plate was connected to the positive supply.

Edison devised a circuit where his modified light bulb effectively replaced the resistor in a DC voltmeter. Edison was awarded a patent for this invention in 1884. Since there was no apparent practical use for such a device at the time, the patent application was most likely simply a precaution in case someone else did find a use for the so-called Edison effect.

About 20 years later, John Ambrose Fleming (scientific adviser to the Marconi Company and former Edison employee) realized that the Edison effect could be used as a precision radio detector. Fleming patented the first true thermionic diode, the Fleming valve, in Britain on November 16, 1904 (followed by U.S. Patent 803,684 in November 1905).

Solid-state Diodes

In 1874 German scientist Karl Ferdinand Braun discovered the "unilateral conduction" of crystals. Braun patented the crystal rectifier in 1899. Copper oxide and selenium rectifiers were developed for power applications in the 1930s.

Indian scientist Jagadish Chandra Bose was the first to use a crystal for detecting radio waves in 1894. He also worked with microwaves in the centimeter and also the millimeter range. The crystal detector was developed into a practical device for wireless telegraphy by Greenleaf Whittier Pickard, who invented a silicon crystal detector in 1903 and received a patent for it on November 20, 1906. Other experimenters tried a variety of other substances, of which the most widely used was the mineral galena (lead sulfide). Other substances offered slightly better performance, but galena was most widely used because it had the advantage of being cheap and easy to obtain. The crystal detector in these early crystal radio sets consisted of an adjustable wire point-contact (the so-called "cat's whisker"), which could be manually moved over the face of the crystal in order to obtain optimum signal. This troublesome device was superseded by thermionic diodes by the 1920s, but after high purity semiconductor materials became available, the crystal detector returned to dominant use with the advent of inexpensive fixed-germanium diodes in the 1950s. Bell Labs also developed a germanium diode for microwave reception, and AT&T used these in their microwave towers that criss-crossed the nation starting in the late 1940s, carrying telephone and network television signals. Bell Labs did not develop a satisfactory thermionic diode for microwave reception.

Etymology

At the time of their invention, such devices were known as rectifiers. In 1919, the year tetrodes were invented, William Henry Eccles coined the term *diode* from the Greek roots *di*, meaning "two", and *ode*, meaning "path". (However, the word *diode* itself, as well as *triode, tetrode, penthode, hexode*, was already in use as a term of multiplex telegraphy.

Rectifiers

Although all diodes *rectify*, the term 'rectifier' is normally reserved for higher currents and voltages than would normally be found in the rectification of lower power signals; examples include:

- power supply (*half-wave* or *full-wave* or *bridge*) rectifiers
- CRT (especially TV) Extra-high voltage flyback, "damper" or "booster" diodes such as the 6AU4GTA.

Thermionic Diodes

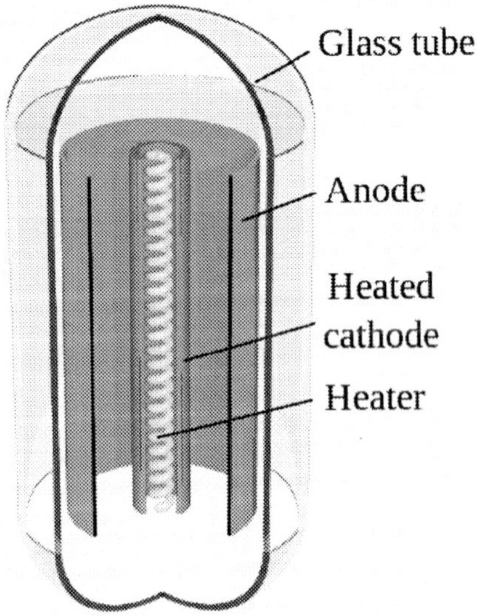

Figure: *Diode vacuum tube construction*

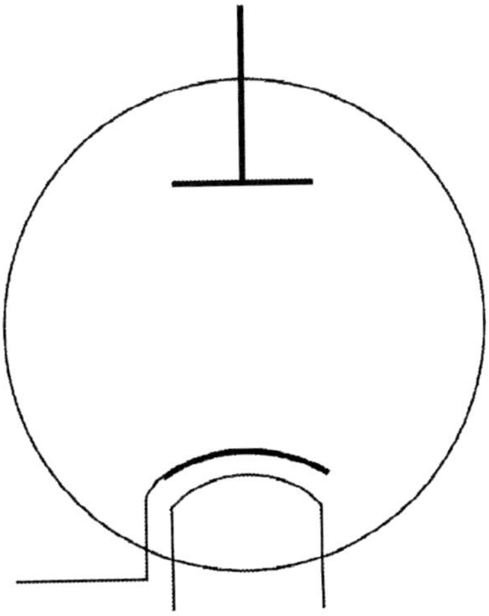

Figure: *The symbol for an indirect heated vacuum-tube diode. From top to bottom, the components are the anode, the cathode, and the heater filament.*

A thermionic diode is a thermionic-valve device (also known as a vacuum tube, tube, or valve), consisting of a sealed evacuated glass envelope containing two electrodes: a cathode heated by a filament, and a plate (anode). Early examples were fairly similar in appearance to incandescent light bulbs.

In operation, a separate current through the filament (heater), a high resistance wire made of nichrome, heats the cathode red hot (800-1000° C), causing it to release electrons into the vacuum, a process called thermionic emission. The cathode is coated with oxides of alkaline earth metals such as barium and strontium oxides, which have a low work function, to increase the number of electrons emitted. (Some valves use *direct heating*, in which a tungsten filament acts as both heater and cathode.) The alternating voltage to be rectified is applied between the cathode and the concentric plate electrode. When the plate has a positive voltage with respect to the cathode, it electrostatically attracts the electrons from the cathode, so a current of electrons flows through the tube from cathode to plate. However when the polarity is reversed and the plate has a negative voltage, no current flows, because the cathode electrons are not attracted to it. The unheated plate does not emit any electrons itself. So current can only flow through the tube in one direction, from cathode to plate.

In a mercury-arc valve, an arc forms between a refractory conductive anode and a pool of liquid mercury acting as cathode. Such units were made with ratings up to hundreds of kilowatts, and were important in the development of HVDC power transmission. Some types of smaller thermionic rectifiers sometimes had mercury vapor fill to reduce their forward voltage drop and to increase current rating over thermionic hard-vacuum devices.

Throughout the vacuum tube era, valve diodes were used in analog signal applications and as rectifiers in DC power supplies in consumer electronics such as radios, televisions, and sound systems. They were replaced in power supplies beginning in the 1940s by selenium rectifiers and then by semiconductor diodes by the 1960s. Today they are still used in a few high power applications where their ability to withstand transients and their robustness gives them an advantage over semiconductor devices. The recent (2012) resurgence of interest among audiophiles and recording studios in old valve audio gear such as guitar amplifiers and home audio systems has provided a market for the legacy consumer diode valves.

Semiconductor Diodes

Point-contact Diodes: A point-contact diode works the same as the junction diodes described below, but their construction is simpler. A block of n-type semiconductor is built, and a conducting sharp-point contact made with some group-3 metal is placed in contact with the semiconductor. Some metal migrates into the semiconductor to make a small region of p-type semiconductor near the contact. The long-popular 1N34 germanium version is still used in radio receivers as a detector and occasionally in specialized analog electronics.

Junction Diodes

Most diodes today are silicon junction diodes. A junction is formed between the p and n regions which is also called a depletion region.

p–n junction Diode

A p–n junction diode is made of a crystal of semiconductor. Impurities are added to it to create a region on one side that contains negative charge carriers (electrons), called n-type semiconductor, and a region on the other side that contains positive charge carriers (holes), called p-type semiconductor. When two materials i.e. n-type and p-type are attached together, a momentary flow of electrons occur from n to p side resulting in a third region where no charge carriers are present. It is called Depletion region due to the absence of charge carriers (electrons and holes in this case). The diode's terminals are attached to each of these regions. The boundary between these two regions, called a p–n junction, is where the action of the diode takes place. The crystal allows electrons to flow from the N-type side (called the cathode) to the P-type side (called the anode), but not in the opposite direction.

Schottky Diode

Another type of junction diode, the Schottky diode, is formed from a metal–semiconductor junction rather than a p–n junction, which reduces capacitance and increases switching speed.

Current–voltage Characteristic

A semiconductor diode's behaviour in a circuit is given by its current–voltage characteristic, or I–V graph. The shape of the curve is determined by the transport of charge carriers through the so-called *depletion layer* or *depletion region* that exists at the p–n junction between differing semiconductors. When a p–n junction is first created, conduction-band (mobile) electrons from the N-doped region diffuse into the P-doped region where there is a large population of holes (vacant places for electrons) with which the electrons "recombine". When a mobile electron recombines with a hole, both hole and electron vanish, leaving behind an immobile positively charged donor (dopant) on the N side and negatively charged acceptor (dopant) on the P side. The region around the p–n junction becomes depleted of charge carriers and thus behaves as an insulator.

However, the width of the depletion region (called the depletion width) cannot grow without limit. For each electron–hole pair that recombines, a positively charged dopant ion is left behind in the N-doped region, and a negatively charged dopant ion is left behind in the P-doped region. As recombination proceeds more ions are created, an increasing electric field develops through the depletion zone that acts to slow and then finally stop recombination. At this point, there is a "built-in" potential across the depletion zone.

If an external voltage is placed across the diode with the same polarity as the built-in potential, the depletion zone continues to act as an insulator, preventing any significant electric current flow (unless electron/hole pairs are actively being created in the junction by, for instance, light). This is the *reverse bias* phenomenon. However, if the polarity of the external voltage opposes the built-in potential, recombination can once again proceed, resulting in substantial electric current through the p–n junction (i.e. substantial numbers of electrons and holes recombine at the junction). For silicon diodes, the built-in potential is approximately 0.7 V (0.3 V for Germanium and 0.2 V for Schottky). Thus, if an external current is passed through the diode, about 0.7 V will be developed across the diode such that the P-doped region is positive with respect to the N-doped region and the diode is said to be "turned on" as it has a *forward bias.*

A diode's *I–V characteristic* can be approximated by four regions of operation.

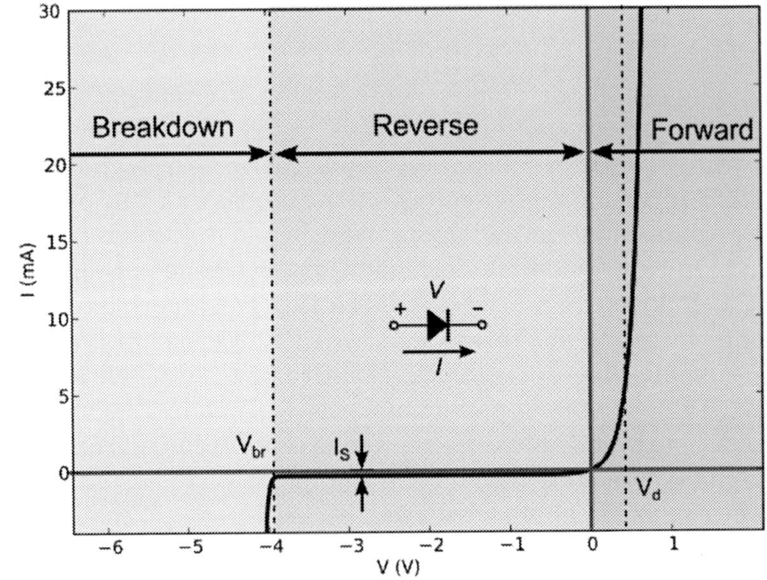

Figure: *I–V characteristics of a p–n junction diode.*

At very large reverse bias, beyond the peak inverse voltage or PIV, a process called reverse breakdown occurs that causes a large increase in current (i.e., a large number of electrons and holes are created at, and move away from the p–n junction) that usually damages the device permanently. The avalanche diode is deliberately designed for use in the avalanche region. In the Zener diode, the concept of PIV is not applicable. A Zener diode contains a heavily doped p–n junction allowing electrons to tunnel from the valence band of the p-type material to the conduction band of the n-type material, such that the reverse voltage is "clamped" to a known value (called the *Zener voltage*), and avalanche does not occur. Both devices, however, do have a limit to the maximum current and power in the clamped reverse-voltage region. Also, following the end of forward conduction in any diode, there is reverse current for a short time. The device does not attain its full blocking capability until the reverse current ceases.

The second region, at reverse biases more positive than the PIV, has only a very small reverse saturation current. In the reverse bias region for a normal P–N rectifier diode, the current through the device is very low (in the μA range). However, this is temperature dependent, and at sufficiently high temperatures, a substantial amount of reverse current can be observed (mA or more).

The third region is forward but small bias, where only a small forward current is conducted.

As the potential difference is increased above an arbitrarily defined "cut-in voltage" or "on-voltage" or "diode forward voltage drop (V_d)", the diode current becomes appreciable (the level of current considered "appreciable" and the value of cut-in voltage depends on the application), and the diode presents a very low resistance. The current–voltage curve is exponential. In a normal silicon diode at rated currents, the arbitrary cut-in voltage is defined as 0.6 to 0.7 volts. The value is different for other diode types—Schottky diodes can be rated as low as 0.2 V, Germanium diodes 0.25 to 0.3 V, and red or blue light-emitting diodes (LEDs) can have values of 1.4 V and 4.0 V respectively.

At higher currents the forward voltage drop of the diode increases. A drop of 1 V to 1.5 V is typical at full rated current for power diodes.

Shockley Diode Equation

The *Shockley ideal diode equation* or the *diode law* (named after transistor co-inventor William Bradford Shockley) gives the I–V characteristic of an ideal diode in either forward or reverse bias (or no bias). The *Shockley ideal diode equation* is below, where n, the ideality factor, is equal to 1 :

$$I = I_S\left(e^{V_D/(nV_T)} - 1\right),$$

where

I is the diode current,

I_S is the reverse bias saturation current (or scale current),

V_D is the voltage across the diode,

V_T is the thermal voltage, and

n is the *ideality factor*, also known as the *quality factor* or sometimes *emission coefficient*. The ideality factor n typically varies from 1 to 2 (though can in some cases be higher), depending on the fabrication process and semiconductor material and in many cases is assumed to be approximately equal to 1 (thus the notation n is omitted). The ideality factor does not form part of the *Shockley ideal diode equation*, and was added to account for imperfect junctions as observed in real transistors. By setting n = 1 above, the equation reduces to the *Shockley ideal diode equation*. The thermal voltage V_T is approximately 25.85 mV at 300 K, a temperature close to "room temperature" commonly used in device simulation software. At any temperature it is a known constant defined by:

$$V_T = \frac{kT}{q},$$

where k is the Boltzmann constant, T is the absolute temperature of the p–n junction, and q is the magnitude of charge of an electron (the elementary charge).

The reverse saturation current, I_S, is not constant for a given device, but varies with temperature; usually more significantly than V_T, so that V_D typically decreases as T increases.

The *Shockley ideal diode equation* or the *diode law* is derived with the assumption that the only processes giving rise to the current in the diode are drift (due to electrical field), diffusion, and thermal recombination–generation (R–G) (this equation is derived by setting n = 1 above). It also assumes that the R–G current in the depletion region is insignificant. This means that the *Shockley ideal diode equation* doesn't account for the processes involved in reverse breakdown and photon-assisted R–G. Additionally, it doesn't describe the "levelling off" of the I–V curve at high forward bias due to internal resistance. Introducing the ideality factor, n, accounts for recombination and generation of carriers.

Under *reverse bias* voltages the exponential in the diode equation is negligible, and the current is a constant (negative) reverse current value of $-I_S$. The reverse *breakdown region* is not modelled by the Shockley diode equation. For even rather small *forward bias* voltages the exponential is very large because the thermal voltage is very small, so the subtracted '1' in the diode equation is negligible and the forward diode current is often approximated as

$$I = I_S e^{V_D/(nV_T)}$$

The use of the diode equation in circuit problems is illustrated in the article on diode modelling.

Small-signal Behaviour

For circuit design, a small-signal model of the diode behaviour often proves useful. A specific example of diode modelling is discussed in the article on small-signal circuits.

Reverse-recovery Effect

Following the end of forward conduction in a p–n type diode, a reverse current flows for a short time. The device does not attain its blocking capability until the mobile charge in the junction is depleted.

The effect can be significant when switching large currents very quickly. A certain amount of "reverse recovery time" t_r (on the order of tens of nanoseconds to a few microseconds) may be required to remove the reverse recovery charge Q_r from the diode. During this recovery time, the diode can actually conduct in the reverse direction. In certain real-world cases it can be important to consider the losses incurred by this non-ideal diode effect. However, when the slew rate of the current is not so severe (e.g. Line frequency) the effect can be safely ignored. For most applications, the effect is also negligible for Schottky diodes.

The reverse current ceases abruptly when the stored charge is depleted; this abrupt stop is exploited in step recovery diodes for generation of extremely short pulses.

Types of Semiconductor Diode

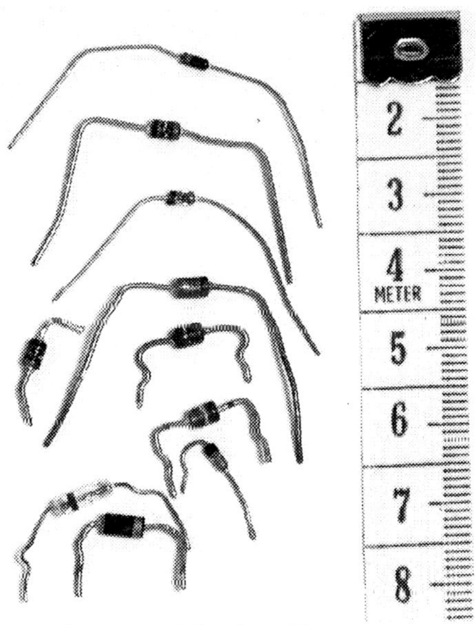

Figure: *Several types of diodes. The scale is centimeters.*

There are several types of p–n junction diodes, which either emphasize a different physical aspect of a diode often by geometric scaling, doping level, choosing the right electrodes, are just an application of a diode in a special circuit, or are really different devices like the Gunn and laser diode and the MOSFET:

Normal (p–n) diodes, which operate as described above, are usually made of doped silicon or, more rarely, germanium. Before the development of silicon power rectifier diodes, cuprous oxide and later selenium was used; its low efficiency gave it a much higher forward voltage drop (typically 1.4 to 1.7 V per "cell", with multiple cells stacked to increase the peak inverse voltage rating in high voltage rectifiers), and required a large heat sink (often an extension of the diode's metal substrate), much larger than a silicon diode of the same current ratings would require. The vast majority of all diodes are the p–n diodes found in CMOS integrated circuits, which include two diodes per pin and many other internal diodes.

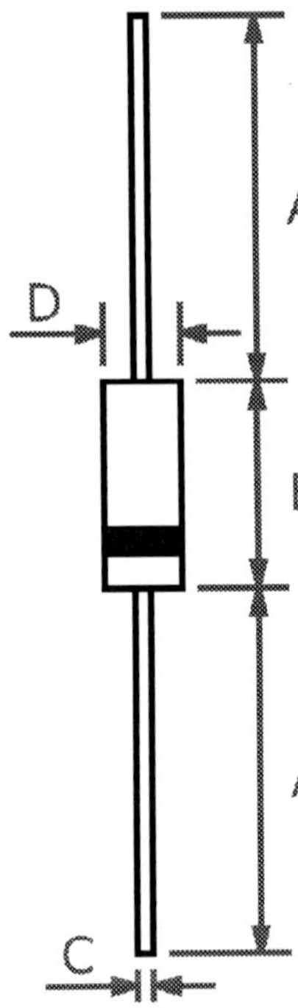

Figure: *Typical datasheet drawing showing the dimensions of a DO-41 diode package.*

Avalanche Diodes: Diodes that conduct in the reverse direction when the reverse bias voltage exceeds the breakdown voltage. These are electrically very similar to Zener diodes, and are often mistakenly called Zener diodes, but break down by a different mechanism, the *avalanche effect*. This occurs when the reverse electric field across the p–n junction causes a wave of ionization, reminiscent of an avalanche, leading to a large current. Avalanche diodes are designed to break down at a well-defined reverse voltage without being destroyed. The difference between the avalanche diode (which has a reverse breakdown above about 6.2 V) and the Zener is that the channel length of the former exceeds the mean free path of the electrons, so there are collisions between them on the way out. The only practical difference is that the two types have temperature coefficients of opposite polarities.

Cat's whisker or Crystal Diodes: These are a type of point-contact diode. The cat's whisker diode consists of a thin or sharpened metal wire pressed against a semiconducting crystal, typically galena or a piece of coal. The wire forms the anode and the crystal forms the cathode. Cat's whisker diodes were also called crystal diodes and found application in crystal radio receivers. Cat's whisker diodes are generally obsolete, but may be available from a few manufacturers.

Constant Current Diodes: These are actually a JFET with the gate shorted to the source, and function like a two-terminal current-limiter analog to the Zener diode, which is limiting voltage. They allow a current through them to rise to a certain value, and then level off at a specific value. Also called *CLDs*, *constant-current diodes*, *diode-connected transistors*, or *current-regulating diodes*.

Esaki or Tunnel Diodes: These have a region of operation showing negative resistance caused by quantum tunneling, allowing amplification of signals and very simple bistable circuits. Due to the high carrier concentration, tunnel diodes are very fast, may be used at low (mK) temperatures, high magnetic fields, and in high radiation environments. Because of these properties, they are often used in spacecraft.

Gunn Diodes: These are similar to tunnel diodes in that they are made of materials such as GaAs or InP that exhibit a region of negative differential resistance. With appropriate biasing, dipole domains form and travel across the diode, allowing high frequency microwave oscillators to be built.

Light-emitting Diodes (LEDs): In a diode formed from a direct band-gap semiconductor, such as gallium arsenide, carriers that cross the junction emit photons when they recombine with the majority carrier on the other side. Depending on the material, wavelengths (or colours) from the infrared to the near ultraviolet may be produced. The forward potential of these diodes depends on the wavelength of the emitted photons: 2.1 V corresponds to red, 4.0 V to violet. The first

LEDs were red and yellow, and higher-frequency diodes have been developed over time. All LEDs produce incoherent, narrow-spectrum light; "white" LEDs are actually combinations of three LEDs of a different colour, or a blue LED with a yellow scintillator coating. LEDs can also be used as low-efficiency photodiodes in signal applications. An LED may be paired with a photodiode or phototransistor in the same package, to form an opto-isolator.

Laser Diodes: When an LED-like structure is contained in a resonant cavity formed by polishing the parallel end faces, a laser can be formed. Laser diodes are commonly used in optical storage devices and for high speed optical communication.

Thermal Diodes: This term is used both for conventional p–n diodes used to monitor temperature due to their varying forward voltage with temperature, and for Peltier heat pumps for thermoelectric heating and cooling. Peltier heat pumps may be made from semiconductor, though they do not have any rectifying junctions, they use the differing behaviour of charge carriers in N and P type semiconductor to move heat.

Photodiodes: All semiconductors are subject to optical charge carrier generation. This is typically an undesired effect, so most semiconductors are packaged in light blocking material. Photodiodes are intended to sense light (photodetector), so they are packaged in materials that allow light to pass, and are usually PIN (the kind of diode most sensitive to light). A photodiode can be used in solar cells, in photometry, or in optical communications. Multiple photodiodes may be packaged in a single device, either as a linear array or as a two-dimensional array. These arrays should not be confused with charge-coupled devices.

PIN Diodes: A PIN diode has a central un-doped, or *intrinsic*, layer, forming a p-type/intrinsic/n-type structure. They are used as radio frequency switches and attenuators. They are also used as large volume ionizing radiation detectors and as photodetectors. PIN diodes are also used in power electronics, as their central layer can withstand high voltages. Furthermore, the PIN structure can be found in many power semiconductor devices, such as IGBTs, power MOSFETs, and thyristors.

Schottky Diodes: Schottky diodes are constructed from a metal to semiconductor contact. They have a lower forward voltage drop than p–n junction diodes. Their forward voltage drop at forward currents of about 1 mA is in the range 0.15 V to 0.45 V, which makes them useful in voltage clamping applications and prevention of transistor saturation. They can also be used as low loss rectifiers, although their reverse leakage current is in general higher than that of other diodes. Schottky diodes are majority carrier devices and so do not suffer from minority carrier storage problems that slow down many other diodes—so they have a faster reverse recovery than p–n junction diodes. They also tend to have much lower junction capacitance than p–n diodes, which provides for high switching speeds and their use in high-speed circuitry and RF devices such as switched-mode power supply, mixers, and detectors.

Super Barrier Diodes: Super barrier diodes are rectifier diodes that incorporate the low forward voltage drop of the Schottky diode with the surge-handling capability and low reverse leakage current of a normal p–n junction diode.

Gold-doped Diodes: As a dopant, gold (or platinum) acts as recombination centres, which helps a fast recombination of minority carriers. This allows the diode to operate at signal frequencies, at the expense of a higher forward voltage drop. Gold-doped diodes are faster than other p–n diodes (but not as fast as Schottky diodes). They also have less reverse-current leakage than Schottky diodes (but not as good as other p–n diodes). A typical example is the 1N914.

Snap-off or Step Recovery Diodes: The term *step recovery* relates to the form of the reverse recovery characteristic of these devices. After a forward current has been passing in an SRD and the current is interrupted or reversed, the reverse conduction will cease very abruptly (as in a step waveform). SRDs can, therefore, provide very fast voltage transitions by the very sudden disappearance of the charge carriers.

Stabistors or Forward Reference Diodes: The term *stabistor* refers to a special type of diodes featuring extremely stable forward voltage characteristics. These devices are specially designed for low-voltage stabilization applications requiring a guaranteed voltage over a wide current range and highly stable over temperature.

Transient Voltage Suppression Diode (TVS): These are avalanche diodes designed specifically to protect other semiconductor devices from high-voltage transients. Their p–n junctions have a much larger cross-sectional area than those of a normal diode, allowing them to conduct large currents to ground without sustaining damage.

Varicap or Varactor Diodes: These are used as voltage-controlled capacitors. These are important in PLL (phase-locked loop) and FLL (frequency-locked loop) circuits, allowing tuning circuits, such as those in television receivers, to lock quickly. They also enabled tunable oscillators in early discrete tuning of radios, where a cheap and stable, but fixed-frequency, crystal oscillator provided the reference frequency for a voltage-controlled oscillator.

Zener Diodes: Diodes that can be made to conduct backward, correctly termed reverse breakdown diodes. This effect, called Zener breakdown, occurs at a precisely defined voltage, allowing the diode to be used as a precision voltage reference.

The term Zener diode is colloquially applied to several types of breakdown diodes, but strictly speaking Zener diodes have a breakdown voltage of below 5 volts, whilst those above that value are usually avalanche diodes. In practical voltage reference circuits, Zener and switching diodes are connected in series and opposite directions to balance the temperature coefficient to near-zero. Some devices labelled as high-voltage Zener diodes are actually avalanche diodes.

Two (equivalent) Zeners in series and in reverse order, in the same package, constitute a transient absorber (or Transorb, a registered trademark). The Zener diode is named for Dr. Clarence Melvin Zener of Carnegie Mellon University, inventor of the device.

Other uses for semiconductor diodes include sensing temperature, and computing analog logarithms.

Numbering and Coding Schemes

There are a number of common, standard and manufacturer-driven numbering and coding schemes for diodes; the two most common being the EIA/JEDEC standard and the European Pro Electron standard:

Eia/Jedec

The standardized 1N-series numbering *EIA370* system was introduced in the US by EIA/JEDEC (Joint Electron Device Engineering Council) about 1960. Among the most popular in this series were: 1N34A/1N270 (Germanium signal), 1N914/1N4148 (Silicon signal), 1N4001-1N4007 (Silicon 1A power rectifier) and 1N54xx (Silicon 3A power rectifier)

JIS

The JIS semiconductor designation system has all semiconductor diode designations starting with "1S".

Pro Electron

The European Pro Electron coding system for active components was introduced in 1966 and comprises two letters followed by the part code. The first letter represents the semiconductor material used for the component (A = Germanium and B = Silicon) and the second letter represents the general function of the part (for diodes: A = low-power/ signal, B = Variable capacitance, X = Multiplier, Y = Rectifier and Z = Voltage reference), for example:

- AA-series germanium low-power/signal diodes (e.g.: AA119)
- BA-series silicon low-power/signal diodes (e.g.: BAT18 Silicon RF Switching Diode)
- BY-series silicon rectifier diodes (e.g.: BY127 1250V, 1A rectifier diode)
- BZ-series silicon Zener diodes (e.g.: BZY88C4V7 4.7V Zener diode)

Other common numbering / coding systems (generally manufacturer-driven) include:

- GD-series germanium diodes (e.g.: GD9) – this is a very old coding system
- OA-series germanium diodes (e.g.: OA47) – a coding sequence developed by Mullard, a UK company

As well as these common codes, many manufacturers or organisations have their own systems too – for example:

- HP diode 1901-0044 = JEDEC 1N4148
- UK military diode CV448 = Mullard type OA81 = GEC type GEX23

Related Devices

- Rectifier
- Transistor
- Thyristor or silicon controlled rectifier (SCR)
- TRIAC
- Diac
- Varistor.

In optics, an equivalent device for the diode but with laser light would be the Optical isolator, also known as an Optical Diode, that allows light to only pass in one direction. It uses a Faraday rotator as the main component.

Applications

Radio Demodulation: The first use for the diode was the demodulation of amplitude modulated (AM) radio broadcasts. The history of this discovery is treated in depth in the radio article. In summary, an AM signal consists of alternating positive and negative peaks of a radio carrier wave, whose amplitude or envelope is proportional to the original audio signal. The diode (originally a crystal diode) rectifies the AM radio frequency signal, leaving only the positive peaks of the carrier wave. The audio is then extracted from the rectified carrier wave using a simple filter and fed into an audio amplifier or transducer, which generates sound waves.

Power Conversion

Rectifiers are constructed from diodes, where they are used to convert alternating current (AC) electricity into direct current (DC). Automotive alternators are a common example, where the diode, which rectifies the AC into DC, provides better performance than the commutator or earlier, dynamo. Similarly, diodes are also used in *Cockcroft–Walton voltage multipliers* to convert AC into higher DC voltages.

Over-voltage Protection

Diodes are frequently used to conduct damaging high voltages away from sensitive electronic devices. They are usually reverse-biased (non-conducting) under normal circumstances. When the voltage rises above the normal range, the diodes become forward-biased (conducting). For example, diodes are used in (stepper motor and H-bridge) motor controller and relay circuits to de-energize coils rapidly without the damaging voltage spikes that would otherwise occur. (Any diode used in such an application is called a flyback diode). Many integrated circuits also incorporate diodes on the connection pins to prevent external voltages from damaging their sensitive transistors. Specialized diodes are used to protect from over-voltages at higher power.

Logic Gates

Diodes can be combined with other components to construct AND and OR logic gates. This is referred to as diode logic.

Ionizing Radiation Detectors

In addition to light, mentioned above, semiconductor diodes are sensitive to more energetic radiation. In electronics, cosmic rays and other sources of ionizing radiation cause noise pulses and single and multiple bit errors. This effect is sometimes exploited by particle detectors to detect radiation. A single particle of radiation, with thousands or millions of electron volts of energy, generates many charge carrier pairs, as its energy is deposited in the semiconductor material. If the depletion layer is large enough to catch the whole shower or to stop a heavy particle, a fairly accurate measurement of the particle's energy can be made, simply by measuring the charge conducted and without the complexity of a magnetic spectrometer, etc. These semiconductor radiation detectors need efficient and uniform charge collection and low leakage current. They are often cooled by liquid nitrogen. For longer-range (about a centimetre) particles, they need a very large depletion depth and large area. For short-range particles, they need any contact or un-depleted semiconductor on at least one surface to be very thin. The back-bias voltages are near breakdown (around a thousand volts per centimetre). Germanium and silicon are common materials. Some of these detectors sense position as well as energy. They have a finite life, especially when detecting heavy particles, because of radiation damage. Silicon and germanium are quite different in their ability to convert gamma rays to electron showers.

Semiconductor detectors for high-energy particles are used in large numbers. Because of energy loss fluctuations, accurate measurement of the energy deposited is of less use.

Temperature Measurements

A diode can be used as a temperature measuring device, since the forward voltage drop across the diode depends on temperature, as in a silicon bandgap temperature sensor. From the Shockley ideal diode equation given above, it might *appear* that the voltage has a *positive* temperature coefficient (at a constant current), but usually the variation of the reverse saturation current term is more significant than the variation in the thermal voltage term. Most diodes therefore have a *negative* temperature coefficient, typically “2 mV/ÚC for silicon diodes at room temperature. This is approximately linear for temperatures above about 20 kelvins. Some graphs are given for: 1N400x series, and CY7 cryogenic temperature sensor.

Current Steering

Diodes will prevent currents in unintended directions. To supply power to an electrical circuit during a power failure, the circuit can draw current from a battery. An uninterruptible power supply may use diodes in this way to ensure that current is only drawn from the battery when necessary. Likewise, small boats typically have two circuits each with their own battery/batteries: one used for engine starting; one used for domestics. Normally, both are charged from a single alternator, and a heavy-duty split-charge diode is used to prevent the higher-charge battery (typically the engine battery) from discharging through the lower-charge battery when the alternator is not running.

Diodes are also used in electronic musical keyboards. To reduce the amount of wiring needed in electronic musical keyboards, these instruments often use keyboard matrix circuits. The keyboard controller scans the rows and columns to determine which note the player has pressed. The problem with matrix circuits is that, when several notes are pressed at once, the current can flow backwards through the circuit and trigger "phantom keys" that cause "ghost" notes to play. To avoid triggering unwanted notes, most keyboard matrix circuits have diodes soldered with the switch under each key of the musical keyboard. The same principle is also used for the switch matrix in solid-state pinball machines.

Abbreviations

Diodes are usually referred to as *D* for diode on PCBs. Sometimes the abbreviation *CR* for *crystal rectifier* is used.

Two-terminal Nonlinear Devices

Many other two-terminal nonlinear devices exist, for example a neon lamp has two terminals in a glass envelope and has interesting and useful nonlinear properties. Lamps including arc-discharge lamps, incandescent lamps, fluorescent lamps and mercury vapor lamps have two terminals and display nonlinear current–voltage characteristics.

Semiconductor Diode

The *diode* essentially acts as a one-way switch controlled by voltage. For one polarity of voltage, if the voltage across the diode is greater than a threshold value (often ~0.6 Volts), it conducts current with essentially no resistance. If the voltage is below that value, or has the opposite polarity, the diode acts as an open switch and conducts no current. The *transistor* differs from the other devices we've considered so far in that it has three leads or connections. The voltage applied between two of these leads controls whether current can be conducted between two others. Although the transistor also acts as a switch, it does not merely shift between fully off and fully on. Its importance lies in the fact that a relatively low-power voltage supply can control the flow of a more powerful current over a range of values. It is this switching and decision-making property of diodes and transistors, which makes all of modern day electronics, including computers and telecommunications, possible.

The study of semiconductors and the devices made from them falls under the category of solid state physics. In this experiment, we will work with one useful device: the diode. Although we will discuss the theory briefly, you will mainly be expected to be able to understand how diodes function in circuits.

Energy Bands

The electrons of an isolated atom have discrete allowed energies that we call energy levels. The Pauli Exclusion Principle states that at most two electrons can occupy any allowed energy level. For example, Figure shows schematically the energy levels for a Lithium atom. The vertical scale is associated with increased energy. The dots denote electrons occupying a given level. In order to minimize energy the electrons fill the levels from the bottom up.

When two isolated atoms are brought close together their electric fields interact and cause a splitting of the energy levels. Each original energy level splits into two, one going slightly higher and the other going lower. (A detailed explanation of the cause of this will have to be postponed until a quantum mechanics course.) In a crystalline solid, where many atoms exist close together, this effect is multiplied many times over. Each energy level spreads out over a small continuous range of energies called an energy band.

If the original energy levels were spaced closely together the associated energy bands might overlap, resulting in a larger continuous band. Sometimes, however, the energy levels are spaced far enough apart that the bands don't overlap. This "gap" between bands is what provides the interesting physics of semiconductor devices. Remember that energy level diagrams represent allowed or accessible states that an electron may occupy. A gap between bands indicates a forbidden energy range for electrons.

Conduction and Valence Bands

Conceptually, we can imagine gradually filling up the electronic states with electrons until all of them are accommodated, even though solids aren't really made that way. In order to minimize the energy, the lowest states of the system fill first. Each band can hold 2 electrons for each atom in the crystal, since each band is derived from one atomic energy level for each atom, and each atomic energy level can hold two electrons (one spin up and one spin down). The highest fully occupied band is called the valence band. The next band above that, which may be partially filled, is called the conduction band. For an electron to participate in conduction it must be able to gain energy in small amounts from the applied electric field, i.e., there must be empty levels close in energy to that occupied by the electron. Thus the valence band electrons are immobile and cannot contribute to the conductivity, whereas electrons in the conduction band can contribute to conduction (as the name suggests).

Partially Populated Conduction Band: Conductors

Figure diagrams a material for which the conduction band is partly occupied by electrons. If an external electric field is applied to this material, some of the electrons can gain a small amount of energy and jump to a higher state in the previously unoccupied section of the conduction band. Thus a material of this sort responds to the application of an electric field with a large current flow. This is the typical metallic behaviour.

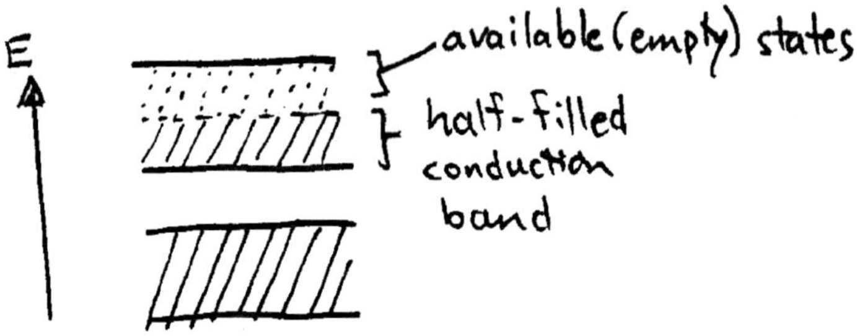

Figure: *The energy bands for a conductor*

Empty Conduction Band: Insulators

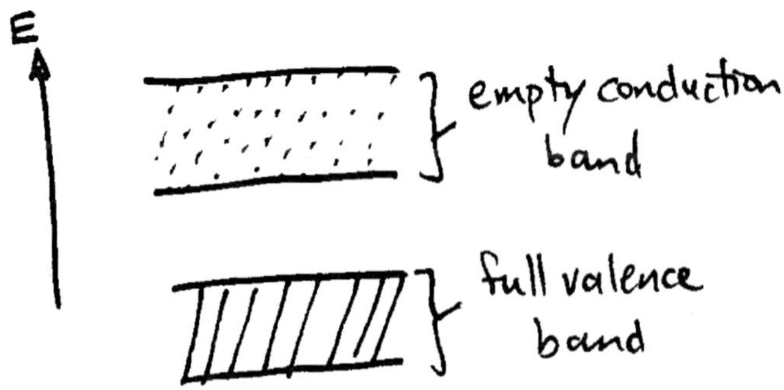

Figure: *The energy bands for an insulator*

If the conduction band is completely empty the material acts like an electrical insulator. Electrons in the valence band cannot gain enough energy to jump over the forbidden region into the conduction band. In addition, the electrons in the valence band cannot move through the solid to create a current because there are no empty states in the valence band for a travelling electron to occupy. Since no current can flow, the material is an insulator.

Semiconductors

Semiconductors are a special case of insulators in which the forbidden region between the valence and conduction bands is relatively small (about 0.5 to 1.5 eV). In this case an extremely small number of electrons are excited across the energy gap by thermal excitation and occupy states in the conduction band.

These excited electrons can respond to the applied fields but since their number is comparatively few, the material as a whole is not a good conductor, so we call it a semiconductor. The conductivity of a semiconductor is very sensitive to temperature since it depends on thermally excited electrons.

Doping

In general, the amount of current that a semiconductor can carry is not enough to make a useful device. Most commercial semiconductors are made by introducing small amounts of impurities to an intrinsic semiconductor (a process called doping). We will use silicon as an example.

Silicon (Si) is an intrinsic semiconductor, but in its natural state it conducts very poorly. Silicon is a group IV element on the periodic chart and has four electrons in its outermost shell. When silicon is doped with arsenic (As), a group V element, the arsenic atoms replace silicon atoms at a small number of points on the crystal lattice. Since arsenic has 5 electrons in its outer shell, it adds a loosely-bound "extra" electron to the crystal. This extra electron (often called a "donor" electron) is easily excited into the conduction band as a freely roaming current carrier.

Table: *Section of Periodic Table*

III	IV	V	
5B	6C	7N	Group III: -1e-: acceptor Ñ p type
13Al	14Si	15P	
31Ga	32Ge	33As	Group V: +1e-: donor Ñ n type
49In	50Sn	51Sb	

Silicon can also be doped with an element from group III of the periodic table, such as gallium (Ga). In this case the impurity has only three electrons in its outermost shell so there is a deficiency of one electron at every point where a gallium atom replaces a silicon atom. This is called an "acceptor site" since the gallium would very much like to have a fourth electron to complete its bonds. The gallium often "steals"an electron from a neighbouring silicon atom leaving a "hole" or empty state in the valence band of the silicon. This "hole" is free to roam around in the valence band and effectively acts as a positive charge carrier.

Holes move through the crystal lattice in the same way spaces between cars move in a traffic jam: the cars (electrons) move forward to fill up the spaces (holes) in front of them, only to create another space behind them. The holes move in the direction opposite the electrons, hence the "effective" positive charge.

Doping with group V elements results in an n-type semiconductor since the charge carriers (electrons) are negative. On the other hand, doping with group III elements produces a p-type semiconductor because the effective charge carriers (holes) are positive.

Diodes

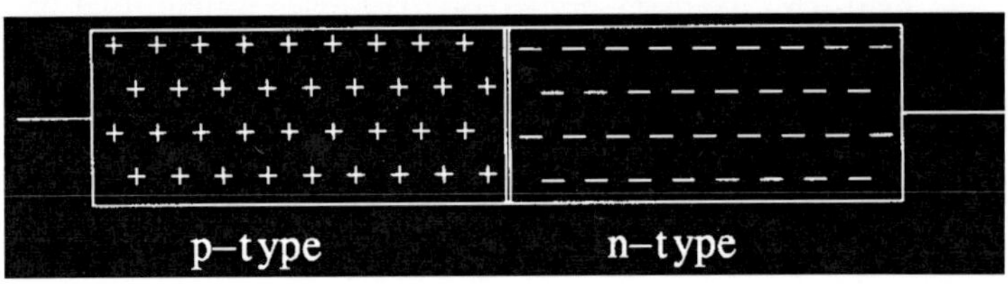

Figure: *distribution of mobile charge carriers in a p-n junction before (spontaneous) charge diffusion*

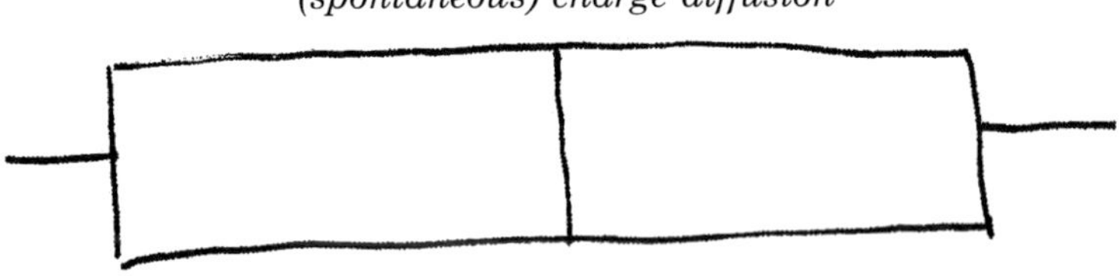

Figure: *distribution of excess charge in a p-n junction before (spontaneous) charge diffusion. This figure shows that each side starts out electrically neutral, because the negative charges due to electrons are exactly balanced by the positively charged nuclei in both the n- and p-type regions.*

Diodes are formed by producing a piece of semiconductor that is p-type at one end and n-type at the other such as Figure. Although electrons and holes are free to roam in each section, the material is electrically neutral.

However, in less than a nanosecond, some of the free electrons will diffuse into the p-type, and an equal number of holes will diffuse into the n-type. As the electrons and holes diffuse across the junction they recombine and "'eliminate" each other. The end result is a lack of mobile charge carriers in the immediate vicinity of the junction. In addition, the region of either side of the junction is no longer electrically neutral so a "built-in" electric field is established. Since the junction region in Figure is devoid of free charge carriers, it will have a low conductivity and high resistance. Another way to see this is that the built-in electric field opposes the motion of any holes from the p-type side which are trying to move to the right and any electrons from the n-type side which are trying to move to the left.

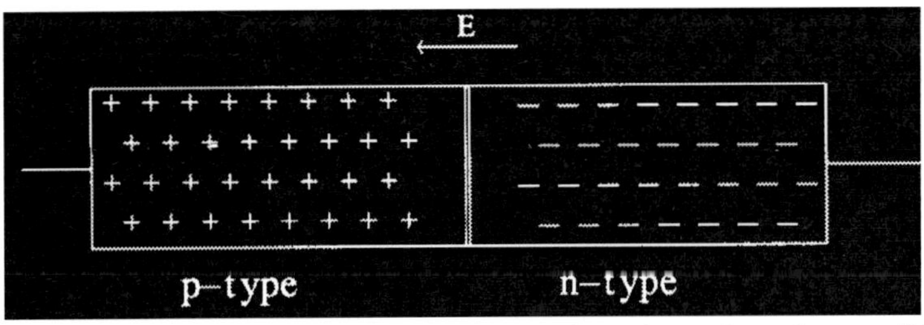

Figure: *Diffusion of mobile charge in a p-n junction after charge diffusion.*

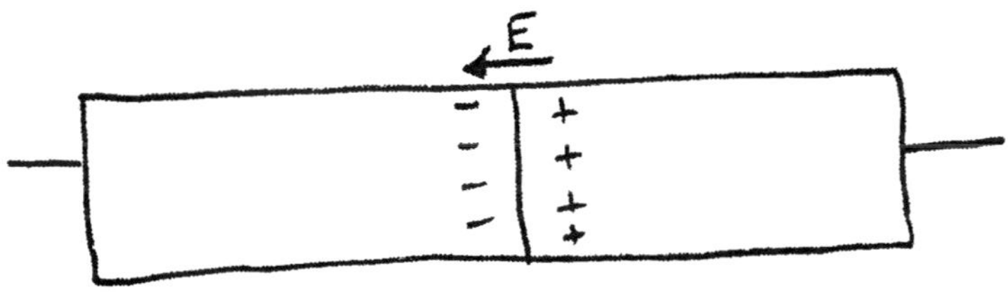

Figure: *Diffusion of excess charge in a p-n junction after charge diffusion. The negative charges which have diffused out of the n-type side into the p-type side leave a net negative charge there, and vice-versa. This charge distribution causes the "built-in" electric field shown near the junction.*

Reverse Biased Junction

Figure shows a p-n junction attached to a voltage source such that the positive terminal is connected to the n-type semiconductor. The electric field due to the applied voltage source adds to the built-in field. (Note that the conductivity of the p- and n-type regions away from the junction is greater than that of the junction region so the potential varies mainly in the proximity of the junction.) Hence, the addition of the second field even further opposes the motion of any holes from the p-type side which are trying to move to the right and any electrons from the n-type side which are trying to move to the left.

So, very little current flows.

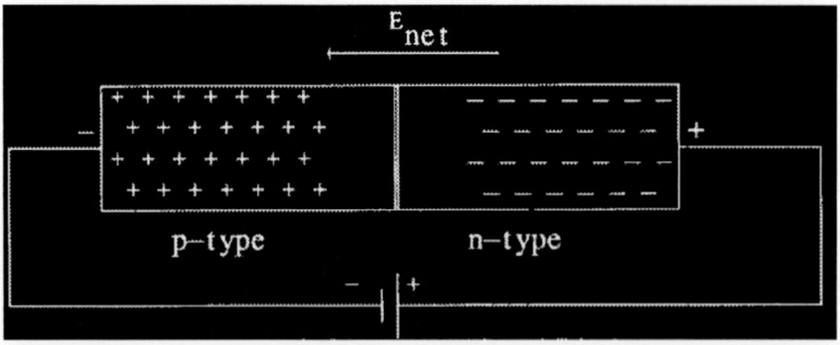

Figure: *Distribution of mobile charges in a reverse biased diode*

Forward Biased Junction

Suppose, on the other hand, that the voltage supply was connected with the positive terminal wired to the p-type semiconductor. The electric field due to the voltage source will now be in the direction opposite to the built-in field. Now, the opposition is reduced holes from the p-type side which are trying to move to the right and electrons from the n-type side which are trying to move to the left. Since these are the directions that the applied voltage is trying to push these charges, current can flow fairly easily, at least once the applied voltage is big enough to mostly overcome the opposition from the built-in field. Thus the p-n junction provides an interesting device that conducts current in only one direction.

In fact, if you literally connected a battery in the forward-bias direction directly across a diode, so much current would flow that the diode would burn out! To avoid this, you would need to add a "current-limiting resistor", as shown below.

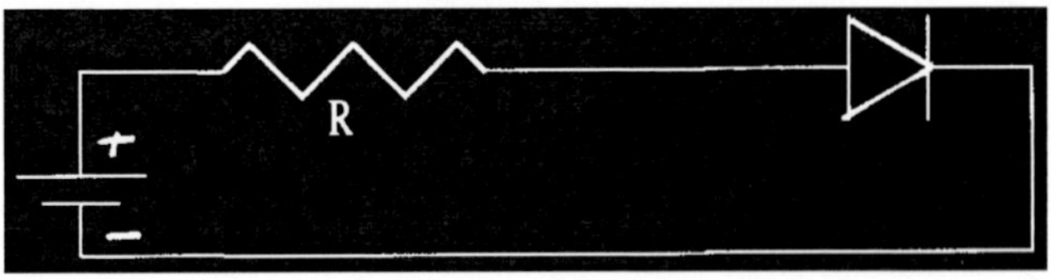

Figure: *Circuit diagram of a forward biased diode, with resistor added to limit the current.*

Resistance Levels

An electron travelling through the wires and loads of the external circuit encounters resistance. Resistance is the hindrance to the flow of charge. For an electron, the journey from terminal to terminal is not a direct route. Rather, it is a zigzag path that results from countless collisions with fixed atoms within the conducting material.

The electrons encounter resistance - a hindrance to their movement. While the electric potential difference established between the two terminals *encourages* the movement of charge, it is resistance that *discourages* it. The rate at which charge flows from terminal to terminal is the result of the combined affect of these two quantities.

Variables Affecting Electrical Resistance

The flow of charge through wires is often compared to the flow of water through pipes. The resistance to the flow of charge in an electric circuit is analogous to the frictional affects between water and the pipe surfaces as well as the resistance offered by obstacles that are present in its path. It is this resistance that hinders the water flow and reduces both its flow rate and its *drift* speed. Like the resistance to water flow, the total amount of resistance to charge flow within a wire of an electric circuit is affected by some clearly identifiable variables.

First, the total length of the wires will affect the amount of resistance. The longer the wire, the more resistance that there will be. There is a direct relationship between the amount of resistance encountered by charge and the length of wire it must traverse. After all, if resistance occurs as the result of collisions between charge carriers and the atoms of the wire, then there is likely to be more collisions in a longer wire. More collisions mean more resistance.

Second, the cross-sectional area of the wires will affect the amount of resistance. Wider wires have a greater cross-sectional area. Water will flow through a wider pipe at a higher rate than it will flow through a narrow pipe. This can be attributed to the lower amount of resistance that is present in the wider pipe. In the same manner, the wider the wire, the less resistance that there will be to the flow of electric charge. When all other variables are the same, charge will flow at higher rates through wider wires with greater cross-sectional areas than through thinner wires.

A third variable that is known to affect the resistance to charge flow is the material that a wire is made of. Not all materials are created equal in terms of their conductive ability. Some materials are better conductors than others and offer less resistance to the flow of charge. Silver is one of the best conductors but is never used in wires of household circuits due to its cost. Copper and aluminum are among the least expensive materials with suitable conducting ability to permit their use in wires of household circuits. The conducting ability of a material is often indicated by its resistivity. The resistivity of a material is dependent upon the material's electronic structure and its temperature. For most (but not all) materials, resistivity increases with increasing temperature. The table below lists resistivity values for various materials at temperatures of 20 degrees Celsius.

Material	*Resistivity (ohm•meter)*
Silver	1.59×10^{-8}
Copper	1.7×10^{-8}
Gold	2.4×10^{-8}
Aluminum	2.8×10^{-8}
Tungsten	5.6×10^{-8}
Iron	10×10^{-8}
Platinum	11×10^{-8}
Lead	22×10^{-8}
Nichrome	150×10^{-8}
Carbon	3.5×10^{-5}
Polystyrene	10^{7} - 10^{11}
Polyethylene	10^{8} - 10^{9}
Glass	10^{10} - 10^{14}
Hard Rubber	10^{13}

There is a broad range of resistivity values for various materials. Those materials with lower resistivities offer less resistance to the flow of charge; they are better conductors. The materials shown in the last four rows of the above table have such high resistivity that they would not even be considered to be conductors.

Diode Equivalent Circuits

An equivalent circuit is a combination of elements properly chosen to best represent the actual terminal characteristics of a device, system, or such in a particular operating region.

In other words, once the equivalent circuit is defined, the device symbol can be removed from a schematic and the equivalent circuit inserted in its place without severely affecting the actual behaviour of the system. The result is often a network that can be solved using traditional circuit analysis techniques.

Piecewise-Linear Equivalent Circuit

One technique for obtaining an equivalent circuit for a diode is to approximate the characteristics of the device by straight-line segments. The resulting equivalent circuit is naturally called the piecewise-linear equivalent circuit. It should be obvious that the straight-line segments do not result in an exact duplication of the actual characteristics, especially in the knee region.

However, the resulting segments are sufficiently close to the actual curve to establish an equivalent circuit that will provide an excellent first approximation to the actual behaviour of the device. For the sloping section of the equivalence the average ac resistance is the resistance level appearing in the equivalent circuit of Figure next to the actual device. In essence, it defines the resistance level of the device when it is in the "on" state. The ideal diode is included to establish that there is only one direction of conduction through the device, and a reverse-bias condition .

Load Line Analysis

A load line is used in graphical analysis of nonlinear electronic circuits, representing the constraint other parts of the circuit place on a non-linear device, like a diode or transistor. It is usually drawn on a graph of the current vs the voltage in the nonlinear device, called the device's characteristic curve. A load line, usually a straight line, represents the response of a linear circuit connected to the nonlinear device in question. The operating point(s) of the circuit are the points where the characteristic curve and the load line intersect; at these points the current and voltage parameters of both parts of the circuit match.

The example at right shows how a load line is used to determine the current and voltage in a simple diode circuit. The nonlinear diode is in series with a linear circuit consisting of a resistor and a voltage source. The graph, representing voltage across the diode V_D versus current I through the diode, is an exponential curve. The load line *(diagonal line)* represents the relationship between current and voltage in the linear part of the circuit. Since the current going through the three elements in series must be the same, and the voltage at the connection of the resistor and diode must be the same, the operating point of the circuit will be at the intersection of the curve with the load line.

In a BJT circuit, the BJT has a different current-voltage (I_C-V_{CE}) characteristic depending on the base current. Placing a series of these curves on the graph shows how the base current will affect the operating point of the circuit.

DC and AC Load Lines

Semiconductor circuits typically have both DC and AC currents in them, with a source of DC current to bias the nonlinear semiconductor to the correct operating point, and the AC signal superimposed on the DC. Load lines can be used separately for both DC and AC analysis. The DC load line is the load line of the DC equivalent circuit, defined by reducing the reactive components to zero (replacing capacitors by open circuits and inductors by closed circuits). It is used to determine the correct DC operating point, often called the Q point.

Once a DC operating point is defined by the DC load line, an AC load line with, in general, a different slope can be drawn, through the DC operating point, to calculate the AC output. Because the impedance of the reactive components will vary with frequency, the slope of the AC load line depends on the frequency of the applied signal. So there are many AC load lines, that vary from the DC load line (at low frequency) to a limiting AC load line, all having a common intersection at the dc operating point. This limiting load line, generally referred to as the *AC load line*, is the load line of the circuit at "infinite frequency", and can be found by replacing capacitors with short circuits, and inductors with open circuits.

Load Lines for Common Configurations

Transistor load line:

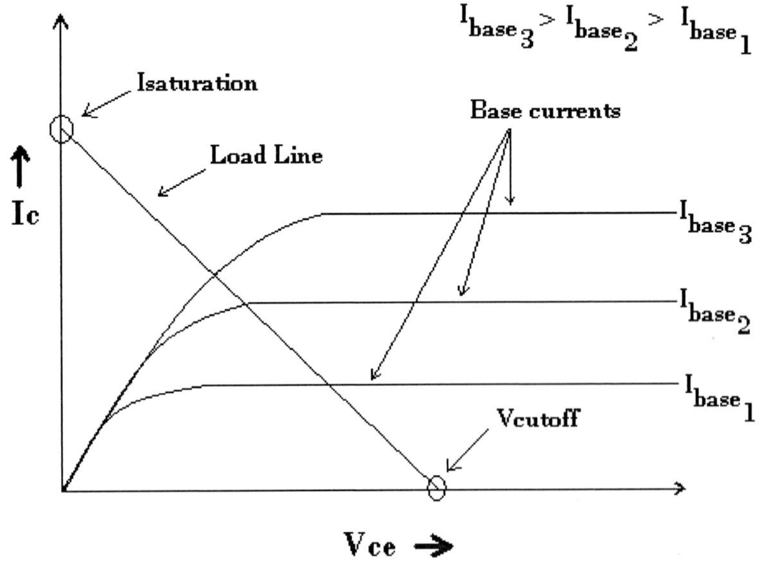

Figure: *Common emitter transistor load line.*

The load line diagram at right is for a transistor connected in a common emitter circuit. It shows the collector current in the transistor I_C versus collector voltage V_{CE} for different values of base current I_{base}. The load line represents a particular value of collector load resistor (R_C). The intersections of the load line with the transistor characteristic curve represent the different values of I_C and V_{CE} at different base currents.

The point on the load line where it intersects the collector current axis is referred to as *saturation point.* At this point, the transistor current is maximum and voltage across collector is minimum, for a given load. For this circuit, $I_{C\text{-}SAT} = V_{CC}/R_C$.

The *cutoff point* is the point where the load line intersects with the collector voltage axis. Here the transistor current is minimum (approximately zero) and emitter is grounded. Hence $V_{CE\text{-}CUTOFF} = V_{cc}$.

The *operating point* of the circuit in this configuration is generally designed to be in the *active region*, approximately between middle of the load line and close to saturation point.

In this region, the collector current is proportional to the base current, and hence useful for amplifier applications. a load line is normally drawn on ic-vce characteristics curves for the transistor used in amplifier circuit.

Half Wave and Full Wave Rectifiers with and Without Filters

Half-wave Rectification

In half wave rectification of a single-phase supply, either the positive or negative half of the AC wave is passed, while the other half is blocked. Because only one half of the input waveform reaches the output, mean voltage is lower. Half-wave rectification requires a single diode in a single-phase supply, or three in a three-phase supply. Rectifiers yield a unidirectional but pulsating direct current; half-wave rectifiers produce far more ripple than full-wave rectifiers, and much more filtering is needed to eliminate harmonics of the AC frequency from the output.

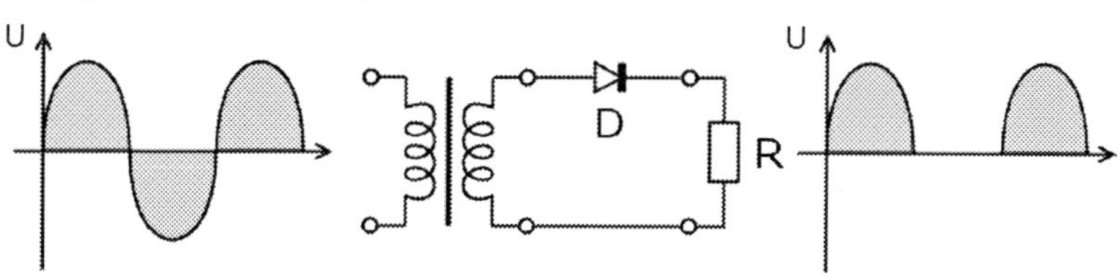

Figure: *Half-wave rectifier*

The no-load output DC voltage of an ideal half wave rectifier is:

$$V_{rms} = \frac{V_{peak}}{2}$$

$$V_{dc} = \frac{V_{peak}}{\pi}$$

Where:

V_{dc}, V_{av} - the DC or average output voltage,

V_{peak}, the peak value of the phase input voltages,

V_{rms}, the root-mean-square value of output voltage.

Full-wave Rectification

A full-wave rectifier converts the whole of the input waveform to one of constant polarity (positive or negative) at its output. Full-wave rectification converts both polarities of the input waveform to pulsating DC (direct current), and yields a higher average output voltage.

Two diodes and a centre tapped transformer, or four diodes in a bridge configuration and any AC source (including a transformer without centre tap), are needed. Single semiconductor diodes, double diodes with common cathode or common anode, and four-diode bridges, are manufactured as single components.

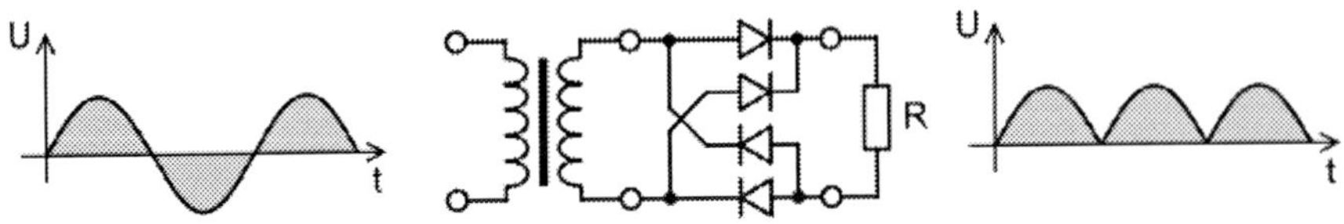

Figure: *Graetz bridge rectifier: a full-wave rectifier using 4 diodes.*

For single-phase AC, if the transformer is centre-tapped, then two diodes back-to-back (cathode-to-cathode or anode-to-anode, depending upon output polarity required) can form a full-wave rectifier. Twice as many turns are required on the transformer secondary to obtain the same output voltage than for a bridge rectifier, but the power rating is unchanged.

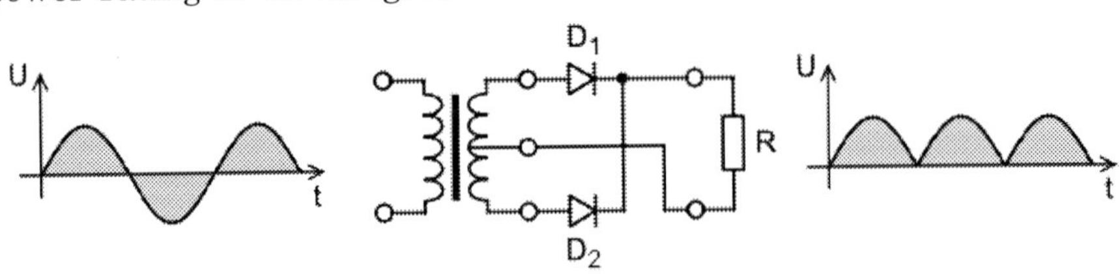

Figure: *Full-wave rectifier using a centre tap transformer and 2 diodes.*

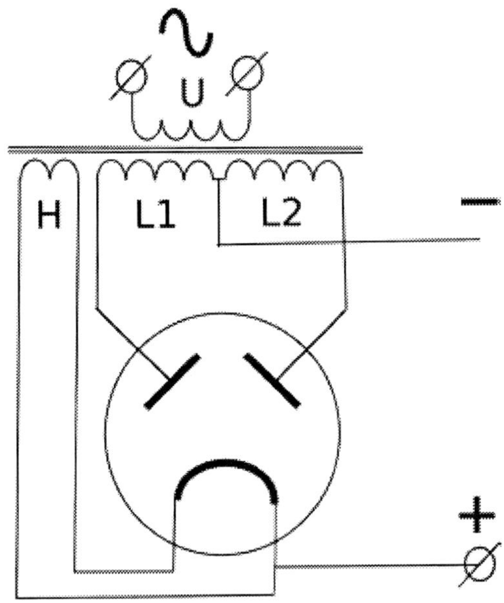

Figure: *Full-wave rectifier, with vacuum tube having two anodes.*

The average and root-mean-square no-load output voltages of an ideal single-phase full-wave rectifier are:

$$V_{dc} = V_{av} = \frac{2V_{peak}}{\pi}$$

$$V_{rms} = \frac{V_{peak}}{\sqrt{2}}$$

Very common double-diode rectifier vacuum tubes contained a single common cathode and two anodes inside a single envelope, achieving full-wave rectification with positive output. The 5U4 and 5Y3 were popular examples of this configuration.

Breakdown Mechanisms

The term electrical breakdown or electric breakdown has several similar but distinctly different meanings. For example, the term can apply to the failure of an electric circuit.

Alternatively, it may refer to a rapid reduction in the resistance of an electrical insulator that can lead to a spark jumping around or through the insulator.

This may be a momentary event (as in an electrostatic discharge), or may lead to a continuous arc discharge if protective devices fail to interrupt the current in a high power circuit.

Electrical System Failure

The most common meaning is related to automobiles and is the failure of an electric circuit or associated device resulting in a loss of vehicle function (a breakdown). Common problems include battery discharge, alternator failure, broken wires, blown fuses, etc.

Failure of Electrical Insulation

The second meaning of the term is more specifically a reference to the breakdown of the insulation of an electrical wire or other electrical component. Such breakdown usually results in a short circuit or a blown fuse. This occurs at the breakdown voltage. Actual insulation breakdown is more generally found in high-voltage applications, where it sometimes causes the opening of a protective circuit breaker. Electrical breakdown is often associated with the failure of solid or liquid insulating materials used inside high voltage transformers or capacitors in the electricity distribution grid. Electrical breakdown can also occur across the insulators that suspend overhead power lines, within underground power cables, or lines arcing to nearby branches of trees. Under sufficient electrical stress, electrical breakdown can occur within solids, liquids, gases or vacuum. However, the specific breakdown mechanisms are significantly different for each, particularly in different kinds of dielectric medium. All this leads to catastrophic failure of the instruments.

Disruptive Devices

A disruptive device is designed to electrically overstress a dielectric beyond its dielectric strength so as to intentionally cause electrical breakdown of the device. This results in the sudden transition of a portion of the dielectric from an insulating state to a highly conductive state. This transition is characterized by the formation of an electric spark (plasma channel), possibly followed by an electric arc through part of the dielectric material. If the dielectric happens to be a solid, permanent physical and chemical changes along the path of the discharge will significantly reduce the material's dielectric strength, and the device can only be used one time. However, if the dielectric material is a fluid or gas, the dielectric can fully recover its insulating properties once current through the plasma channel has been externally interrupted. Commercial Spark gaps use this property to abruptly switch high voltages in pulsed power systems, to provide surge protection for telecommunication and electrical power systems, and ignite fuel via spark plugs in many internal combustion engines.

Mechanism

Electrical breakdown occurs within a gas (or mixture of gases, such as air) when the dielectric strength of the gas(es) is exceeded. Regions of high electrical stress can cause nearby gas to partially ionize and begin conducting. This is done deliberately in low pressure discharges such as in fluorescent lights or in an electrostatic precipitator.

Partial electrical breakdown of the air causes the "fresh air" smell of ozone during thunderstorms or around high-voltage equipment. Although air is normally an excellent insulator, when stressed by a sufficiently high voltage (an electric field strength of about 3 x 10^6V/m), air can begin to break down, becoming partially conductive.

If the voltage is sufficiently high, complete electrical breakdown of the air will culminate in an electrostatic discharge or Electric arc that bridges the entire gap.

While the small sparks generated by static electricity may barely be audible, larger sparks are often accompanied by a loud snap or bang. Lightning is an example of an immense spark that can be many miles long. The colour of the spark depends upon the gases that make up the gaseous media.

If a fuse or circuit breaker fails to interrupt the current through a spark in a power circuit, current may continue, forming a very hot electric arc. The colour of an arc depends primarily upon the conductor materials (as they are vaporized and mix within the hot plasma in the arc).

The free ions in and around the arc recombine to create new chemical compounds (ozone, carbon monoxide, nitrous oxide and other compounds). Ozone is most easily noticed due to its distinct odour. Although sparks and arcs are usually undesirable, they can be useful in everyday applications such as spark plugs for gasoline engines, electrical welding of metals, or for metal melting in an electric arc furnace.

Voltage-current Relation

Before gas breakdown, there is a non-linear relation between voltage and current as shown in the figure. In region 1, there are free ions that can be accelerated by the field and induce a current. These will be saturated after a certain voltage and give a constant current, region 2. Region 3 and 4 are caused by ion avalanche as explained by the Townsend discharge mechanism.

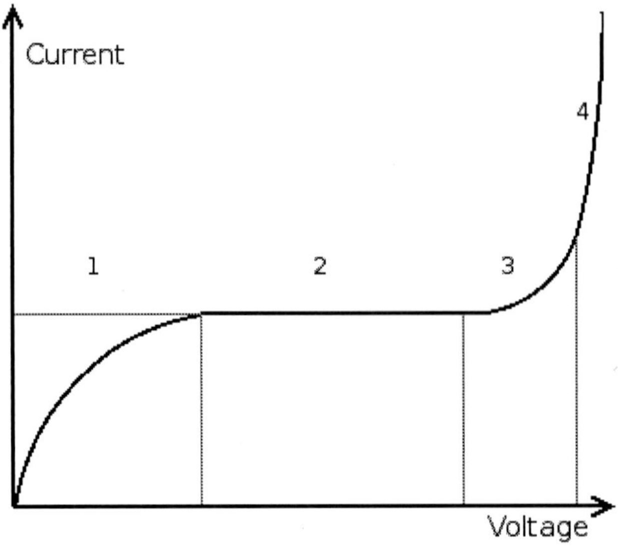

***Figure:** Voltage-current relation before breakdown*

Corona Breakdown

Partial breakdown of the air occurs as a corona discharge on high voltage conductors at points with the highest electrical stress. As the dielectric strength of the material surrounding the conductor determines the maximum strength of the electric field the surrounding material can tolerate before becoming conductive, conductors that consist of sharp points, or balls with small radii, are more prone to causing dielectric breakdown. Corona is sometimes seen as a bluish glow around high voltage wires and heard as a sizzling sound along high voltage power lines. Corona also generates radio frequency noise that can also be heard as "static" or buzzing on radio receivers. Corona can also occur naturally at high points (such as church spires, treetops, or ship masts) during thunderstorms as St. Elmo's Fire. Although corona discharge is usually undesirable, until recently it was essential in the operation of photocopiers (xerography) and laser printers. Many modern copiers and laser printers now charge the photoconductor drum with an electrically conductive roller, reducing undesirable indoor ozone pollution. Additionally, lightning rods use corona discharge to create conductive paths in the air that point towards the rod, deflecting potentially-damaging lightning away from buildings and other structures.

Corona discharge ozone generators have been used for more than 30 years in the water purification process. Ozone is a toxic gas, even more potent than chlorine. In a typical drinking water treatment plant, the ozone gas is dissolved into the filtered water to kill bacteria and viruses. Ozone also removes the bad odours and taste from the water. The main advantage of ozone is that the overdose (residual) decomposes to gaseous oxygen well before the water reaches the consumer. This is in contrast with chlorine which stays in the water and can be tasted by the consumer.

Corona discharges are also used to modify the surface properties of many polymers. An example is the corona treatment of plastic materials which allows paint or ink to adhere properly.

Zener Diode

A Zener diode is a diode which allows current to flow in the forward direction in the same manner as an ideal diode, but will also permit it to flow in the reverse direction when the voltage is above a certain value known as the breakdown voltage, "zener knee voltage", "zener voltage" or "avalanche point".

The device was named after Clarence Zener, who discovered this electrical property. Many diodes described as "zener" diodes rely instead on avalanche breakdown as the mechanism. Both types are used. Common applications include providing a reference voltage for voltage regulators, or to protect other semiconductor devices from momentary voltage pulses.

Operation

A conventional solid-state diode will allow significant current if it is reverse-biased above its reverse breakdown voltage. When the reverse bias breakdown voltage is exceeded, a conventional diode is subject to high current due to avalanche breakdown. Unless this current is limited by circuitry, the diode will be permanently damaged due to overheating. A zener diode exhibits almost the same properties, except the device is specially designed so as to have a reduced breakdown voltage, the so-called zener voltage. By contrast with the conventional device, a reverse-biased zener diode will exhibit a controlled breakdown and allow the current to keep the voltage across the zener diode close to the zener breakdown voltage. For example, a diode with a zener breakdown voltage of 3.2 V will exhibit a voltage drop of very nearly 3.2 V across a wide range of reverse currents. The zener diode is therefore ideal for applications such as the generation of a reference voltage (e.g. for an amplifier stage), or as a voltage stabilizer for low-current applications.

Another mechanism that produces a similar effect is the avalanche effect as in the avalanche diode. The two types of diode are in fact constructed the same way and both effects are present in diodes of this type. In silicon diodes up to about 5.6 volts, the zener effect is the predominant effect and shows a marked negative temperature coefficient. Above 5.6 volts, the avalanche effect becomes predominant and exhibits a positive temperature coefficient.

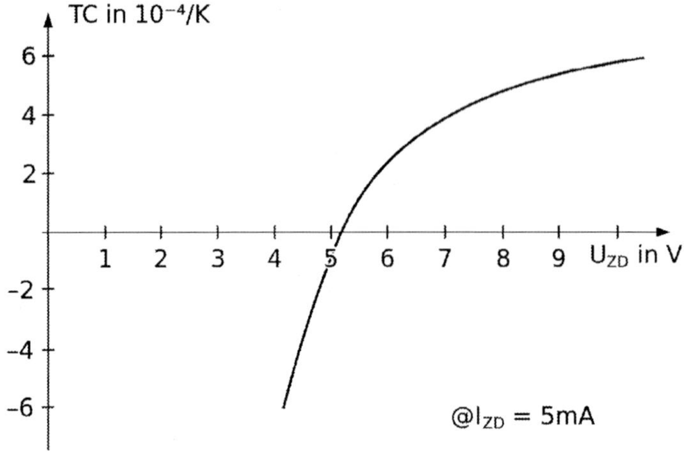

Figure: *TC depending on zener voltage*

In a 5.6 V diode, the two effects occur together and their temperature coefficients nearly cancel each other out, thus the 5.6 V diode is useful in temperature-critical applications. An alternative which is used for voltage references that need to be highly stable over long periods of time is to use a Z-diode with a TC of +2 mV/°C (breakdown voltage 6.2-6.3 V) connected in series with a forward-biased silicon diode (or a transistor B-E junction) manufactured on the same chip. The forward biased diode has a TC of -2 mV/°C, causing the TCs to cancel out.

Modern manufacturing techniques have produced devices with voltages lower than 5.6 V with negligible temperature coefficients, but as higher voltage devices are encountered, the temperature coefficient rises dramatically. A 75 V diode has 10 times the coefficient of a 12 V diode. Zener and avalanche diodes, regardless of breakdown voltage, are usually marketed under the umbrella term of "zener diode".

Construction

The zener diode's operation depends on the heavy doping of its p-n junction. The depletion region formed in the diode is very thin

(<1 μm) and the electric field is consequently very high (about 500 kV/m) even for a small reverse bias voltage of about 5 V, allowing electrons to tunnel from the valence band of the n-type material to the conduction band of the n-type material.

In the atomic scale, this tunnelling corresponds to the transport of valence band electrons into the empty conduction band states; as a result of the reduced barrier between these bands and high electric fields that are induced due to the relatively high levels of dopings on both sides. The breakdown voltage can be controlled quite accurately in the doping process. While tolerances within 0.05% are available, the most widely used tolerances are 5% and 10%. Breakdown voltage for commonly available zener diodes can vary widely from 1.2 volts to 200 volts.

Surface Zeners

The emitter-base junction of a bipolar NPN transistor behaves as a zener diode, with breakdown voltage at about 6.8 V for common bipolar processes and about 10 V for lightly doped base regions in BiCMOS processes. Older processes with poor control of doping characteristics had the variation of Zener voltage up to +-1 V, newer processes using ion implantation can achieve no more than +-0.25 V. The NPN transistor structure can be employed as a *surface zener diode*, with collector and emitter connected together as its cathode and base region as anode.

In this approach the base doping profile usually narrows towards the surface, creating a region with intensified electric field where the avalanche breakdown occurs. The hot carriers produced by acceleration in the intense field sometime shoot into the oxide layer above the junction and become trapped there. The accumulation of trapped charges can then cause Zener walkout, a corresponding change of the Zener voltage of the junction. The same effect can be achieved by radiation damage.

The emitter-base zener diodes can handle only smaller currents as the energy is dissipated in the base depletion region which is very small. Higher amount of dissipated energy (higher current for longer time, or a short very high current spike) will cause thermal damage to the junction and/or its contacts. Partial damage of the junction can shift its Zener voltage. Total destruction of the Zener junction by overheating it and causing migration of metallization across the junction ("spiking") can be used intentionally as a Zener zap antifuse.

Subsurface Zeners

A subsurface Zener diode, also called buried Zener, is a device similar to the surface Zener, but with the avalanche region located deeper in the structure, typically several micrometers below the oxide. The hot carriers then lose energy by collisions with the semiconductor lattice before reaching the oxide layer and cannot be trapped there. The Zener walkout phenomenon therefore does not occur here, and the buried Zeners have voltage constant over their entire lifetime. Most buried Zeners have breakdown voltage of 5-7 volts. Several different junction structures are used.

Uses

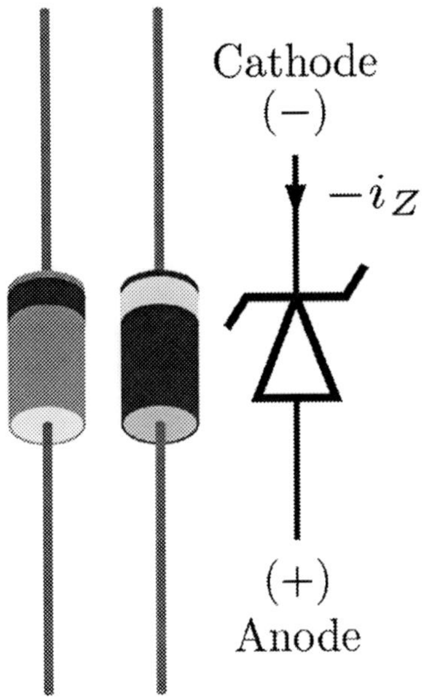

Figure: *Zener diode shown with typical packages.*
Reverse *current* $-i_Z$ *is shown.*

Zener diodes are widely used as voltage references and as shunt regulators to regulate the voltage across small circuits. When connected in parallel with a variable voltage source so that it is reverse biased, a zener diode conducts when the voltage reaches the diode's reverse breakdown voltage. From that point on, the relatively low impedance of the diode keeps the voltage across the diode at that value.

In this circuit, a typical voltage reference or regulator, an input voltage, U_{IN}, is regulated down to a stable output voltage U_{OUT}. The breakdown voltage of diode D is stable over a wide current range and holds U_{OUT} relatively constant even though the input voltage may fluctuate over a fairly wide range. Because of the low impedance of the diode when operated like this, resistor R is used to limit current through the circuit.

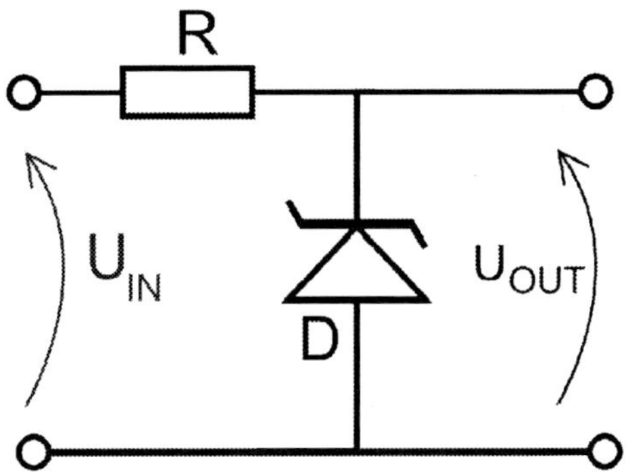

In the case of this simple reference, the current flowing in the diode is determined using Ohm's law and the known voltage drop across the resistor R;

$I_{Diode} = (U_{IN} - U_{OUT}) / R_{\Omega}$

The value of R must satisfy two conditions :

1. R must be small enough that the current through D keeps D in reverse breakdown. The value of this current is given in the data sheet for D. For example, the common BZX79C5V6 device, a 5.6 V 0.5 W zener diode, has a recommended reverse current of 5 mA. If insufficient current exists through D, then U_{OUT} will be unregulated, and less than the nominal breakdown voltage (this differs to voltage regulator tubes where the output voltage will be higher than nominal and could rise as high as U_{IN}). When calculating R, allowance must be made for any current through the external load, not shown in this diagram, connected across U_{OUT}.
2. R must be large enough that the current through D does not destroy the device. If the current through D is I_D, its breakdown voltage V_B and its maximum power dissipation P_{MAX}, then $I_D V_B < P_{MAX}$.

A load may be placed across the diode in this reference circuit, and as long as the zener stays in reverse breakdown, the diode will provide a stable voltage source to the load. Zener diodes in this configuration are often used as stable references for more advanced voltage regulator circuits.

Shunt regulators are simple, but the requirements that the ballast resistor be small enough to avoid excessive voltage drop during worst-case operation (low input voltage concurrent with high load current) tends to leave a lot of current flowing in the diode much of the time, making for a fairly wasteful regulator with high quiescent power dissipation, only suitable for smaller loads.

These devices are also encountered, typically in series with a base-emitter junction, in transistor stages where selective choice of a device centred around the avalanche or zener point can be used to introduce compensating temperature co-efficient balancing of the transistor PN junction. An example of this kind of use would be a DC error amplifier used in a regulated power supply circuit feedback loop system.

Zener diodes are also used in surge protectors to limit transient voltage spikes.

Another notable application of the zener diode is the use of noise caused by its avalanche breakdown in a random number generator.

Opto-Electronic Devices

LEDs

Photo Diode and Applications: A light-emitting diode (LED) is a semiconductor light source. LEDs are used as indicator lamps in many devices and are increasingly used for other lighting. Appearing as practical electronic components in 1962, early LEDs emitted low-intensity red light, but modern versions are available across the visible, ultraviolet, and infrared wavelengths, with very high brightness.

When a light-emitting diode is switched on, electrons are able to recombine with holes within the device, releasing energy in the form of photons. This effect is called electroluminescence and the colour of the light (corresponding to the energy of the photon) is determined by the energy band gap of the semiconductor. An LED is often small in area (less than 1 mm^2), and integrated optical components may be used to shape its radiation pattern. LEDs present many advantages over incandescent light sources including lower energy consumption, longer lifetime, improved physical robustness, smaller size, and faster switching. However, LEDs powerful enough for room lighting are relatively expensive and require more precise current and heat management than compact fluorescent lamp sources of comparable output.

Light-emitting diodes are used in applications as diverse as aviation lighting, digital microscopes, automotive lighting, advertising, general lighting, and traffic signals. LEDs have allowed new text, video displays, and sensors to be developed, while their high switching rates are also useful in advanced communications technology. Infrared LEDs are also used in the remote control units of many commercial products including televisions, DVD players and other domestic appliances. LEDs are also used in seven-segment display.

History

Discoveries and Early Devices: Electroluminescence as a phenomenon was discovered in 1907 by the British experimenter H. J. Round of Marconi Labs, using a crystal of silicon carbide and a cat's-whisker detector. Russian Oleg Vladimirovich Losev reported creation of the first LED in 1927. His research was distributed in Russian, German and British scientific journals, but no practical use was made of the discovery for several decades. Rubin Braunstein of the Radio Corporation of America reported on infrared emission from gallium arsenide (GaAs) and other semiconductor alloys in 1955. Braunstein observed infrared emission generated by simple diode structures using gallium antimonide (GaSb), GaAs, indium phosphide (InP), and silicon-germanium (SiGe) alloys at room temperature and at 77 Kelvin.

In 1961 American experimenters Robert Biard and Gary Pittman, working at Texas Instruments, found that GaAs emitted infrared radiation when electric current was applied and received the patent for the infrared LED.

The first practical visible-spectrum (red) LED was developed in 1962 by Nick Holonyak, Jr., while working at General Electric Company. Holonyak first reported this breakthrough in the journal Applied Physics Letters on the December 1, 1962. Holonyak is seen as the "father of the light-emitting diode". M. George Craford, a former graduate student of Holonyak, invented the first yellow LED and improved the brightness of red and red-orange LEDs by a factor of ten in 1972. In 1976, T. P. Pearsall created the first high-brightness, high-efficiency LEDs for optical fibre telecommunications by inventing new semiconductor materials specifically adapted to optical fibre transmission wavelengths.

Commercial Development

The first commercial LEDs were commonly used as replacements for incandescent and neon indicator lamps, and in seven-segment displays, first in expensive equipment such as laboratory and electronics test equipment, then later in such appliances as TVs, radios, telephones, calculators, and even watches. Until 1968, visible and infrared LEDs were extremely costly, in the order of US$200 per unit, and so had little practical use. The Monsanto Company was the first organization to mass-produce visible LEDs, using gallium arsenide phosphide (GaAsP) in 1968 to produce red LEDs suitable for indicators. Hewlett Packard (HP) introduced LEDs in 1968, initially using GaAsP supplied by Monsanto. These red LEDs were bright enough only for use as indicators, as the light output was not enough to illuminate an area. Readouts in calculators were so small that plastic lenses were built over each digit to make them legible. Later, other colours became widely available and appeared in appliances and equipment. In the 1970s commercially successful LED devices at less than five cents each were produced by Fairchild Optoelectronics. These devices employed compound semiconductor chips fabricated with the planar process invented by Dr. Jean Hoerni at Fairchild Semiconductor. The combination of planar processing for chip fabrication and innovative packaging methods enabled the team at Fairchild led by optoelectronics pioneer Thomas Brandt to achieve the needed cost reductions. These methods continue to be used by LED producers.

As LED materials technology grew more advanced, light output rose, while maintaining efficiency and reliability at acceptable levels. The invention and development of the high-power white-light LED led to use for illumination, and is slowly replacing incandescent and fluorescent lighting.

Most LEDs were made in the very common 5 mm T1¾ and 3 mm T1 packages, but with rising power output, it has grown increasingly necessary to shed excess heat to maintain reliability, so more complex packages have been adapted for efficient heat dissipation. Packages for state-of-the-art high-power LEDs bear little resemblance to early LEDs.

The Blue and White LED

The first high-brightness blue LED was demonstrated by Shuji Nakamura of Nichia Corporation in 1994 and was based on InGaN. Its development built on critical developments in GaN nucleation on sapphire substrates and the demonstration of p-type doping of GaN, developed by Isamu Akasaki and H. Amano in Nagoya. In 1995, Alberto Barbieri at the Cardiff University Laboratory (GB) investigated the efficiency and reliability of high-brightness LEDs and demonstrated a "transparent contact" LED using indium tin oxide (ITO) on (AlGaInP/GaAs). The existence of blue LEDs and high-efficiency LEDs quickly led to the development of the first white LED, which employed a Y 3Al 5O 12:Ce, or "YAG", phosphor coating to mix down-converted yellow light with blue to produce light that appears white. Nakamura was awarded the 2006 Millennium Technology Prize for his invention.

The development of LED technology has caused their efficiency and light output to rise exponentially, with a doubling occurring approximately every 36 months since the 1960s, in a way similar to Moore's law. This trend is generally attributed to the parallel development of other semiconductor technologies and advances in optics and material science, and has been called Haitz's law after Dr. Roland Haitz.

In 2001 and 2002, processes for growing gallium nitride (GaN) LEDs on silicon were successfully demonstrated. In January 2012, Osram demonstrated high-power InGaN LEDs grown on Silicon substrates commercially. It has been speculated that the use of six-inch silicon wafers instead of two-inch sapphire wafers and epitaxy manufacturing processes could reduce production costs by up to 90%.

Technology

Physics: The LED consists of a chip of semiconducting material doped with impurities to create a *p-n junction*. As in other diodes, current flows easily from the p-side, or anode, to the n-side, or cathode, but not in the reverse direction. Charge-carriers—electrons and holes—flow into the junction from electrodes with different voltages. When an electron meets a hole, it falls into a lower energy level, and releases energy in the form of a photon.

The wavelength of the light emitted, and thus its colour depends on the band gap energy of the materials forming the *p n junction*. In silicon or germanium diodes, the electrons and holes recombine by a *non-radiative transition*, which produces no optical emission, because these are indirect band gap materials. The materials used for the LED have a direct band gap with energies corresponding to near-infrared, visible, or near-ultraviolet light.

LED development began with infrared and red devices made with gallium arsenide. Advances in materials science have enabled making devices with ever-shorter wavelengths, emitting light in a variety of colours.

LEDs are usually built on an n-type substrate, with an electrode attached to the p-type layer deposited on its surface. P-type substrates, while less common, occur as well. Many commercial LEDs, especially GaN/InGaN, also use sapphire substrate.

Most materials used for LED production have very high refractive indices. This means that much light will be reflected back into the material at the material/air surface interface. Thus, light extraction in LEDs is an important aspect of LED production, subject to much research and development.

Refractive Index

Bare uncoated semiconductors such as silicon exhibit a very high refractive index relative to open air, which prevents passage of photons at sharp angles relative to the air-contacting surface of the semiconductor. This property affects both the light-emission efficiency of LEDs as well as the light-absorption efficiency of photovoltaic cells. The refractive index of silicon is 3.96 (590 nm), while air is 1.0002926.

In general, a flat-surface uncoated LED semiconductor chip will emit light only perpendicular to the semiconductor's surface, and a few degrees to the side, in a cone shape referred to as the *light cone*, *cone of light*, or the *escape cone*. The maximum angle of incidence is referred to as the critical angle. When this angle is exceeded, photons no longer penetrate the semiconductor but are instead reflected both internally inside the semiconductor crystal and externally off the surface of the crystal as if it were a mirror.

Internal reflections can escape through other crystalline faces, if the incidence angle is low enough and the crystal is sufficiently transparent to not re-absorb the photon emission. But for a simple square LED with 90-degree angled surfaces on all sides, the faces all act as equal angle mirrors. In this case the light can not escape and is lost as waste heat in the crystal.

A convoluted chip surface with angled facets similar to a jewel or fresnel lens can increase light output by allowing light to be emitted perpendicular to the chip surface while far to the sides of the photon emission point.

The ideal shape of a semiconductor with maximum light output would be a microsphere with the photon emission occurring at the exact centre, with electrodes penetrating to the centre to contact at the emission point. All light rays emanating from the centre would be perpendicular to the entire surface of the sphere, resulting in no internal reflections. A hemispherical semiconductor would also work, with the flat back-surface serving as a mirror to back-scattered photons.

Transition Coatings

After doping the wafer, people at the fabrication plant cut the wafer apart into individual dies. Each die is commonly called a chip.

Many LED semiconductor chips are encapsulated or potted in clear or coloured molded plastic shells. The plastic shell has three purposes:

1. Mounting the semiconductor chip in devices is easier to accomplish.
2. The tiny fragile electrical wiring is physically supported and protected from damage.
3. The plastic acts as a refractive intermediary between the relatively high-index semiconductor and low-index open air.

The third feature helps to boost the light emission from the semiconductor by acting as a diffusing lens, allowing light to be emitted at a much higher angle of incidence from the light cone than the bare chip is able to emit alone.

Efficiency and Operational Parameters

Typical indicator LEDs are designed to operate with no more than 30–60 milliwatts (mW) of electrical power. Around 1999, Philips Lumileds introduced power LEDs capable of continuous use at one watt. These LEDs used much larger semiconductor die sizes to handle the large power inputs. Also, the semiconductor dies were mounted onto metal slugs to allow for heat removal from the LED die. LED power densities up to 300W/cm2 have been achieved.

One of the key advantages of LED-based lighting sources is high luminous efficiency. White LEDs quickly matched and overtook the efficacy of standard incandescent lighting systems. In 2002, Lumileds made five-watt LEDs available with a luminous efficacy of 18–22 lumens per watt (lm/W). For comparison, a conventional incandescent light bulb of 60–100 watts emits around 15 lm/W, and standard fluorescent lights emit up to 100 lm/W. A recurring problem is that efficacy falls sharply with rising current. This effect is known as droop and effectively limits the light output of a given LED, raising heating more than light output for higher current.

The mechanism behind droop efficiency loss was identified in 2013 as Auger recombination.

As of 2012, the Lumiled catalogue gives the following as the best efficacy for each colour:

Colour	*Wavelength range (nm)*	*Typical efficacy (lm/W)*
Red	$620 < \lambda < 645$	72
Red-orange	$610 < \lambda < 620$	98
Green	$520 < \lambda < 550$	93
Cyan	$490 < \lambda < 520$	75
Blue	$460 < \lambda < 490$	37

In September 2003, a new type of blue LED was demonstrated by the company Cree Inc. to provide 24 mW at 20 milliamperes (mA). This produced a commercially packaged white light giving 65 lm/W at 20 mA, becoming the brightest white LED commercially available at the time, and more than four times as efficient as standard incandescents. In 2006, they demonstrated a prototype with a record white LED luminous efficacy of 131 lm/W at 20 mA. Nichia Corporation has developed a white LED with luminous efficacy of 150 lm/W at a forward current of 20 mA. Cree's XLamp XM-L LEDs, commercially available in 2011, produce 100 lumens per watt at their full power of 10 watts, and up to 160 lumens/watt at around 2 watts input power. In 2012, Cree announced a white LED giving 254 lumens per watt.

Practical general lighting needs high-power LEDs, of one watt or more. Typical operating currents for such devices begin at 350 mA.

Note that these efficiencies are for the LED chip only, held at low temperature in a lab. Lighting works at higher temperature and with drive circuit losses, so efficiencies are much lower. United States Department of Energy (DOE) testing of commercial LED lamps designed to replace incandescent lamps or CFLs showed that average efficacy was still about 46 lm/W in 2009 (tested performance ranged from 17 lm/W to 79 lm/W).

Cree issued a press release on February 3, 2010 about a laboratory prototype LED achieving 208 lumens per watt at room temperature. The correlated colour temperature was reported to be 4579 K. In December 2012 Cree issued another press release announcing commercial availability of 200 lumens per watt LED at room temperature.

Lifetime and Failure

Solid-state devices such as LEDs are subject to very limited wear and tear if operated at low currents and at low temperatures. Many of the LEDs made in the 1970s and 1980s are still in service today. Typical lifetimes quoted are 25,000 to 100,000 hours, but heat and current settings can extend or shorten this time significantly.

The most common symptom of LED (and diode laser) failure is the gradual lowering of light output and loss of efficiency. Sudden failures, although rare, can occur as well. Early red LEDs were notable for their short service life. With the development of high-power LEDs the devices are subjected to higher junction temperatures and higher current densities than traditional devices. This causes stress on the material and may cause early light-output degradation. To quantitatively classify useful lifetime in a standardized manner it has been suggested to use the terms L70 and L50, which is the time it will take a given LED to reach 70% and 50% light output respectively.

LED performance is temperature dependent. Most manufacturers' published ratings of LEDs are for an operating temperature of 25 °C. LEDs used outdoors, such as traffic signals or in-pavement signal lights, and that are utilized in climates where the temperature within the luminaire gets very hot, could result in low signal intensities or even failure.

LED light output rises at lower temperatures, levelling off, depending on type, at around –30 °C. Thus, LED technology may be a good replacement in uses such as supermarket freezer lighting and will last longer than other technologies. Because LEDs emit less heat than incandescent bulbs, they are an energy-efficient technology for uses such as in freezers and refrigerators. However, because they emit little heat, ice and snow may build up on the LED luminaire in colder climates. Similarly, this lack of waste heat generation has been observed to sometimes cause significant problems with street traffic signals and airport runway lighting in snow-prone areas. In response to this problem, some LED lighting systems have been designed with an added heating circuit at the expense of reduced overall electrical efficiency of the system; additionally, research has been done to develop heat sink technologies that will transfer heat produced within the junction to appropriate areas of the luminaire.

Ultraviolet and blue LEDs

Current bright blue LEDs are based on the wide band gap semiconductors GaN (gallium nitride) and InGaN (indium gallium nitride). They can be added to existing red and green LEDs to produce the impression of white light. Modules combining the three colours are used in big video screens and in adjustable-colour fixtures.

The first blue LEDs using gallium nitride were made in 1971 by Jacques Pankove at RCA Laboratories. These devices had too little light output to be of practical use and research into gallium nitride devices slowed. In August 1989, Cree Inc. introduced the first commercially available blue LED based on the indirect bandgap semiconductor, silicon carbide. SiC LEDs had very low efficiency, no more than about 0.03%, but did emit in the blue portion of the visible light spectrum.

In the late 1980s, key breakthroughs in GaN epitaxial growth and p-type doping ushered in the modern era of GaN-based optoelectronic devices. Building upon this foundation, in 1993 high-brightness blue LEDs were demonstrated. High-brightness blue LEDs invented by Shuji Nakamura of Nichia Corporation using gallium nitride revolutionized LED lighting, making high-power light sources practical.

By the late 1990s, blue LEDs had become widely available. They have an active region consisting of one or more InGaN quantum wells sandwiched between thicker layers of GaN, called cladding layers. By varying the relative In/Ga fraction in the InGaN quantum wells, the light emission can in theory be varied from violet to amber. Aluminium gallium nitride (AlGaN) of varying Al/Ga fraction can be used to manufacture the cladding and quantum well layers for ultraviolet LEDs, but these devices have not yet reached the level of efficiency and technological maturity of InGaN/GaN blue/green devices. If un-alloyed GaN is used in this case to form the active quantum well layers, the device will emit near-ultraviolet light with a peak wavelength centred around 365 nm. Green LEDs manufactured from the InGaN/GaN system are far more efficient and brighter than green LEDs produced with non-nitride material systems, but practical devices still exhibit efficiency too low for high-brightness applications.

With nitrides containing aluminium, most often AlGaN and AlGaInN, even shorter wavelengths are achievable. Ultraviolet LEDs in a range of wavelengths are becoming available on the market. Near-UV emitters at wavelengths around 375–395 nm are already cheap and often encountered, for example, as black light lamp replacements for inspection of anti-counterfeiting UV watermarks in some documents and paper currencies. Shorter-wavelength diodes, while substantially more expensive, are commercially available for wavelengths down to 240 nm. As the photosensitivity of microorganisms approximately matches the absorption spectrum of DNA, with a peak at about 260 nm, UV LED emitting at 250–270 nm are to be expected in prospective disinfection and sterilization devices. Recent research has shown that commercially available UVA LEDs (365 nm) are already effective disinfection and sterilization devices.

Deep-UV wavelengths were obtained in laboratories using aluminium nitride (210 nm), boron nitride (215 nm) and diamond (235 nm).

White Light

There are two primary ways of producing white light-emitting diodes (WLEDs), LEDs that generate high-intensity white light. One is to use individual LEDs that emit three primary colours—red, green, and blue—and then mix all the colours to form white light. The other is to use a phosphor material to convert monochromatic light from a blue or UV LED to broad-spectrum white light, much in the same way a fluorescent light bulb works.

Because of metamerism, it is possible to have quite different spectra that appear white. But the appearance of some objects may be more sensitive to details of the light spectrum illuminating them.

RGB Systems

White light can be formed by mixing differently coloured lights; the most common method is to use red, green, and blue (RGB). Hence the method is called multi-colour white LEDs (sometimes referred to as RGB LEDs). Because these need electronic circuits to control the blending and diffusion of different colours, and because the individual colour LEDs typically have slightly different emission patterns (leading to variation of the colour depending on direction) even if they are made as a single unit, these are seldom used to produce white lighting. Nevertheless, this method is particularly interesting in many uses because of the flexibility of mixing different colours, and, in principle, this mechanism also has higher quantum efficiency in producing white light.

There are several types of multi-colour white LEDs: di-, tri-, and tetrachromatic white LEDs. Several key factors that play among these different methods, include colour stability, colour rendering capability, and luminous efficacy. Often, higher efficiency will mean lower colour rendering, presenting a trade-off between the luminous efficiency and colour rendering. For example, the dichromatic white LEDs have the best luminous efficacy (120 lm/W), but the lowest colour rendering capability. However, although tetrachromatic white LEDs have excellent colour rendering capability, they often have poor luminous efficiency. Trichromatic white LEDs are in between, having both good luminous efficacy (>70 lm/W) and fair colour rendering capability.

One of the challenges is the development of more efficient green LEDs. The theoretical maximum for green LEDs is 683 lumens per watt but today few green LEDs exceed even 100 lumens per watt. The blue and red LEDs get closer to their theoretical limits.

Multi-colour LEDs offer not merely another means to form white light but a new means to form light of different colours. Most perceivable colours can be formed by mixing different amounts of three primary colours. This allows precise dynamic colour control. As more effort is devoted to investigating this method, multi-colour LEDs should have profound influence on the fundamental method that we use to produce and control light colour. However, before this type of LED can play a role on the market, several technical problems must be solved. These include that this type of LED's emission power decays exponentially with rising temperature, resulting in a substantial change in colour stability. Such problems inhibit and may preclude industrial use. Thus, many new package designs aimed at solving this problem have been proposed and their results are now being reproduced by researchers and scientists.

CCT dimming for LED technology is regarded as a difficult task, since binning, age and temperature drift effects of LEDs change the actual colour value output. Feedback loop systems are used for example with colour sensors, to actively monitor and control the colour output of multiple colour mixing LEDs.

Phosphor-based LEDs

This method involves coating LEDs of one colour (mostly blue LEDs made of InGaN) with phosphors of different colours to form white light; the resultant LEDs are called phosphor-based white LEDs. A fraction of the blue light undergoes the Stokes shift being transformed from shorter wavelengths to longer. Depending on the colour of the original LED, phosphors of different colours can be employed. If several phosphor layers of distinct colours are applied, the emitted spectrum is broadened, effectively raising the colour rendering index (CRI) value of a given LED.

Phosphor-based LED efficiency losses are due to the heat loss from the Stokes shift and also other phosphor-related degradation issues. Their efficiencies compared to normal LEDs depend on the spectral distribution of the resultant light output and the original wavelength of the LED itself. For example, the efficiency of a typical YAG yellow phosphor based white LED ranges from 3 to 5 times the efficiency of the original blue LED because of the greater luminous efficacy of yellow compared to blue light. Due to the simplicity of manufacturing the phosphor method is still the most popular method for making high-intensity white LEDs. The design and production of a light source or light fixture using a monochrome emitter with phosphor conversion is simpler and cheaper than a complex RGB system, and the majority of high-intensity white LEDs presently on the market are manufactured using phosphor light conversion.

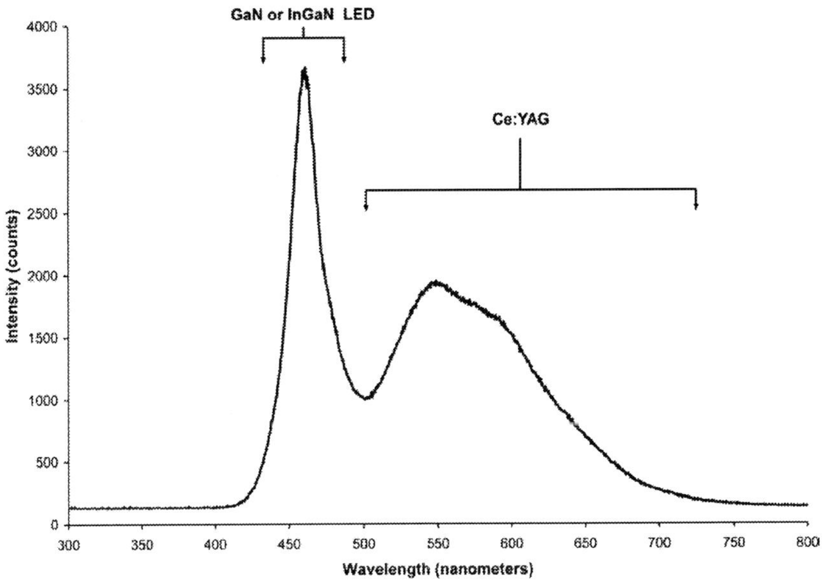

Figure: *Spectrum of a "white" LED clearly showing blue light directly emitted by the GaN-based LED (peak at about 465 nm) and the more broadband Stokes-shifted light emitted by the Ce:*$^{3+}$*YAG phosphor, which emits at roughly 500–700 nm.*

Among the challenges being faced to improve the efficiency of LED-based white light sources is the development of more efficient phosphors. Today the most efficient yellow phosphor is still the YAG phosphor, with less than 10% Stoke shift loss. Losses attributable to internal optical losses due to re-absorption in the LED chip and in the LED packaging itself account typically for another 10% to 30% of efficiency loss.

Currently, in the area of phosphor LED development, much effort is being spent on optimizing these devices to higher light output and higher operation temperatures. For instance, the efficiency can be raised by adapting better package design or by using a more suitable type of phosphor. Conformal coating process is frequently used to address the issue of varying phosphor thickness.

The phosphor-based white LEDs encapsulate InGaN blue LEDs inside phosphor coated epoxy. A common yellow phosphor material is cerium-doped yttrium aluminium garnet (Ce^{3+}:YAG).

White LEDs can also be made by coating near-ultraviolet (NUV) LEDs with a mixture of high-efficiency europium-based phosphors that emit red and blue, plus copper and aluminium-doped zinc sulfide (ZnS:Cu, Al) that emits green. This is a method analogous to the way fluorescent lamps work. This method is less efficient than blue LEDs with YAG:Ce phosphor, as the Stokes shift is larger, so more energy is converted to heat, but yields light with better spectral characteristics, which render colour better. Due to the higher radiative output of the ultraviolet LEDs than of the blue ones, both methods offer comparable brightness. A concern is that UV light may leak from a malfunctioning light source and cause harm to human eyes or skin.

Other white LEDs

Another method used to produce experimental white light LEDs used no phosphors at all and was based on homoepitaxially grown zinc selenide (ZnSe) on a ZnSe substrate that simultaneously emitted blue light from its active region and yellow light from the substrate.

Organic Light-emitting Diodes (OLEDs)

In an organic light-emitting diode (OLED), the electroluminescent material comprising the emissive layer of the diode is an organic compound. The organic material is electrically conductive due to the delocalization of pi electrons caused by conjugation over all or part of the molecule, and the material therefore functions as an organic semiconductor. The organic materials can be small organic molecules in a crystalline phase, or polymers.

The potential advantages of OLEDs include thin, low-cost displays with a low driving voltage, wide viewing angle, and high contrast and colour gamut. Polymer LEDs have the added benefit of printable and flexible displays. OLEDs have been used to make visual displays for portable electronic devices such as cellphones, digital cameras, and MP3 players while possible future uses include lighting and televisions.

Quantum dot LEDs (Experimental)

Quantum dots (QD) are semiconductor nanocrystals that possess unique optical properties. Their emission colour can be tuned from the visible throughout the infrared spectrum. This allows quantum dot LEDs to create almost any colour on the CIE diagram. This provides more colour options and better colour rendering than white LEDs. Quantum dot LEDs are available in the same package types as traditional phosphor-based LEDs. There are two types of schemes for QD excitation. One uses photo excitation with a primary light source LED (typically blue or UV LEDs are used). The other is direct electrical excitation first demonstrated by Alivisatos et al.

One example of the photo-excitation scheme is a method developed by Michael Bowers, at Vanderbilt University in Nashville, involving coating a blue LED with quantum dots that glow white in response to the blue light from the LED. This method emits a warm, yellowish-white light similar to that made by incandescent bulbs. Quantum dots are also being considered for use in white light-emitting diodes in liquid crystal display (LCD) televisions.

The major difficulty in using quantum dots-based LEDs is the insufficient stability of QDs under prolonged irradiation. In February 2011 scientists at PlasmaChem GmbH could synthesize quantum dots for LED applications and build a light converter on their basis, which could efficiently convert light from blue to any other colour for many hundred hours. Such QDs can be used to emit visible or near infrared light of any wavelength being excited by light with a shorter wavelength.

Figure: *LEDs are produced in a variety of shapes and sizes. The colour of the plastic lens is often the same as the actual colour of light emitted, but not always. For instance, purple plastic is often used for infrared LEDs, and most blue devices have colourless housings. Modern high power LEDs such as those used for lighting and backlighting are generally found in surface-mount technology (SMT) packages*

The structure of QD-LEDs used for the electrical-excitation scheme is similar to basic design of OLED. A layer of quantum dots is sandwiched between layers of electron-transporting and hole-transporting materials. An applied electric field causes electrons and holes to move into the quantum dot layer and recombine forming an exciton that excites a QD. This scheme is commonly studied for quantum dot display. The tunability of emission wavelengths and narrow bandwidth is also beneficial as excitation sources for fluorescence imaging. Fluorescence near-field scanning optical microscopy (NSOM) utilizing an integrated QD-LED has been demonstrated.

In February 2008, a luminous efficacy of 300 lumens of visible light per watt of radiation (not per electrical watt) and warm-light emission was achieved by using nanocrystals.

The main types of LEDs are miniature, high power devices and custom designs such as alphanumeric or multi-colour.

Miniature

These are mostly single-die LEDs used as indicators, and they come in various sizes from 2 mm to 8 mm, through-hole and surface mount packages. They usually do not use a separate heat sink. Typical current ratings ranges from around 1 mA to above 20 mA. The small size sets a natural upper boundary on power consumption due to heat caused by the high current density and need for a heat sink.

Common package shapes include round, with a domed or flat top, rectangular with a flat top (as used in bar-graph displays), and triangular or square with a flat top. The encapsulation may also be clear or tinted to improve contrast and viewing angle.

There are three main categories of miniature single die LEDs:

- Low-current: typically rated for 2 mA at around 2 V (approximately 4 mW consumption).
- Standard: 20 mA LEDs (ranging from approximately 40 mW to 90 mW) at around:

1.9 to 2.1 V for red, orange and yellow,

3.0 to 3.4 V for green and blue,

2.9 to 4.2 V for violet, pink, purple and white.

- Ultra-high-output: 20 mA at approximately 2 V or 4–5 V, designed for viewing in direct sunlight.

5 V and 12 V LEDs are ordinary miniature LEDs that incorporate a suitable series resistor for direct connection to a 5 V or 12 V supply.

Mid-range

Medium-power LEDs are often through-hole-mounted and mostly utilized when an output of just a few lumen is needed. They sometimes have the diode mounted to four leads (two cathode leads, two anode leads) for better heat conduction and carry an integrated lens. An example of this is the Superflux package, from Philips Lumileds. These LEDs are most commonly used in light panels, emergency lighting, and automotive tail-lights. Due to the larger amount of metal in the LED, they are able to handle higher currents (around 100 mA). The higher current allows for the higher light output required for tail-lights and emergency lighting.

High-power

High-power LEDs (HPLED) can be driven at currents from hundreds of mA to more than an ampere, compared with the tens of mA for other LEDs. Some can emit over a thousand lumens. LED power densities up to 300W/cm2 have been achieved. Since overheating is destructive, the HPLEDs must be mounted on a heat sink to allow for heat dissipation. If the heat from a HPLED is not removed, the device will fail in seconds. One HPLED can often replace an incandescent bulb in a flashlight, or be set in an array to form a powerful LED lamp.

Some well-known HPLEDs in this category are the Nichia 19 series, Lumileds Rebel Led, Osram Opto Semiconductors Golden Dragon, and Cree X-lamp. As of September 2009, some HPLEDs manufactured by Cree Inc. now exceed 105 lm/W (e.g. the XLamp XP-G LED chip emitting Cool White light) and are being sold in lamps intended to replace incandescent, halogen, and even fluorescent lights, as LEDs grow more cost competitive.

The impact of Haitz's law which describes the exponential rise in light output of LEDs over time can be readily seen in year over year increases in lumen output and efficiency. For example, the CREE XP-G series LED achieved 105 lm/W in 2009, while Nichia released the 19 series with a typical efficacy of 140 lm/W in 2010.

LEDs have been developed by Seoul Semiconductor that can operate on AC power without the need for a DC converter. For each half-cycle, part of the LED emits light and part is dark, and this is reversed during the next half-cycle. The efficacy of this type of HPLED is typically 40 lm/W. A large number of LED elements in series may be able to operate directly from line voltage. In 2009, Seoul Semiconductor released a high DC voltage LED capable of being driven from AC power with a simple controlling circuit. The low-power dissipation of these LEDs affords them more flexibility than the original AC LED design.

Application-specific Variations

- *Flashing LEDs* are used as attention seeking indicators without requiring external electronics. Flashing LEDs resemble standard LEDs but they contain an integrated multivibrator circuit that causes the LED to flash with a typical period of one second. In diffused lens LEDs this is visible as a small black dot. Most flashing LEDs emit light of one colour, but more sophisticated devices can flash between multiple colours and even fade through a colour sequence using RGB colour mixing.
- *Bi-colour LEDs* are two different LED emitters in one case. There are two types of these. One type consists of two dies connected to the same two leads antiparallel to each other. Current flow in one direction emits one colour, and current in the opposite direction emits the other colour. The other type consists of two dies with separate leads for both dies and another lead for common anode or cathode, so that they can be controlled independently.
- *Tri-colour LEDs* are three different LED emitters in one case. Each emitter is connected to a separate lead so they can be controlled independently. A four-lead arrangement is typical with one common lead (anode or cathode) and an additional lead for each colour.
- *RGB LEDs* are Tri-colour LEDs with red, green, and blue emitters, in general using a four-wire connection with one common lead (anode or cathode). These LEDs can have either common positive or common negative leads. Others however, have only two leads (positive and negative) and have a built in tiny electronic control unit.
- *Alphanumeric LED displays* are available in seven-segment, starburst and dot-matrix format. Seven-segment displays handle all numbers and a limited set of letters. Starburst displays can display all letters. Dot-matrix displays typically use 5x7 pixels per character. Seven-segment LED displays were in widespread use in the 1970s and 1980s, but rising use of liquid crystal displays, with their lower power needs and greater display flexibility, has reduced the popularity of numeric and alphanumeric LED displays.

Considerations for Use

Power Sources: The current/voltage characteristic of an LED is similar to other diodes, in that the current is dependent exponentially on the voltage. This means that a small change in voltage can cause a large change in current. If the maximum voltage rating is exceeded by a small amount, the current rating may be exceeded by a large amount, potentially damaging or destroying the LED. The typical solution is to use constant-current power supplies, or driving the LED at a voltage much below the maximum rating. Since most common power sources (batteries, mains) are constant-voltage sources, most LED fixtures must include a power converter, at least a current-limiting resistor. However, the high resistance of 3 V coin cells combined with the high differential resistance of nitride-based LEDs makes it possible to power such an LED from such a coin cell without an external resistor.

Electrical Polarity

As with all diodes, current flows easily from p-type to n-type material. However, no current flows and no light is emitted if a small voltage is applied in the reverse direction. If the reverse voltage grows large enough to exceed the breakdown voltage, a large current flows and the LED may be damaged. If the reverse current is sufficiently limited to avoid damage, the reverse-conducting LED is a useful noise diode.

Safety and Health

The vast majority of devices containing LEDs are "safe under all conditions of normal use", and so are classified as "Class 1 LED product"/ "LED Klasse 1". At present, only a few LEDs—extremely bright LEDs that also have a tightly focused viewing angle of 8° or less—could, in theory, cause temporary blindness, and so are classified as "Class 2". The Opinion of the French Agency for Food, Environmental and Occupational Health & Safety (ANSES) of 2010, on the health issues concerning LEDs, suggested banning public use of lamps which were in the moderate Risk Group 2, especially those with a high blue component in places frequented by children. In general, laser safety regulations—and the "Class 1", "Class 2", etc. system—also apply to LEDs.

While LEDs have the advantage over fluorescent lamps that they do not contain mercury, they may contain other hazardous metals such as lead and arsenic. A study published in 2011 states: "According to federal standards, LEDs are not hazardous except for low-intensity red LEDs, which leached Pb [lead] at levels exceeding regulatory limits (186 mg/L; regulatory limit: 5). However, according to California regulations, excessive levels of copper (up to 3892 mg/kg; limit: 2500), lead (up to 8103 mg/kg; limit: 1000), nickel (up to 4797 mg/kg; limit: 2000), or silver (up to 721 mg/kg; limit: 500) render all except low-intensity yellow LEDs hazardous."

Advantages

- *Efficiency:* LEDs emit more light per watt than incandescent light bulbs. The efficiency of LED lighting fixtures is not affected by shape and size, unlike fluorescent light bulbs or tubes.
- *Colour:* LEDs can emit light of an intended colour without using any colour filters as traditional lighting methods need. This is more efficient and can lower initial costs.
- *Size:* LEDs can be very small (smaller than 2 mm^2) and are easily attached to printed circuit boards.
- *On/Off time:* LEDs light up very quickly. A typical red indicator LED will achieve full brightness in under a microsecond. LEDs used in communications devices can have even faster response times.
- *Cycling:* LEDs are ideal for uses subject to frequent on-off cycling, unlike fluorescent lamps that fail faster when cycled often, or HID lamps that require a long time before restarting.
- *Dimming:* LEDs can very easily be dimmed either by pulse-width modulation or lowering the forward current. This pulse-width modulation is why LED lights viewed on camera, particularly headlights on cars, appear to be flashing or flickering. This is a type of Stroboscopic effect.
- *Cool light:* In contrast to most light sources, LEDs radiate very little heat in the form of IR that can cause damage to sensitive objects or fabrics. Wasted energy is dispersed as heat through the base of the LED.
- *Slow failure:* LEDs mostly fail by dimming over time, rather than the abrupt failure of incandescent bulbs.
- *Lifetime:* LEDs can have a relatively long useful life. One report estimates 35,000 to 50,000 hours of useful life, though time to complete failure may be longer. Fluorescent tubes typically are rated at about 10,000 to 15,000 hours, depending partly on the conditions of use, and incandescent light bulbs at 1,000 to 2,000 hours. Several DOE demonstrations have shown that reduced maintenance costs from this extended lifetime, rather than energy savings, is the primary factor in determining the payback period for an LED product.

- *Shock resistance:* LEDs, being solid-state components, are difficult to damage with external shock, unlike fluorescent and incandescent bulbs, which are fragile.
- *Focus:* The solid package of the LED can be designed to focus its light. Incandescent and fluorescent sources often require an external reflector to collect light and direct it in a usable manner. For larger LED packages total internal reflection (TIR) lenses are often used to the same effect. However, when large quantities of light is needed many light sources are usually deployed, which are difficult to focus or collimate towards the same target.

Disadvantages

- *High initial price:* LEDs are currently more expensive, price per lumen, on an initial capital cost basis, than most conventional lighting technologies. As of 2010, the cost per thousand lumens (kilolumen) was about \$18. The price is expected to reach \$2/kilolumen by 2015. The additional expense partially stems from the relatively low lumen output and the drive circuitry and power supplies needed.
- *Temperature dependence:* LED performance largely depends on the ambient temperature of the operating environment – or "thermal management" properties. Over-driving an LED in high ambient temperatures may result in overheating the LED package, eventually leading to device failure. An adequate heat sink is needed to maintain long life. This is especially important in automotive, medical, and military uses where devices must operate over a wide range of temperatures, which require low failure rates.
- *Voltage sensitivity:* LEDs must be supplied with the voltage above the threshold and a current below the rating. This can involve series resistors or current-regulated power supplies.
- *Light quality:* Most cool-white LEDs have spectra that differ significantly from a black body radiator like the sun or an incandescent light. The spike at 460 nm and dip at 500 nm can cause the colour of objects to be perceived differently under cool-white LED illumination than sunlight or incandescent sources, due to metamerism, red surfaces being rendered particularly badly by typical phosphor-based cool-white LEDs. However, the colour rendering properties of common fluorescent lamps are often inferior to what is now available in state-of-art white LEDs.

- *Area light source:* Single LEDs do not approximate a point source of light giving a spherical light distribution, but rather a lambertian distribution. So LEDs are difficult to apply to uses needing a spherical light field, however different fields of light can be manipulated by the application of different optics or "lenses". LEDs cannot provide divergence below a few degrees. In contrast, lasers can emit beams with divergences of 0.2 degrees or less.
- *Electrical polarity:* Unlike incandescent light bulbs, which illuminate regardless of the electrical polarity, LEDs will only light with correct electrical polarity. To automatically match source polarity to LED devices, rectifiers can be used.
- *Blue hazard:* There is a concern that blue LEDs and cool-white LEDs are now capable of exceeding safe limits of the so-called blue-light hazard as defined in eye safety specifications such as ANSI/IESNA RP-27.1–05: Recommended Practice for Photobiological Safety for Lamp and Lamp Systems.
- *Blue pollution:* Because cool-white LEDs with high colour temperature emit proportionally more blue light than conventional outdoor light sources such as high-pressure sodium vapor lamps, the strong wavelength dependence of Rayleigh scattering means that cool-white LEDs can cause more light pollution than other light sources. The International Dark-Sky Association discourages using white light sources with correlated colour temperature above 3,000 K.
- *Droop:* The efficiency of conventional InGaN based LEDs decreases as one increases current above a given level.

Applications

LED uses fall into four major categories:

- Visual signals where light goes more or less directly from the source to the human eye, to convey a message or meaning.
- Illumination where light is reflected from objects to give visual response of these objects.
- Measuring and interacting with processes involving no human vision.
- Narrow band light sensors where LEDs operate in a reverse-bias mode and respond to incident light, instead of emitting light.

Indicators and Signs

The low energy consumption, low maintenance and small size of LEDs has led to uses as status indicators and displays on a variety of equipment and installations. Large-area LED displays are used as stadium displays and as dynamic decorative displays. Thin, lightweight message displays are used at airports and railway stations, and as destination displays for trains, buses, trams, and ferries.

One-colour light is well suited for traffic lights and signals, exit signs, emergency vehicle lighting, ships' navigation lights or lanterns (chromacity and luminance standards being set under the Convention on the International Regulations for Preventing Collisions at Sea 1972, Annex I and the CIE) and LED-based Christmas lights. In cold climates, LED traffic lights may remain snow covered. Red or yellow LEDs are used in indicator and alphanumeric displays in environments where night vision must be retained: aircraft cockpits, submarine and ship bridges, astronomy observatories, and in the field, e.g. night time animal watching and military field use.

Figure: *Automotive applications for LEDs continue to grow*

Because of their long life and fast switching times, LEDs have been used in brake lights for cars' high-mounted brake lights, trucks, and buses, and in turn signals for some time, but many vehicles now use LEDs for their rear light clusters. The use in brakes improves safety, due to a great reduction in the time needed to light fully, or faster rise time, up to 0.5 second faster than an incandescent bulb. This gives drivers behind more time to react. It is reported that at normal highway speeds, this equals one car length equivalent in increased time to react. In a dual intensity circuit (rear markers and brakes) if the LEDs are not pulsed at a fast enough frequency, they can create a phantom array, where ghost images of the LED will appear if the eyes quickly scan across the array. White LED headlamps are starting to be used. Using LEDs has styling advantages because LEDs can form much thinner lights than incandescent lamps with parabolic reflectors.

Due to the relative cheapness of low output LEDs, they are also used in many temporary uses such as glowsticks, throwies, and the photonic textile Lumalive. Artists have also used LEDs for LED art.

Weather/all-hazards radio receivers with Specific Area Message Encoding (SAME) have three LEDs: red for warnings, orange for watches, and yellow for advisories & statements whenever issued.

Lighting

With the development of high-efficiency and high-power LEDs, it has become possible to use LEDs in lighting and illumination. Replacement light bulbs have been made, as well as dedicated fixtures and LED lamps. To encourage the shift to very high efficiency lighting, the US Department of Energy has created the L Prize competition. The Philips Lighting North America LED bulb won the first competition on August 3, 2011 after successfully completing 18 months of intensive field, lab, and product testing.

LEDs are used as street lights and in other architectural lighting where colour changing is used. The mechanical robustness and long lifetime is used in automotive lighting on cars, motorcycles, and bicycle lights.

LED street lights are employed on poles and in parking garages. In 2007, the Italian village Torraca was the first place to convert its entire illumination system to LEDs.

LEDs are used in aviation lighting. Airbus has used LED lighting in their Airbus A320 Enhanced since 2007, and Boeing plans its use in the 787. LEDs are also being used now in airport and heliport lighting. LED airport fixtures currently include medium-intensity runway lights, runway centreline lights, taxiway centreline and edge lights, guidance signs, and obstruction lighting.

LEDs are also suitable for backlighting for LCD televisions and lightweight laptop displays and light source for DLP projectors. RGB LEDs raise the colour gamut by as much as 45%. Screens for TV and computer displays can be made thinner using LEDs for backlighting.

LEDs are used increasingly in aquarium lights. In particular for reef aquariums, LED lights provide an efficient light source with less heat output to help maintain optimal aquarium temperatures. LED-based aquarium fixtures also have the advantage of being manually adjustable to emit a specific colour-spectrum for ideal colouration of corals, fish, and invertebrates while optimizing photosynthetically active radiation (PAR), which raises growth and sustainability of photosynthetic life such as corals, anemones, clams, and macroalgae. These fixtures can be electronically programmed to simulate various lighting conditions throughout the day, reflecting phases of the sun and moon for a dynamic reef experience.

LED fixtures typically cost up to five times as much as similarly rated fluorescent or high-intensity discharge lighting designed for reef aquariums and are not as high output to date.

The lack of IR or heat radiation makes LEDs ideal for stage lights using banks of RGB LEDs that can easily change colour and decrease heating from traditional stage lighting, as well as medical lighting where IR-radiation can be harmful.

In energy conservation, the lower heat output of LEDs also means air conditioning (cooling) systems have less heat to dispose of, reducing carbon dioxide emissions.

LEDs are small, durable and need little power, so they are used in hand held devices such as flashlights. LED strobe lights or camera flashes operate at a safe, low voltage, instead of the 250+ volts commonly found in xenon flashlamp-based lighting. This is especially useful in cameras on mobile phones, where space is at a premium and bulky voltage-raising circuitry is undesirable.

LEDs are used for infrared illumination in night vision uses including security cameras. A ring of LEDs around a video camera, aimed forward into a retroreflective background, allows chroma keying in video productions.

LEDs are used in mining operations, as cap lamps to provide light for miners. Research has been done to improve LEDs for mining, to reduce glare and to increase illumination, reducing risk of injury for the miners.

LEDs are now used commonly in all market areas from commercial to home use: standard lighting, AV, stage, theatrical, architectural, and public installations, and wherever artificial light is used.

LEDs are increasingly finding uses in medical and educational applications, for example as mood enhancement, and new technologies such as AmBX, exploiting LED versatility. NASA has even sponsored research for the use of LEDs to promote health for astronauts.

Smart Lighting

Light can be used to transmit broadband data, which is already implemented in IrDA standards using infrared LEDs. Because LEDs can cycle on and off millions of times per second, they can be wireless transmitters and access points for data transport. Lasers can also be modulated in this manner.

Sustainable Lighting

Efficient lighting is needed for sustainable architecture. In 2009, a typical 13-watt LED lamp emitted 450 to 650 lumens, which is equivalent to a standard 40-watt incandescent bulb. In 2011, LEDs have become more efficient, so that a 6-watt LED can easily achieve the same results. A standard 40-watt incandescent bulb has an expected lifespan of 1,000 hours, whereas an LED can continue to operate with reduced efficiency for more than 50,000 hours, 50 times longer than the incandescent bulb.

Energy Consumption

In the US, one kilowatt-hour of electricity will cause 1.34 pounds (610 g) of CO

2 emission. Assuming the average light bulb is on for 10 hours a day, one 40-watt incandescent bulb will cause 196 pounds (89 kg) of CO

2 emission per year. The 6-watt LED equivalent will only cause 30 pounds (14 kg) of CO

2 over the same time span. A building's carbon footprint from lighting can be reduced by 85% by exchanging all incandescent bulbs for new LEDs.

Economically Sustainable

LED light bulbs could be a cost-effective option for lighting a home or office space because of their very long lifetimes. Consumer use of LEDs as a replacement for conventional lighting system is currently hampered by the high cost and low efficiency of available products. 2009 DOE testing results showed an average efficacy of 35 lm/W, below that of typical CFLs, and as low as 9 lm/W, worse than standard incandescents. However, as of 2011, there are LED bulbs available as efficient as 150 lm/W and even inexpensive low-end models typically exceed 50 lm/W. The high initial cost of commercial LED bulbs is due to the expensive sapphire substrate, which is key to the production process. The sapphire apparatus must be coupled with a mirror-like collector to reflect light that would otherwise be wasted.

Low Maintenance Lighting

Most of LED lamps or luminaires can achieve 50,000 hours life time (running 10 hours/day, 7 days/week, they theoretically should be able to last for 14 years), but very rare manufacturers can promise a 5 year warranty of their products. The main reason that the product's life time has been shortened is the heat issue. Using high efficiency LED module (better thermal management) and well designed heat sink (efficient cooling material/structure/device) for the LED unit will significantly enhance the its performance (longer life and lower failure rate). Here is an example of how premium LED luminaires help to significantly save energy costs and maintenance costs at the same time. UK's biggest retrofit LED deal

Other Applications

The light from LEDs can be modulated very quickly so they are used extensively in optical fibre and free space optics communications. This includes remote controls, such as for TVs, VCRs, and LED Computers, where infrared LEDs are often used. Opto-isolators use an LED combined with a photodiode or phototransistor to provide a signal path with electrical isolation between two circuits. This is especially useful in medical equipment where the signals from a low-voltage sensor circuit (usually battery-powered) in contact with a living organism must be electrically isolated from any possible electrical failure in a recording or monitoring device operating at potentially dangerous voltages. An optoisolator also allows information to be transferred between circuits not sharing a common ground potential.

Many sensor systems rely on light as the signal source. LEDs are often ideal as a light source due to the requirements of the sensors. LEDs are used as movement sensors, for example in optical computer mice. The Nintendo Wii's sensor bar uses infrared LEDs. Pulse oximeters use them for measuring oxygen saturation. Some flatbed scanners use arrays of RGB LEDs rather than the typical cold-cathode fluorescent lamp as the light source. Having independent control of three illuminated colours allows the scanner to calibrate itself for more accurate colour balance, and there is no need for warm-up. Further, its sensors only need be monochromatic, since at any one time the page being scanned is only lit by one colour of light. Touch sensing: Since LEDs can also be used as photodiodes, they can be used for both photo emission and detection. This could be used, for example, in a touch-sensing screen that registers reflected light from a finger or stylus.

Many materials and biological systems are sensitive to or dependent on light. Grow lights use LEDs to increase photosynthesis in plants and bacteria and viruses can be removed from water and other substances using UV LEDs for sterilization. Other uses are as UV curing devices for some ink and coating methods, and in LED printers.

Plant growers are interested in LEDs because they are more energy-efficient, emit less heat (can damage plants close to hot lamps), and can provide the optimum light frequency for plant growth and bloom periods compared to currently used grow lights: HPS (high-pressure sodium), metal-halide or CFL/low-energy. However, LEDs have not replaced these grow lights due to higher price. As mass production and LED kits develop, the LED products will become cheaper.

LEDs have also been used as a medium-quality voltage reference in electronic circuits. The forward voltage drop (e.g., about 1.7 V for a normal red LED) can be used instead of a Zener diode in low-voltage regulators. Red LEDs have the flattest *I*/*V* curve above the knee. Nitride-based LEDs have a fairly steep *I*/*V* curve and are useless for this purpose. Although LED forward voltage is far more current-dependent than a good Zener, Zener diodes are not widely available below voltages of about 3 V.

LEDs are an essential part of the illumination system in digital microscopes, such as those which can be directly attached to the USB port of a computer. Such USB microscopes and endoscopes enable images to be recorded and stored directly on the computer without use of a separate camera.

Light Sources for Machine Vision Systems

Machine vision systems often require bright and homogeneous illumination, so features of interest are easier to process. LEDs are often used for this purpose, and this is likely to remain one of their major uses until price drops low enough to make signalling and illumination uses more widespread. Barcode scanners are the most common example of machine vision, and many low cost ones use red LEDs instead of lasers. Optical computer mice are also another example of LEDs in machine vision, as it is used to provide an even light source on the surface for the miniature camera within the mouse. LEDs constitute a nearly ideal light source for machine vision systems for several reasons:

The size of the illuminated field is usually comparatively small and machine vision systems are often quite expensive, so the cost of the light source is usually a minor concern. However, it might not be easy to replace a broken light source placed within complex machinery, and here the long service life of LEDs is a benefit.

LED elements tend to be small and can be placed with high density over flat or even-shaped substrates (PCBs etc.) so that bright and homogeneous sources that direct light from tightly controlled directions on inspected parts can be designed. This can often be obtained with small, low-cost lenses and diffusers, helping to achieve high light densities with control over lighting levels and homogeneity. LED sources can be shaped in several configurations (spot lights for reflective illumination; ring lights for coaxial illumination; back lights for contour illumination; linear assemblies; flat, large format panels; dome sources for diffused, omnidirectional illumination).

LEDs can be easily strobed (in the microsecond range and below) and synchronized with imaging. High-power LEDs are available allowing well-lit images even with very short light pulses. This is often used to obtain crisp and sharp "still" images of quickly moving parts.

LEDs come in several different colours and wavelengths, allowing easy use of the best colour for each need, where different colour may provide better visibility of features of interest. Having a precisely known spectrum allows tightly matched filters to be used to separate informative bandwidth or to reduce disturbing effects of ambient light. LEDs usually operate at comparatively low working temperatures, simplifying heat management and dissipation. This allows using plastic lenses, filters, and diffusers. Waterproof units can also easily be designed, allowing use in harsh or wet environments (food, beverage, oil industries).

Photo Diode

A photodiode is a type of photodetector capable of converting light into either current or voltage, depending upon the mode of operation. The common, traditional solar cell used to generate electric solar power is a large area photodiode.

Photodiodes are similar to regular semiconductor diodes except that they may be either exposed (to detect vacuum UV or X-rays) or packaged with a window or optical fibre connection to allow light to reach the sensitive part of the device. Many diodes designed for use specifically as a photodiode use a PIN junction rather than a p-n junction, to increase the speed of response. A photodiode is designed to operate in reverse bias.

Principle of Operation

A photodiode is a p-n junction or PIN structure. When a photon of sufficient energy strikes the diode, it excites an electron, thereby creating a *free electron* (and a positively charged electron *hole*). This mechanism is also known as the inner photoelectric effect.

If the absorption occurs in the junction's depletion region, or one diffusion length away from it, these carriers are swept from the junction by the built-in electric field of the depletion region.

Thus holes move toward the anode, and electrons toward the cathode, and a photocurrent is produced. The total current through the photodiode is the sum of the dark current (current that flows with or without light) and the photocurrent, so the dark current must be minimized to maximize the sensitivity of the device.

Photovoltaic Mode

When used in zero bias or *photovoltaic mode*, the flow of photocurrent out of the device is restricted and a voltage builds up. This mode exploits the photovoltaic effect, which is the basis for solar cells – a traditional solar cell is just a large area photodiode.

Photoconductive Mode

In this mode the diode is often reverse biased (with the cathode driven positive with respect to the anode). Compared to forward bias, this dramatically reduces the response time at the expense of increased noise, because it increases the width of the depletion layer, which decreases the junction's capacitance.

The reverse bias induces only a small amount of current (known as saturation or dark current) along its direction while the photocurrent remains virtually the same. For a given spectral distribution, the photocurrent is linearly proportional to the illuminance (and to the irradiance).

Although this mode is faster, the photoconductive mode tends to exhibit more electronic noise. The leakage current of a good PIN diode is so low (<1 nA) that the Johnson–Nyquist noise of the load resistance in a typical circuit often dominates.

Other modes of Operation

Avalanche photodiodes have a similar structure to regular photodiodes, but they are operated with much higher reverse bias. This allows each *photo-generated* carrier to be multiplied by avalanche breakdown, resulting in internal gain within the photodiode, which increases the effective *responsivity* of the device.

A phototransistor is in essence a bipolar transistor encased in a transparent case so that light can reach the *base-collector junction.* It was invented by Dr. John N. Shive (more famous for his wave machine) at Bell Labs in 1948, but it wasn't announced until 1950. The electrons that are generated by photons in the base-collector junction are injected into the base, and this photodiode current is amplified by the transistor's current gain β (or h_{fe}).

If the emitter is left unconnected, the phototransistor becomes a photodiode. While phototransistors have a higher responsivity for light they are not able to detect low levels of light any better than photodiodes. Phototransistors also have significantly longer response times.

Materials

The material used to make a photodiode is critical to defining its properties, because only photons with sufficient energy to excite electrons across the material's bandgap will produce significant photocurrents.

Materials commonly used to produce photodiodes include:

Material	*Electromagnetic spectrum wavelength range (nm)*
Silicon	190–1100
Germanium	400–1700
Indium gallium arsenide	800–2600
Lead(II) sulfide	<1000–3500

Because of their greater bandgap, silicon-based photodiodes generate less noise than germanium-based photodiodes.

Unwanted Photodiodes

Any p-n junction, if illuminated, is potentially a photodiode. Semiconductor devices such as transistors and ICs contain p-n junctions, and will not function correctly if they are illuminated by unwanted electromagnetic radiation (light) of wavelength suitable to produce a photocurrent; this is avoided by encapsulating devices in opaque housings. If these housings are not completely opaque to high-energy radiation (ultraviolet, X-rays, gamma rays), transistors and ICs can malfunction due to induced photo-currents. Plastic cases are more vulnerable than metal ones.

Features

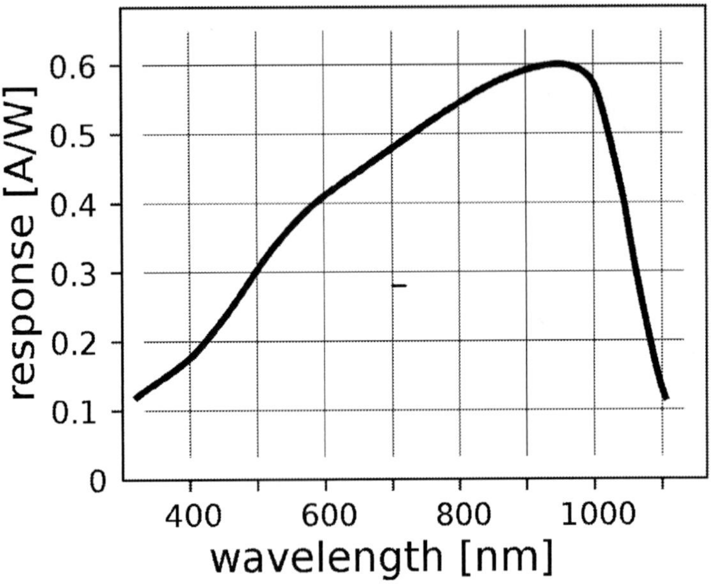

Figure: *Response of a silicon photo diode vs wavelength of the incident light*

Critical performance parameters of a photodiode include:

Responsivity: The Spectral responsivity is a ratio of the generated photocurrent to incident light power, expressed in A/W when used in photoconductive mode. The wavelength-dependence may also be expressed as a *Quantum efficiency*, or the ratio of the number of photogenerated carriers to incident photons, a unitless quantity.

Dark Current: The current through the photodiode in the absence of light, when it is operated in photoconductive mode. The dark current includes photocurrent generated by background radiation and the saturation current of the semiconductor junction. Dark current must be accounted for by calibration if a photodiode is used to make an accurate optical power measurement, and it is also a source of noise when a photodiode is used in an optical communication system.

Response Time: A photon absorbed by the semiconducting material will generate an electron-hole pair which will in turn start moving in the material under the effect of the electric field and thus generate a current. The finite duration of this current is known as the transit-time spread and can be evaluated by using Ramo's theorem. One can also show with this theorem that the total charge generated in the external circuit is well e and not 2e as might seem by the presence of the two carriers. Indeed the integral of the current due to both electron and hole over time must be equal to e. The resistance and capacitance of the photodiode and the external circuitry give rise to another response time known as RC time constant $\tau = RC$. This combination of R and C integrates the photoresponse over time and thus lengthens the impulse response of the photodiode. When used in an optical communication system, the response time determines the bandwidth available for signal modulation and thus data transmission.

Noise-equivalent Power: (NEP) The minimum input optical power to generate photocurrent, equal to the rms noise current in a 1 hertz bandwidth. NEP is essentially the minimum detectable power. The related characteristic detectivity (*D*) is the inverse of NEP, 1/NEP. There is also the specific detectivity ($D^{\star}$) which is the detectivity multiplied by the square root of the area (*A*) of the photodetector, ($D^{\star} = D\sqrt{A}$) for a 1 Hz bandwidth. The specific detectivity allows different systems to be compared independent of sensor area and system bandwidth; a higher detectivity value indicates a low-noise device or system. Although it is traditional to give ($D^{\star}$) in many catalogues as a measure of the diode's quality, in practice, it is hardly ever the key parameter.

When a photodiode is used in an optical communication system, all these parameters contribute to the *sensitivity* of the optical receiver, which is the minimum input power required for the receiver to achieve a specified *bit error rate.*

Applications

P-N photodiodes are used in similar applications to other photodetectors, such as photoconductors, charge-coupled devices, and photomultiplier tubes. They may be used to generate an output which is dependent upon the illumination (analog; for measurement and the like), or to change the state of circuitry (digital; either for control and switching, or digital signal processing).

Photodiodes are used in consumer electronics devices such as compact disc players, smoke detectors, and the receivers for infrared remote control devices used to control equipment from televisions to air conditioners. For many applications either photodiodes or photoconductors may be used. Either type of photosensor may be used for light measurement, as in camera light metres, or to respond to light levels, as in switching on street lighting after dark.

Photosensors of all types may be used to respond to incident light, or to a source of light which is part of the same circuit or system. A photodiode is often combined into a single component with an emitter of light, usually a light-emitting diode (LED), either to detect the presence of a mechanical obstruction to the beam (slotted optical switch), or to couple two digital or analog circuits while maintaining extremely high electrical isolation between them, often for safety (optocoupler).

Photodiodes are often used for accurate measurement of light intensity in science and industry. They generally have a more linear response than photoconductors.

They are also widely used in various medical applications, such as detectors for computed tomography (coupled with scintillators), instruments to analyze samples (immunoassay), and pulse oximeters.

PIN diodes are much faster and more sensitive than p-n junction diodes, and hence are often used for optical communications and in lighting regulation.

P-N photodiodes are not used to measure extremely low light intensities. Instead, if high sensitivity is needed, avalanche photodiodes, intensified charge-coupled devices or photomultiplier tubes are used for applications such as astronomy, spectroscopy, night vision equipment and laser rangefinding.

Comparison with Photomultipliers

Advantages compared to photomultipliers:

1. Excellent linearity of output current as a function of incident light
2. Spectral response from 190 nm to 1100 nm (silicon), longer wavelengths with other semiconductor materials
3. Low noise
4. Ruggedized to mechanical stress
5. Low cost
6. Compact and light weight
7. Long lifetime

8. High quantum efficiency, typically 60-80%
9. No high voltage required

Disadvantages compared to photomultipliers:

1. Small area
2. No internal gain (except avalanche photodiodes, but their gain is typically 10^2–10^3 compared to up to 10^8 for the photomultiplier)
3. Much lower overall sensitivity
4. Photon counting only possible with specially designed, usually cooled photodiodes, with special electronic circuits
5. Response time for many designs is slower

Photodiode Array

Figure: *A 2 x 2 cm photodiode array chip with more than 200 diodes*

A one-dimensional array of hundreds or thousands of photodiodes can be used as a position sensor, for example as part of an angle sensor. One advantage of photodiode arrays (PDAs) is that they allow for high speed parallel read out since the driving electronics may not be built in like a traditional CMOS or CCD sensor.

Silicon Controlled Rectifier

A silicon-controlled rectifier (or semiconductor-controlled rectifier) is a four-layer solid state current controlling device. The name "silicon controlled rectifier" or SCR is General Electric's trade name for a type of thyristor. The SCR was developed by a team of power engineers led by Robert N. Hall and commercialized by Frank W. "Bill" Gutzwiller in 1957.

Some sources define silicon controlled rectifiers and thyristors as synonymous, other sources define silicon controlled rectifiers as a subset of a larger family of devices with at least four layers of alternating N and P-type material, this entire family being referred to as thyristors.

SCRs are unidirectional devices (i.e. can conduct current only in one direction) as opposed to TRIACs which are bidirectional (i.e. current can flow through them in either direction). SCRs can be triggered normally only by currents going into the gate as opposed to TRIACs which can be triggered normally by either a positive or a negative current applied to its gate electrode.

Modes of Operation

This device is generally used in switching applications. In the normal "off" state, the device restricts current to the leakage current. When the gate-to-cathode voltage exceeds a certain threshold, the device turns "on" and conducts current. The device will remain in the "on" state even after gate current is removed as long as current through the device remains above the holding current. Once current falls below the holding current for an appropriate period of time, the device will switch "off". If the gate is pulsed and the current through the device is below the latching current, the device will remain in the "off" state. If the applied voltage increases rapidly enough, capacitive coupling may induce enough charge into the gate to trigger the device into the "on" state; this is referred to as "dv/dt triggering." This is usually prevented by limiting the rate of voltage rise across the device, perhaps by using a snubber. "dv/dt triggering" may not switch the SCR into full conduction rapidly, and the partially triggered SCR may dissipate more power than is usual, possibly harming the device.

SCRs can also be triggered by increasing the forward voltage beyond their rated breakdown voltage (also called as break over voltage), but again, this does not rapidly switch the entire device into conduction and so may be harmful; therefore this mode of operation is also usually avoided.

Reverse Bias

SCR are available with or without reverse blocking capability. Reverse blocking capability adds to the forward voltage drop because of the need to have a long, low doped P1 region. Usually, the reverse blocking voltage rating and forward blocking voltage rating are the same. The typical application for reverse blocking SCR is in current source inverters.

SCR incapable of blocking reverse voltage are known as asymmetrical SCR, abbreviated ASCR. They typically have a reverse breakdown rating in the 10's of volts. ASCR are used where either a reverse conducting diode is applied in parallel (for example, in voltage source inverters) or where reverse voltage would never occur (for example, in switching power supplies or DC traction choppers).

Asymmetrical SCR can be fabricated with a reverse conducting diode in the same package. These are known as RCT, for reverse conducting thyristor.

Thyristor Turn on Methods

1. forward voltage triggering
2. gate triggering
3. dv/dt triggering
4. temperature triggering
5. light triggering

Forward voltage triggering occurs when the anode-cathode forward voltage is increased with the gate circuit opened. This is known as avalanche breakdown, during which junction j2 will breakdown. At sufficient voltages, the thyristor changes to its on state with low voltage drop and large forward current. In this case, J1 and J3 are already forward biased.

Application of SCRs

SCRs are mainly used in devices where the control of high power, possibly coupled with high voltage, is demanded. Their operation makes them suitable for use in medium to high-voltage AC power control applications, such as lamp dimming, regulators and motor control.

SCRs and similar devices are used for rectification of high power AC in high-voltage direct current power transmission. They are also used in the control of welding machines, mainly MTAW and GTAW processes.

Chapter

2

Transistor Characteristics

Introduction

A transistor is a semiconductor device used to amplify and switch electronic signals and electrical power. It is composed of semiconductor material with at least three terminals for connection to an external circuit. A voltage or current applied to one pair of the transistor's terminals changes the current through another pair of terminals. Because the controlled (output) power can be higher than the controlling (input) power, a transistor can amplify a signal. Today, some transistors are packaged individually, but many more are found embedded in integrated circuits.

The transistor is the fundamental building block of modern electronic devices, and is ubiquitous in modern electronic systems. Following its development in the early 1950s, the transistor revolutionized the field of electronics, and paved the way for smaller and cheaper radios, calculators, and computers, among other things.

History

The thermionic triode, a vacuum tube invented in 1907, propelled the electronics age forward, enabling amplified radio technology and long-distance telephony. The triode, however, was a fragile device that consumed a lot of power. Physicist Julius Edgar Lilienfeld filed a patent for a field-effect transistor (FET) in Canada in 1925, which was intended to be a solid-state replacement for the triode. Lilienfeld also filed identical patents in the United States in 1926 and 1928. However, Lilienfeld did not publish any research articles about his devices nor did his patents cite any specific examples of a working prototype. Because the production of high-quality semiconductor materials was still decades away, Lilienfeld's solid-state amplifier ideas would not have found practical use in the 1920s and 1930s, even if such a device had been built. In 1934, German inventor Oskar Heil patented a similar device.

From November 17, 1947 to December 23, 1947, John Bardeen and Walter Brattain at AT&T's Bell Labs in the United States, performed experiments and observed that when two gold point contacts were applied to a crystal of germanium, a signal was produced with the output power greater than the input. Solid State Physics Group leader William Shockley saw the potential in this, and over the next few months worked to greatly expand the knowledge of semiconductors. The term *transistor* was coined by John R. Pierce as a portmanteau of the term "transfer resistor". According to Lillian Hoddeson and Vicki Daitch, authors of a biography of John Bardeen, Shockley had proposed that Bell Labs' first patent for a transistor should be based on the field-effect and that he be named as the inventor. Having unearthed Lilienfeld's patents that went into obscurity years earlier, lawyers at Bell Labs advised against Shockley's proposal because the idea of a field-effect transistor that used an electric field as a "grid" was not new. Instead, what Bardeen, Brattain, and Shockley invented in 1947 was the first point-contact transistor. In acknowledgement of this accomplishment, Shockley, Bardeen, and Brattain were jointly awarded the 1956 Nobel Prize in Physics "for their researches on semiconductors and their discovery of the transistor effect."

In 1948, the point-contact transistor was independently invented by German physicists Herbert Mataré and Heinrich Welker while working at the Compagnie des Freins et Signaux, a Westinghouse subsidiary located in Paris. Mataré had previous experience in developing crystal rectifiers from silicon and germanium in the German radar effort during World War II. Using this knowledge, he began researching the phenomenon of "interference" in 1947. By witnessing currents flowing through point-contacts, similar to what Bardeen and Brattain had accomplished earlier in December 1947, Mataré by June 1948, was able to produce consistent results by using samples of germanium produced by Welker. Realizing that Bell Labs' scientists had already invented the transistor before them, the company rushed to get its "transistron" into production for amplified use in France's telephone network.

The first high-frequency transistor was the surface-barrier germanium transistor developed by Philco in 1953, capable of operating up to 60 MHz. These were made by etching depressions into an N-type germanium base from both sides with jets of indium sulphate until it was a few ten-thousandths of an inch thick. Indium electroplated into the depressions formed the collector and emitter. The first all-transistor car radio, which was produced in 1955 by Chrysler and Philco, used these transistors in its circuitry and also they were the first suitable for high-speed computers.

The first working silicon transistor was developed at Bell Labs on January 26, 1954 by Morris Tanenbaum. The first commercial silicon transistor was produced by Texas Instruments in 1954. This was the work of Gordon Teal, an expert in growing crystals of high purity, who had previously worked at Bell Labs. The first MOS transistor actually built was by Kahng and Atalla at Bell Labs in 1960.

In March 2013 American researchers at Stanford University announced they had built a transistor out of DNA and RNA molecules.

Importance

The transistor is the key active component in practically all modern electronics. Many consider it to be one of the greatest inventions of the 20th century. Its importance in today's society rests on its ability to be mass-produced using a highly automated process (semiconductor device fabrication) that achieves astonishingly low per-transistor costs. The invention of the first transistor at Bell Labs was named an IEEE Milestone in 2009.

Although several companies each produce over a billion individually packaged (known as *discrete*) transistors every year, the vast majority of transistors are now produced in integrated circuits (often shortened to *IC*, *microchips* or simply *chips*), along with diodes, resistors, capacitors and other electronic components, to produce complete electronic circuits. A logic gate consists of up to about twenty transistors whereas an advanced microprocessor, as of 2009, can use as many as 3 billion transistors (MOSFETs). "About 60 million transistors were built in 2002 ... for [each] man, woman, and child on Earth."

The transistor's low cost, flexibility, and reliability have made it a ubiquitous device. Transistorized mechatronic circuits have replaced electromechanical devices in controlling appliances and machinery. It is often easier and cheaper to use a standard microcontroller and write a computer programme to carry out a control function than to design an equivalent mechanical control function.

Simplified Operation

The essential usefulness of a transistor comes from its ability to use a small signal applied between one pair of its terminals to control a much larger signal at another pair of terminals. This property is called gain. A transistor can control its output in proportion to the input signal; that is, it can act as an amplifier. Alternatively, the transistor can be used to turn current on or off in a circuit as an electrically controlled switch, where the amount of current is determined by other circuit elements.

There are two types of transistors, which have slight differences in how they are used in a circuit. A *bipolar transistor* has terminals labelled base, collector, and emitter. A small current at the base terminal (that is, flowing between the base and the emitter) can control or switch a much larger current between the collector and emitter terminals. For a *field-effect transistor*, the terminals are labelled gate, source, and drain, and a voltage at the gate can control a current between source and drain.

The image to the right represents a typical bipolar transistor in a circuit. Charge will flow between emitter and collector terminals depending on the current in the base. Because internally the base and emitter connections behave like a semiconductor diode, a voltage drop develops between base and emitter while the base current exists. The amount of this voltage depends on the material the transistor is made from, and is referred to as V_{BE}.

Transistor as a Switch

Transistors are commonly used as electronic switches, both for high-power applications such as switched-mode power supplies and for low-power applications such as logic gates.

In a grounded-emitter transistor circuit, such as the light-switch circuit shown, as the base voltage rises, the emitter and collector currents rise exponentially. The collector voltage drops because of reduced resistance from collector to emitter. If the voltage difference between the collector and emitter were zero (or near zero), the collector current would be limited only by the load resistance (light bulb) and the supply voltage. This is called *saturation* because current is flowing from collector to emitter freely. When saturated the switch is said to be *on*.

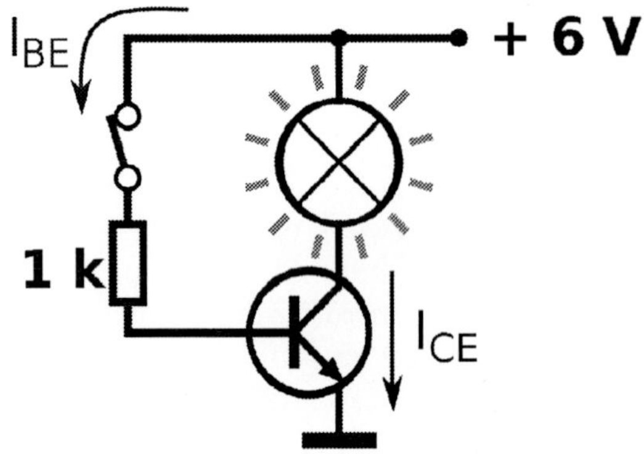

Figure: *BJT used as an electronic switch, in grounded-emitter configuration.*

Providing sufficient base drive current is a key problem in the use of bipolar transistors as switches. The transistor provides current gain, allowing a relatively large current in the collector to be switched by a much smaller current into the base terminal. The ratio of these currents varies depending on the type of transistor, and even for a particular type, varies depending on the collector current. In the example light-switch circuit shown, the resistor is chosen to provide enough base current to ensure the transistor will be saturated.

In any switching circuit, values of input voltage would be chosen such that the output is either completely off, or completely on. The transistor is acting as a switch, and this type of operation is common in digital circuits where only "on" and "off" values are relevant.

Transistor as an Amplifier

The common-emitter amplifier is designed so that a small change in voltage (V_{in}) changes the small current through the base of the transistor; the transistor's current amplification combined with the properties of the circuit mean that small swings in V_{in} produce large changes in V_{out}.

Various configurations of single transistor amplifier are possible, with some providing current gain, some voltage gain, and some both.

From mobile phones to televisions, vast numbers of products include amplifiers for sound reproduction, radio transmission, and signal processing. The first discrete-transistor audio amplifiers barely supplied a few hundred milliwatts, but power and audio fidelity gradually increased as better transistors became available and amplifier architecture evolved.

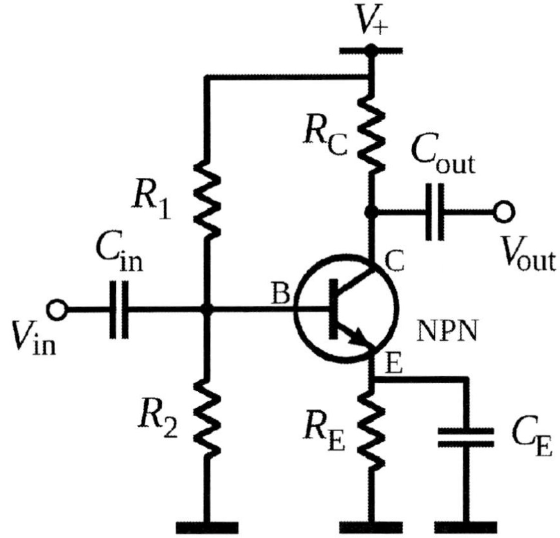

Figure: *Amplifier circuit, common-emitter configuration with a voltage-divider bias circuit.*

Modern transistor audio amplifiers of up to a few hundred watts are common and relatively inexpensive.

Comparison with Vacuum Tubes

Prior to the development of transistors, vacuum (electron) tubes (or in the UK "thermionic valves" or just "valves") were the main active components in electronic equipment.

Advantages

The key advantages that have allowed transistors to replace their vacuum tube predecessors in most applications are

- No power consumption by a cathode heater.
- Small size and minimal weight, allowing the development of miniaturized electronic devices.
- Low operating voltages compatible with batteries of only a few cells.
- No warm-up period for cathode heaters required after power application.
- Lower power dissipation and generally greater energy efficiency.
- Higher reliability and greater physical ruggedness.

- Extremely long life. Some transistorized devices have been in service for more than 50 years.
- Complementary devices available, facilitating the design of complementary-symmetry circuits, something not possible with vacuum tubes.
- Insensitivity to mechanical shock and vibration, thus avoiding the problem of microphonics in audio applications.

Limitations

- Silicon transistors can age and fail.
- High-power, high-frequency operation, such as that used in over-the-air television broadcasting, is better achieved in vacuum tubes due to improved electron mobility in a vacuum.
- Solid-state devices are more vulnerable to Electrostatic discharge in handling and operation
- A vacuum tube momentarily overloaded will just get a little hotter; solid-state devices have less mass to absorb the heat due to overloads, in proportion to their rating
- Sensitivity to radiation and cosmic rays (special radiation hardened chips are used for spacecraft devices).
- Vacuum tubes create a distortion, the so-called tube sound, that some people find to be more tolerable to the ear.

Types

Transistors are categorized by

- Semiconductor material (date first used): the metalloids germanium (1947) and silicon (1954)— in amorphous, polycrystalline and monocrystalline form; the compounds gallium arsenide (1966) and silicon carbide (1997), the alloy silicon-germanium (1989), the allotrope of carbon graphene (research ongoing since 2004), etc.
- Structure: BJT, JFET, IGFET (MOSFET), IGBT, "other types"
- Electrical polarity (positive and negative): n–p–n, p–n–p (BJTs); n-channel, p-channel (FETs)
- Maximum power rating: low, medium, high
- Maximum operating frequency: low, medium, high, radio (RF), microwave frequency (the maximum effective frequency of a transistor is denoted by the term f_T, an abbreviation for transition frequency—the frequency of transition is the frequency at which the transistor yields unity gain)

- Application: switch, general purpose, audio, high voltage, super-beta, matched pair
- Physical packaging: through-hole metal, through-hole plastic, surface mount, ball grid array, power modules
- Amplification factor h_{fe} or β_F (transistor beta)

Thus, a particular transistor may be described as *silicon, surface-mount, BJT, n–p–n, low-power, high-frequency switch.*

Bipolar Junction Transistor (BJT)

Bipolar transistors are so named because they conduct by using both majority and minority carriers. The bipolar junction transistor, the first type of transistor to be mass-produced, is a combination of two junction diodes, and is formed of either a thin layer of p-type semiconductor sandwiched between two n-type semiconductors (an n–p–n transistor), or a thin layer of n-type semiconductor sandwiched between two p-type semiconductors (a p–n–p transistor). This construction produces two p–n junctions: a base–emitter junction and a base–collector junction, separated by a thin region of semiconductor known as the base region (two junction diodes wired together without sharing an intervening semiconducting region will not make a transistor).

BJTs have three terminals, corresponding to the three layers of semiconductor—an *emitter*, a *base*, and a *collector*. They are useful in amplifiers because the currents at the emitter and collector are controllable by a relatively small base current." In an n–p–n transistor operating in the active region, the emitter–base junction is forward biased (electrons and holes recombine at the junction), and electrons are injected into the base region. Because the base is narrow, most of these electrons will diffuse into the reverse-biased (electrons and holes are formed at, and move away from the junction) base–collector junction and be swept into the collector; perhaps one-hundredth of the electrons will recombine in the base, which is the dominant mechanism in the base current. By controlling the number of electrons that can leave the base, the number of electrons entering the collector can be controlled. Collector current is approximately β (common-emitter current gain) times the base current. It is typically greater than 100 for small-signal transistors but can be smaller in transistors designed for high-power applications.

Unlike the field-effect transistor, the BJT is a low–input-impedance device. Also, as the base–emitter voltage (V_{be}) is increased the base–

emitter current and hence the collector–emitter current (I_{ce}) increase exponentially according to the Shockley diode model and the Ebers-Moll model. Because of this exponential relationship, the BJT has a higher transconductance than the FET.

Bipolar transistors can be made to conduct by exposure to light, because absorption of photons in the base region generates a photocurrent that acts as a base current; the collector current is approximately β times the photocurrent. Devices designed for this purpose have a transparent window in the package and are called phototransistors.

Field-effect Transistor (FET)

The *field-effect transistor*, sometimes called a *unipolar transistor*, uses either electrons (in *n-channel FET*) or holes (in *p-channel FET*) for conduction. The four terminals of the FET are named *source*, *gate*, *drain*, and *body* (*substrate*). On most FETs, the body is connected to the source inside the package, and this will be assumed for the following description.

In a FET, the drain-to-source current flows via a conducting channel that connects the *source* region to the *drain* region. The conductivity is varied by the electric field that is produced when a voltage is applied between the gate and source terminals; hence the current flowing between the drain and source is controlled by the voltage applied between the gate and source. As the gate–source voltage (V_{gs}) is increased, the drain–source current (I_{ds}) increases exponentially for V_{gs} below threshold, and then at a roughly quadratic rate ($I_{ds} \propto (V_{gs} - V_T)^2$) (where V_T is the threshold voltage at which drain current begins) in the "space-charge-limited" region above threshold. A quadratic behaviour is not observed in modern devices, for example, at the 65 nm technology node.

For low noise at narrow bandwidth the higher input resistance of the FET is advantageous.

FETs are divided into two families: *junction FET* (JFET) and *insulated gate FET* (IGFET). The IGFET is more commonly known as a *metal–oxide–semiconductor FET* (MOSFET), reflecting its original construction from layers of metal (the gate), oxide (the insulation), and semiconductor. Unlike IGFETs, the JFET gate forms a p–n diode with the channel which lies between the source and drain. Functionally, this makes the n-channel JFET the solid-state equivalent of the vacuum tube triode which, similarly, forms a diode between its grid and cathode. Also, both devices operate in the *depletion mode*, they both have a high input impedance, and they both conduct current under the control of an input voltage.

Metal–semiconductor FETs (MESFETs) are JFETs in which the reverse biased p–n junction is replaced by a metal–semiconductor junction. These, and the HEMTs (high-electron-mobility transistors, or HFETs), in which a two-dimensional electron gas with very high carrier mobility is used for charge transport, are especially suitable for use at very high frequencies (microwave frequencies; several GHz).

Unlike bipolar transistors, FETs do not inherently amplify a photocurrent. Nevertheless, there are ways to use them, especially JFETs, as light-sensitive devices, by exploiting the photocurrents in channel–gate or channel–body junctions.

FETs are further divided into *depletion-mode* and *enhancement-mode* types, depending on whether the channel is turned on or off with zero gate-to-source voltage. For enhancement mode, the channel is off at zero bias, and a gate potential can "enhance" the conduction. For the depletion mode, the channel is on at zero bias, and a gate potential (of the opposite polarity) can "deplete" the channel, reducing conduction. For either mode, a more positive gate voltage corresponds to a higher current for n-channel devices and a lower current for p-channel devices. Nearly all JFETs are depletion-mode because the diode junctions would forward bias and conduct if they were enhancement-mode devices; most IGFETs are enhancement-mode types.

Usage of Bipolar and Field-effect Transistors

The bipolar junction transistor (BJT) was the most commonly used transistor in the 1960s and 70s. Even after MOSFETs became widely available, the BJT remained the transistor of choice for many analog circuits such as amplifiers because of their greater linearity and ease of manufacture. In integrated circuits, the desirable properties of MOSFETs allowed them to capture nearly all market share for digital circuits. Discrete MOSFETs can be applied in transistor applications, including analog circuits, voltage regulators, amplifiers, power transmitters and motor drivers.

Other Transistor Types

- Bipolar junction transistor
  - Heterojunction bipolar transistor, up to several hundred GHz, common in modern ultrafast and RF circuits
  - Schottky transistor

- Avalanche transistor
- Darlington transistors are two BJTs connected together to provide a high current gain equal to the product of the current gains of the two transistors.
- Insulated-gate bipolar transistors (IGBTs) use a medium-power IGFET, similarly connected to a power BJT, to give a high input impedance. Power diodes are often connected between certain terminals depending on specific use. IGBTs are particularly suitable for heavy-duty industrial applications.

  The Asea Brown Boveri (ABB) *5SNA2400E170100* illustrates just how far power semiconductor technology has advanced. Intended for three-phase power supplies, this device houses three n–p–n IGBTs in a case measuring 38 by 140 by 190 mm and weighing 1.5 kg. Each IGBT is rated at 1,700 volts and can handle 2,400 amperes.
- Photo transistor
- Multiple-emitter transistor, used in transistor–transistor logic
- Multiple-base transistor, used to amplify very-low-level signals in noisy environments such as the pickup of a record player or radio front ends. Effectively, it is a very large number of transistors in parallel where, at the output, the signal is added constructively, but random noise is added only stochastically.

• Field-effect transistor
  - Carbon nanotube field-effect transistor (CNFET)
  - JFET, where the gate is insulated by a reverse-biased p–n junction
  - MESFET, similar to JFET with a Schottky junction instead of a p–n junction
    - High-electron-mobility transistor (HEMT, HFET, MODFET)
  - MOSFET, where the gate is insulated by a shallow layer of insulator
  - Inverted-T field-effect transistor (ITFET)
  - FinFET, source/drain region shapes fins on the silicon surface.

- o FREDFET, fast-reverse epitaxial diode field-effect transistor
- o Thin-film transistor, in LCDs.
- o Organic field-effect transistor (OFET), in which the semiconductor is an organic compound
- o Ballistic transistor
- o Floating-gate transistor, for non-volatile storage.
- o FETs used to sense environment
  - Ion-sensitive field effect transistor (IFSET), to measure ion concentrations in solution.
  - EOSFET, electrolyte-oxide-semiconductor field-effect transistor (Neurochip)
  - DNAFET, deoxyribonucleic acid field-effect transistor

- Diffusion transistor, formed by diffusing dopants into semiconductor substrate; can be both BJT and FET
- Unijunction transistors can be used as simple pulse generators. They comprise a main body of either P-type or N-type semiconductor with ohmic contacts at each end (terminals *Base1* and *Base2*). A junction with the opposite semiconductor type is formed at a point along the length of the body for the third terminal (*Emitter*).
- Single-electron transistors (SET) consist of a gate island between two tunnelling junctions. The tunnelling current is controlled by a voltage applied to the gate through a capacitor.
- Nanofluidic transistor, controls the movement of ions through sub-microscopic, water-filled channels.
- Multigate devices
  - o Tetrode transistor
  - o Pentode transistor
  - o Trigate transistors (Prototype by Intel)
  - o Dual-gate FETs have a single channel with two gates in cascode; a configuration optimized for *high-frequency amplifiers*, *mixers*, and oscillators.
- Junctionless nanowire transistor (JNT), developed at Tyndall National Institute in Ireland, was the first transistor successfully fabricated without junctions. (Even MOSFETs have junctions, although its gate is electrically insulated from the region the gate controls.) Junctions are difficult and

expensive to fabricate, and, because they are a significant source of current leakage, they waste significant power and generate significant waste heat. Eliminating them held the promise of cheaper and denser microchips. The JNT uses a simple nanowire of silicon surrounded by an electrically isolated "wedding ring" that acts to gate the flow of electrons through the wire. This method has been described as akin to squeezing a garden hose to gate the flow of water through the hose. The nanowire is heavily n-doped, making it an excellent conductor. Crucially the gate, comprising silicon, is heavily p-doped; and its presence depletes the underlying silicon nanowire thereby preventing carrier flow past the gate.

- Vacuum-channel transistor: In 2012, NASA and the National Nanofab Centre in South Korea were reported to have built a prototype vacuum-channel transistor in only 150 nanometers in size, can be manufactured cheaply using standard silicon semiconductor processing, can operate at high speeds even in hostile environments, and could consume just as much power as a standard transistor.

Part Numbering Standards / Specifications

The types of some transistors can be parsed from the part number. There are three major semiconductor naming standards; in each the alphanumeric prefix provides clues to type of the device.

Japanese Industrial Standard (JIS)

***Table:** JIS Transistor Prefix Table*

Prefix	*Type of transistor*
2SA	high-frequency p–n–p BJTs
2SB	audio-frequency p–n–p BJTs
2SC	high-frequency n–p–n BJTs
2SD	audio-frequency n–p–n BJTs
2SJ	P-channel FETs (both JFETs and MOSFETs)
2SK	N-channel FETs (both JFETs and MOSFETs)

The *JIS-C-7012* specification for transistor part numbers starts with "2S", e.g. 2SD965, but sometimes the "2S" prefix is not marked on the package – a 2SD965 might only be marked "D965"; a 2SC1815 might be listed by a supplier as simply "C1815". This series sometimes has suffixes (such as "R", "O", "BL"... standing for "Red", "Orange", "Blue" etc.) to denote variants, such as tighter h_{FE} (gain) groupings.

European Electronic Component Manufacturers Association (EECA)

The Pro Electron standard, the European Electronic Component Manufacturers Association part numbering scheme, begins with two letters: the first gives the semiconductor type (A for germanium, B for silicon, and C for materials like GaAs); the second letter denotes the intended use (A for diode, C for general-purpose transistor, etc.). A 3-digit sequence number (or one letter then 2 digits, for industrial types) follows. With early devices this indicated the case type. Suffixes may be used, with a letter (e.g. "C" often means high h_{FE}, such as in: BC549C) or other codes may follow to show gain (e.g. BC327-25) or voltage rating (e.g. BUK854-800A). The more common prefixes are:

Table: *Pro Electron / EECA Transistor Prefix Table*

Prefix class	*Type and usage*	*Example*	*Equivalent*	*Reference*
AC	Germanium small-signal AF transistor	AC126	NTE102A	Datasheet
AD	Germanium AF power transistor	AD133	NTE179	Datasheet
AF	Germanium small-signal RF transistor	AF117	NTE160	Datasheet
AL	Germanium RF power transistor	ALZ10	NTE100	Datasheet
AS	Germanium switching transistor	ASY28	NTE101	Datasheet
AU	Germanium power switching transistor	AU103	NTE127	Datasheet
BC	Silicon, small-signal transistor ("general purpose")	BC548	2N3904	Datasheet
BD	Silicon, power transistor	BD139	NTE375	Datasheet
BF	Silicon, RF (high frequency) BJT or FET	BF245	NTE133	Datasheet
BS	Silicon, switching transistor (BJT or MOSFET)	BS170	2N7000	Datasheet
BL	Silicon, high frequency, high power (for transmitters)	BLW60	NTE325	Datasheet
BU	Silicon, high voltage (for CRT horizontal deflection circuits)	BU2520A	NTE2354	Datasheet
CF	Gallium Arsenide small-signal Microwave transistor (MESFET)	CF739	—	Datasheet
CL	Gallium Arsenide Microwave power transistor (FET)	CLY10	—	Datasheet

Joint Electron Devices Engineering Council (JEDEC)

The JEDEC *EIA370* transistor device numbers usually start with "2N", indicating a three-terminal device (dual-gate field-effect transistors are four-terminal devices, so begin with 3N), then a 2, 3 or 4-digit sequential number with no significance as to device properties (although early devices with low numbers tend to be germanium). For example 2N3055 is a silicon n–p–n power transistor, 2N1301 is a p–n–p germanium switching transistor. A letter suffix (such as "A") is sometimes used to indicate a newer variant, but rarely gain groupings.

Proprietary

Manufacturers of devices may have their own proprietary numbering system, for example CK722. Note that a manufacturer's prefix (like "MPF" in MPF102, which originally would denote a Motorola FET) now is an unreliable indicator of who made the device. Some proprietary naming schemes adopt parts of other naming schemes, for example a PN2222A is a (possibly Fairchild Semiconductor) 2N2222A in a plastic case (but a PN108 is a plastic version of a BC108, not a 2N108, while the PN100 is unrelated to other xx100 devices).

Military part numbers sometimes are assigned their own codes, such as the British Military CV Naming System. Manufacturers buying large numbers of similar parts may have them supplied with "house numbers", identifying a particular purchasing specification and not necessarily a device with a standardized registered number. For example, an HP part 1854,0053 is a (JEDEC) 2N2218 transistor which is also assigned the CV number: CV7763

Naming Problems

With so many independent naming schemes, and the abbreviation of part numbers when printed on the devices, ambiguity sometimes occurs. For example two different devices may be marked "J176" (one the J176 low-power Junction FET, the other the higher-powered MOSFET 2SJ176).

As older "through-hole" transistors are given surface-mount packaged counterparts, they tend to be assigned many different part numbers because manufacturers have their own systems to cope with the variety in pinout arrangements and options for dual or matched n–p–n+p–n–p devices in one pack. So even when the original device (such as a 2N3904) may have been assigned by a standards authority, and well known by engineers over the years, the new versions are far from standardized in their naming.

Construction

Semiconductor material:

Semiconductor material characteristics

Semiconductor material	*Junction forwar dvoltage V @ 25 °C*	*Electron mobilitym²/(V·s) @ 25 °C*	*Hole mobilitym²/(V·s) @ 25 °C*	*Max.junction temp.°C*
Ge	0.27	0.39	0.19	70 to 100
Si	0.71	0.14	0.05	150 to 200
GaAs	1.03	0.85	0.05	150 to 200
Al-Si junction	0.3	—	—	150 to 200

The first BJTs were made from germanium (Ge). Silicon (Si) types currently predominate but certain advanced microwave and high-performance versions now employ the *compound semiconductor* material gallium arsenide (GaAs) and the *semiconductor alloy* silicon germanium (SiGe). Single element semiconductor material (Ge and Si) is described as *elemental*.

Rough parameters for the most common semiconductor materials used to make transistors; these parameters will vary with increase in temperature, electric field, impurity level, strain, and sundry other factors.

The *junction forward voltage* is the voltage applied to the emitter–base junction of a BJT in order to make the base conduct a specified current. The current increases exponentially as the junction forward voltage is increased. The values given in the table are typical for a current of 1 mA (the same values apply to semiconductor diodes). The lower the junction forward voltage the better, as this means that less power is required to "drive" the transistor. The junction forward voltage for a given current decreases with increase in temperature. For a typical silicon junction the change is "2.1 mV/°C. In some circuits special compensating elements (sensistors) must be used to compensate for such changes.

The density of mobile carriers in the channel of a MOSFET is a function of the electric field forming the channel and of various other phenomena such as the impurity level in the channel. Some impurities, called dopants, are introduced deliberately in making a MOSFET, to control the MOSFET electrical behaviour.

The *electron mobility* and *hole mobility* columns show the average speed that electrons and holes diffuse through the semiconductor material with an electric field of 1 volt per metre applied across the material. In general, the higher the electron mobility the faster the transistor can operate. The table indicates that Ge is a better material than Si in this respect. However, Ge has four major shortcomings compared to silicon and gallium arsenide:

- Its maximum temperature is limited;
- it has relatively high leakage current;
- it cannot withstand high voltages;
- it is less suitable for fabricating integrated circuits.

Because the electron mobility is higher than the hole mobility for all semiconductor materials, a given bipolar n–p–n transistor tends to be swifter than an equivalent p–n–p transistor. GaAs has the highest electron mobility of the three semiconductors. It is for this reason that GaAs is used in high-frequency applications. A relatively recent FET development, the *high-electron-mobility transistor* (HEMT), has a heterostructure (junction between different semiconductor materials) of aluminium gallium arsenide (AlGaAs)-gallium arsenide (GaAs) which has twice the electron mobility of a GaAs-metal barrier junction. Because of their high speed and low noise, HEMTs are used in satellite receivers working at frequencies around 12 GHz.

Max. junction temperature values represent a cross section taken from various manufacturers' data sheets. This temperature should not be exceeded or the transistor may be damaged.

Al–Si junction refers to the high-speed (aluminum–silicon) metal–semiconductor barrier diode, commonly known as a Schottky diode. This is included in the table because some silicon power IGFETs have a *parasitic* reverse Schottky diode formed between the source and drain as part of the fabrication process. This diode can be a nuisance, but sometimes it is used in the circuit.

Packaging

Discrete transistors are individually packaged transistors. Transistors come in many different semiconductor packages. The two main categories are *through-hole* (or *leaded*), and *surface-mount*, also known as *surface-mount device* (SMD). The *ball grid array* (BGA) is the latest surface-mount package (currently only for large integrated circuits). It has solder "balls" on the underside in place of leads. Because they are smaller and have shorter interconnections, SMDs have better high-frequency characteristics but lower power rating. Transistor packages are made of glass, metal, ceramic, or plastic. The package often dictates the power rating and frequency characteristics. Power transistors have larger packages that can be clamped to heat sinks for enhanced cooling. Additionally, most power transistors have the collector or drain physically connected to the metal enclosure. At the other extreme, some surface-mount *microwave* transistors are as small as grains of sand.

Often a given transistor type is available in several packages. Transistor packages are mainly standardized, but the assignment of a transistor's functions to the terminals is not: other transistor types can assign other functions to the package's terminals. Even for the same transistor type the terminal assignment can vary (normally indicated by a suffix letter to the part number, q.e. BC212L and BC212K).

Bipolar Junction Transistor

A bipolar junction transistor (BJT or bipolar transistor) is a type of transistor that relies on the contact of two types of semiconductor for its operation. BJTs can be used as amplifiers, switches, or in oscillators. BJTs can be found either as individual discrete components, or in large numbers as parts of integrated circuits.

Bipolar transistors are so named because their operation involves both electrons and holes. These two kinds of charge carriers are characteristic of the two kinds of doped semiconductor material. In contrast, unipolar transistors such as the field-effect transistors have only one kind of charge carrier.

Charge flow in a BJT is due to bidirectional diffusion of charge carriers across a junction between two regions of different charge concentrations. The regions of a BJT are called *emitter*, *collector*, and *base*. A discrete transistor has three leads for connection to these regions. By design, most of the BJT collector current is due to the flow of charges injected from a high-concentration emitter into the base where there are minority carriers that diffuse toward the collector, and so BJTs are classified as minority-carrier devices.

Introduction

BJTs come in two types, or polarities, known as PNP and NPN based on the doping types of the three main terminal regions. An NPN transistor comprises two semiconductor junctions that share a thin p-doped anode region, and a PNP transistor comprises two semiconductor junctions that share a thin n-doped cathode region.

In typical operation, the base–emitter junction is forward biased, which means that the p-doped side of the junction is at a more positive potential than the n-doped side, and the base–collector junction is reverse biased. In an NPN transistor, when positive bias is applied to the base–emitter junction, the equilibrium is disturbed between the thermally generated carriers and the repelling electric field of the n-doped emitter depletion region. This allows thermally excited electrons to inject from the emitter into the base region. These electrons diffuse through the base from the region of high concentration near the emitter towards the region of low concentration near the collector. The electrons in the base are called *minority carriers* because the base is doped p-type, which makes holes the *majority carrier* in the base.

To minimize the percentage of carriers that recombine before reaching the collector–base junction, the transistor's base region must be thin enough that carriers can diffuse across it in much less time than the semiconductor's minority carrier lifetime. In particular, the thickness of the base must be much less than the diffusion length of the electrons. The collector–base junction is reverse-biased, and so little electron injection occurs from the collector to the base, but electrons that diffuse through the base towards the collector are swept into the collector by the electric field in the depletion region of the collector–base junction. The thin *shared* base and asymmetric collector–emitter doping is what differentiates a bipolar transistor from two *separate* and oppositely biased diodes connected in series.

Voltage, current, and charge control

The collector–emitter current can be viewed as being controlled by the base–emitter current (current control), or by the base–emitter voltage (voltage control). These views are related by the current–voltage relation of the base–emitter junction, which is just the usual exponential current–voltage curve of a p-n junction (diode).

The physical explanation for collector current is the amount of minority carriers in the base region. Due to low level injection (in which there are much fewer excess carriers than normal majority carriers) the ambipolar transport rates (in which the excess majority and minority carriers flow at the same rate) is in effect determined by the excess minority carriers.

Detailed transistor models of transistor action, such as the Gummel–Poon model, account for the distribution of this charge explicitly to explain transistor behaviour more exactly. The charge-control view easily handles phototransistors, where minority carriers in the base region are created by the absorption of photons, and handles the dynamics of turn-off, or recovery time, which depends on charge in the base region recombining. However, because base charge is not a signal that is visible at the terminals, the current- and voltage-control views are generally used in circuit design and analysis.

In analog circuit design, the current-control view is sometimes used because it is approximately linear. That is, the collector current is approximately β_F times the base current. Some basic circuits can be designed by assuming that the emitter–base voltage is approximately constant, and that collector current is beta times the base current. However, to accurately and reliably design production BJT circuits, the voltage-control (for example, Ebers–Moll) model is required. The voltage-control model requires an exponential function to be taken into account, but when it is linearized such that the transistor can be modelled as a transconductance, as in the Ebers–Moll model, design for circuits such as differential amplifiers again becomes a mostly linear problem, so the voltage-control view is often preferred. For translinear circuits, in which the exponential I–V curve is key to the operation, the transistors are usually modelled as voltage controlled with transconductance proportional to collector current. In general, transistor level circuit design is performed using SPICE or a comparable analog circuit simulator, so model complexity is usually not of much concern to the designer.

Turn-on, Turn-off, and Storage Delay

The Bipolar transistor exhibits a few delay characteristics when turning on and off. Most transistors, and especially power transistors, exhibit long base-storage times that limit maximum frequency of operation in switching applications. One method for reducing this storage time is by using a Baker clamp.

Transistor Parameters: Alpha (α) and Beta (β)

The proportion of electrons able to cross the base and reach the collector is a measure of the BJT efficiency. The heavy doping of the emitter region and light doping of the base region causes many more electrons to be injected from the emitter into the base than holes to be injected from the base into the emitter. The *common-emitter current gain* is represented by β_F or h_{FE}; it is approximately the ratio of the DC collector current to the DC base current in forward-active region. It is typically greater than 100 for small-signal transistors but can be smaller in transistors designed for high-power applications. Another important parameter is the *common-base current gain*, α_F. The common-base current gain is approximately the gain of current from emitter to collector in the forward-active region. This ratio usually has a value close to unity; between 0.98 and 0.998. It is less than unity due to recombination of charge carriers as they cross the base region. Alpha and beta are more precisely related by the following identities (NPN transistor):

$$\alpha_F = \frac{I_C}{I_E}$$

$$\beta_F = \frac{I_C}{I_B}$$

$$\beta_F = \frac{\alpha_F}{1-\alpha_F} \Leftrightarrow \alpha_F = \frac{\beta_F}{\beta_F+1}$$

Structure

A BJT consists of three differently doped semiconductor regions, the *emitter* region, the *base* region and the *collector* region. These regions are, respectively, *p* type, *n* type and *p* type in a PNP transistor, and *n* type, *p* type and *n* type in an NPN transistor. Each semiconductor region is connected to a terminal, appropriately labelled: *emitter* (E), *base* (B) and *collector* (C).

The *base* is physically located between the *emitter* and the *collector* and is made from lightly doped, high resistivity material. The collector surrounds the emitter region, making it almost impossible for the electrons injected into the base region to escape being collected, thus making the resulting value of α very close to unity, and so, giving the transistor a large β. A cross section view of a BJT indicates that the collector–base junction has a much larger area than the emitter–base junction.

The bipolar junction transistor, unlike other transistors, is usually not a symmetrical device. This means that interchanging the collector and the emitter makes the transistor leave the forward active mode and start to operate in reverse mode. Because the transistor's internal structure is usually optimized for forward-mode operation, interchanging the collector and the emitter makes the values of α and β in reverse operation much smaller than those in forward operation; often the α of the reverse mode is lower than 0.5. The lack of symmetry is primarily due to the doping ratios of the emitter and the collector. The emitter is heavily doped, while the collector is lightly doped, allowing a large reverse bias voltage to be applied before the collector–base junction breaks down. The collector–base junction is reverse biased in normal operation. The reason the emitter is heavily doped is to increase the emitter injection efficiency: the ratio of carriers injected by the emitter to those injected by the base. For high current gain, most of the carriers injected into the emitter–base junction must come from the emitter.

The low-performance "lateral" bipolar transistors sometimes used in CMOS processes are sometimes designed symmetrically, that is, with no difference between forward and backward operation.

Small changes in the voltage applied across the base–emitter terminals causes the current that flows between the *emitter* and the *collector* to change significantly. This effect can be used to amplify the input voltage or current. BJTs can be thought of as voltage-controlled current sources, but are more simply characterized as current-controlled current sources, or current amplifiers, due to the low impedance at the base.

Early transistors were made from germanium but most modern BJTs are made from silicon. A significant minority are also now made from gallium arsenide, especially for very high speed applications.

NPN

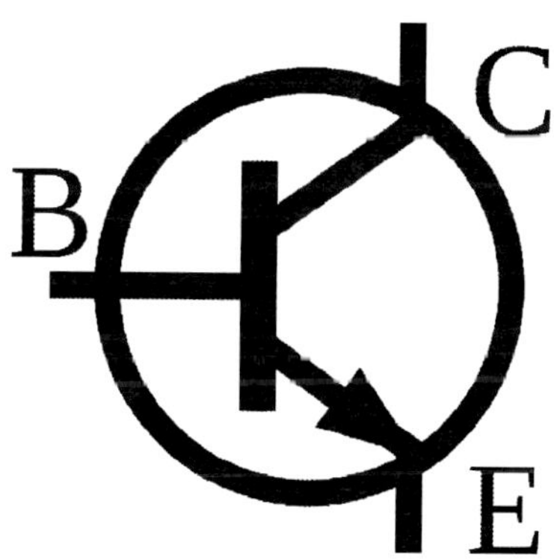

Figure: *The symbol of an NPN BJT. The symbol is "not pointing in."*

NPN is one of the two types of bipolar transistors, consisting of a layer of P-doped semiconductor (the "base") between two N-doped layers. A small current entering the base is amplified to produce a large collector and emitter current. That is, when there is a positive potential difference measured from the emitter of an NPN transistor to its base (i.e., when the base is high relative to the emitter) as well as positive potential difference measured from the base to the collector, the transistor becomes active. In this "on" state, current flows between the collector and emitter of the transistor. Most of the current is carried by electrons moving from emitter to collector as minority carriers in the P-type base region. To allow for greater current and faster operation, most bipolar transistors used today are NPN because electron mobility is higher than hole mobility.

A mnemonic device for the NPN transistor symbol is "*n*ot *p*ointing i*n*," based on the arrows in the symbol and the letters in the name.

PNP

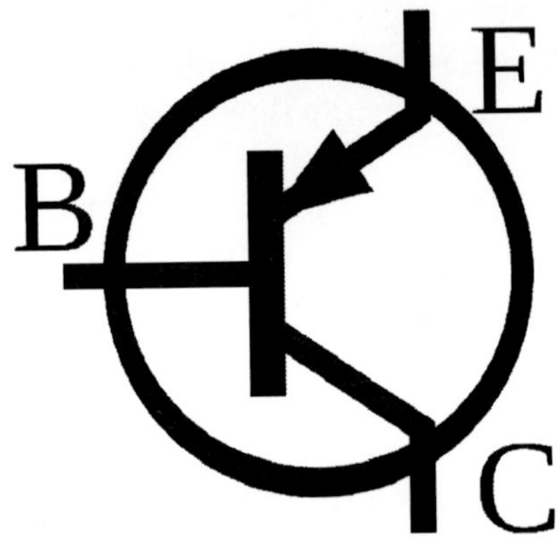

Figure: *The symbol of a PNP BJT. The symbol "points in proudly."*

The other type of BJT is the PNP, consisting of a layer of N-doped semiconductor between two layers of P-doped material. A small current leaving the base is amplified in the collector output. That is, a PNP transistor is "on" when its base is pulled low relative to the emitter.

The arrows in the NPN and PNP transistor symbols are on the emitter legs and point in the direction of the conventional current flow when the device is in forward active mode. A mnemonic device for the PNP transistor symbol is "*p*ointing i*n* (*p*roudly/*p*ermanently)," based on the arrows in the symbol and the letters in the name.

Heterojunction Bipolar Transistor

The heterojunction bipolar transistor (HBT) is an improvement of the BJT that can handle signals of very high frequencies up to several hundred GHz. It is common in modern ultrafast circuits, mostly RF systems. Heterojunction transistors have different semiconductors for the elements of the transistor. Usually the emitter is composed of a larger bandgap material than the base. The figure shows that this difference in bandgap allows the barrier for holes to inject backward into the base, denoted in figure as $\Delta\varphi_p$, to be made large, while the barrier for electrons to inject into the base $\Delta\varphi_n$ is made low. This barrier arrangement helps reduce minority carrier injection from the base when the emitter-base junction is under forward bias, and thus reduces base current and increases emitter injection efficiency.

The improved injection of carriers into the base allows the base to have a higher doping level, resulting in lower resistance to access the base electrode.

In the more traditional BJT, also referred to as homojunction BJT, the efficiency of carrier injection from the emitter to the base is primarily determined by the doping ratio between the emitter and base, which means the base must be lightly doped to obtain high injection efficiency, making its resistance relatively high. In addition, higher doping in the base can improve figures of merit like the Early voltage by lessening base narrowing.

The grading of composition in the base, for example, by progressively increasing the amount of germanium in a SiGe transistor, causes a gradient in bandgap in the neutral base, denoted in the figure by $\Delta\varphi_G$, providing a "built-in" field that assists electron transport across the base. That drift component of transport aids the normal diffusive transport, increasing the frequency response of the transistor by shortening the transit time across the base.

Two commonly used HBTs are silicon–germanium and aluminium gallium arsenide, though a wide variety of semiconductors may be used for the HBT structure. HBT structures are usually grown by epitaxy techniques like MOCVD and MBE.

Regions of Operation

Applied voltages	*B-E Junction Bias (NPN)*	*B-C Junction Bias (NPN)*	*Mode (NPN)*
E < B < C	Forward	Reverse	Forward-active
E < B > C	Forward	Forward	Saturation
E > B < C	Reverse	Reverse	Cut-off
E > B > C	Reverse	Forward	Reverse-active
Applied voltages	B-E Junction Bias (PNP)	B-C Junction Bias (PNP)	Mode (PNP)
E < B < C	Reverse	Forward	Reverse-active
E < B > C	Reverse	Reverse	Cut-off
E > B < C	Forward	Forward	Saturation
E > B > C	Forward	Reverse	Forward-active

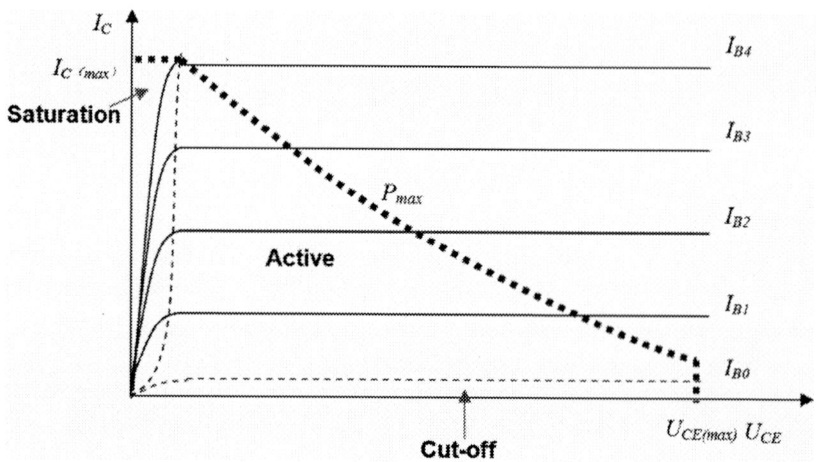

Figure: *The relationship between* I_C, U_{CE} *and* I_B.

Bipolar transistors have five distinct regions of operation, defined by BJT junction biases.

The modes of operation can be described in terms of the applied voltages (this description applies to NPN transistors; polarities are reversed for PNP transistors):

- *Forward-active:* base higher than emitter, collector higher than base (in this mode the collector current is proportional to base current by β_F).
- *Saturation:* base higher than emitter, but collector is not higher than base.
- *Cut-Off:* base lower than emitter, but collector is higher than base. It means the transistor is not letting conventional current go through from collector to emitter.
- *Reverse-active:* base lower than emitter, collector lower than base: reverse conventional current goes through transistor.

In Terms of Junction Biasing: ('reverse biased base–collector junction' means Vbc < 0 for NPN, opposite for PNP)

- *Forward-active (or simply, active):* The base–emitter junction is forward biased and the base–collector junction is reverse biased. Most bipolar transistors are designed to afford the greatest common-emitter current gain, β_F, in forward-active mode. If this is the case, the collector–emitter current is approximately proportional to the base current, but many times larger, for small base current variations.

- Reverse-active (or inverse-active or inverted): By reversing the biasing conditions of the forward-active region, a bipolar transistor goes into reverse-active mode. In this mode, the emitter and collector regions switch roles. Because most BJTs are designed to maximize current gain in forward-active mode, the β_F in inverted mode is several (2–3 for the ordinary germanium transistor) times smaller. This transistor mode is seldom used, usually being considered only for failsafe conditions and some types of bipolar logic. The reverse bias breakdown voltage to the base may be an order of magnitude lower in this region.
- Saturation: With both junctions forward-biased, a BJT is in saturation mode and facilitates high current conduction from the emitter to the collector (or the other direction in the case of NPN, with negatively charged carriers flowing from emitter to collector). This mode corresponds to a logical "on", or a closed switch.
- Cutoff: In cutoff, biasing conditions opposite of saturation (both junctions reverse biased) are present. There is very little current, which corresponds to a logical "off", or an open switch.
- Avalanche breakdown region

Although these regions are well defined for sufficiently large applied voltage, they overlap somewhat for small (less than a few hundred millivolts) biases. For example, in the typical grounded-emitter configuration of an NPN BJT used as a pulldown switch in digital logic, the "off" state never involves a reverse-biased junction because the base voltage never goes below ground; nevertheless the forward bias is close enough to zero that essentially no current flows, so this end of the forward active region can be regarded as the cutoff region.

Active-mode NPN Transistors in Circuits

The diagram shows a schematic representation of an NPN transistor connected to two voltage sources. To make the transistor conduct appreciable current (on the order of 1 mA) from C to E, V_{BE} must be above a minimum value sometimes referred to as the cut-in voltage. The cut-in voltage is usually about 650 mV for silicon BJTs at room temperature but can be different depending on the type of transistor and its biasing. This applied voltage causes the lower P-N junction to 'turn on', allowing a flow of electrons from the emitter into the base. In active mode, the electric field existing between base and collector (caused by V_{CE}) will cause the majority of these electrons to cross the upper P-N junction into the collector to form the collector current I_C. The remainder of the electrons recombine with holes, the majority carriers in the base, making a current through the base connection to form the base current, I_B. As shown in the diagram, the emitter current, I_E, is the total transistor current, which is the sum of the other terminal currents, (i.e., $I_E = I_B + I_C$).

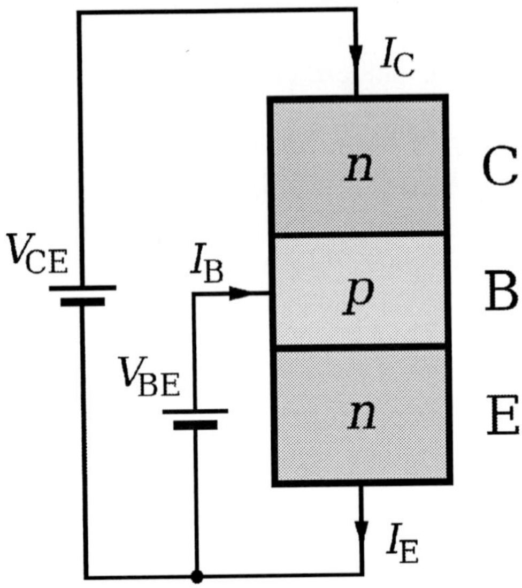

Figure: *Structure and use of NPN transistor. Arrow according to schematic.*

In the diagram, the arrows representing current point in the direction of conventional current – the flow of electrons is in the opposite direction of the arrows because electrons carry negative electric charge. In active mode, the ratio of the collector current to the base current is called the *DC current gain*. This gain is usually 100 or more, but robust circuit designs do not depend on the exact value. The value of this gain for DC signals is referred to as h_{FE}, and the value of this gain for small signals is referred to as h_{fe}. That is, when a small change in the currents occurs, and sufficient time has passed for the new condition to reach a steady state h_{fe} is the ratio of the change in collector current to the change in base current. The symbol β is used for both h_{FE} and h_{fe}.

The emitter current is related to V_{BE} exponentially. At room temperature, an increase in V_{BE} by approximately 60 mV increases the emitter current by a factor of 10. Because the base current is approximately proportional to the collector and emitter currents, they vary in the same way.

Active-mode PNP Transistors in Circuits

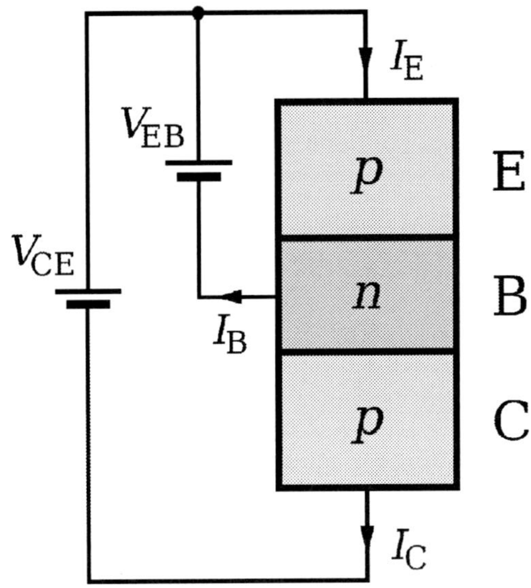

Figure: *Structure and use of PNP transistor.*

The diagram shows a schematic representation of a PNP transistor connected to two voltage sources. To make the transistor conduct appreciable current (on the order of 1 mA) from E to C, V_{EB} must be above a minimum value sometimes referred to as the cut-in voltage. The cut-in voltage is usually about 650 mV for silicon BJTs at room temperature but can be different depending on the type of transistor and its biasing. This applied voltage causes the upper P-N junction to 'turn-on' allowing a flow of holes from the emitter into the base. In active mode, the electric field existing between the emitter and the collector (caused by V_{CE}) causes the majority of these holes to cross the lower p-n junction into the collector to form the collector current I_C. The remainder of the holes recombine with electrons, the majority carriers in the base, making a current through the base connection to form the base current, I_B. As shown in the diagram, the emitter current, I_E, is the total transistor current, which is the sum of the other terminal currents (i.e., $I_E = I_B + I_C$).

In the diagram, the arrows representing current point in the direction of conventional current – the flow of holes is in the same direction of the arrows because holes carry positive electric charge. In active mode, the ratio of the collector current to the base current is called the *DC current gain*. This gain is usually 100 or more, but robust circuit designs do not depend on the exact value. The value of this gain for DC signals is referred to as h_{FE}, and the value of this gain for AC signals is referred to as h_{fe}. However, when there is no particular frequency range of interest, the symbol $\beta$is used.

It should also be noted that the emitter current is related to V_{EB} exponentially. At room temperature, an increase in V_{EB} by approximately 60 mV increases the emitter current by a factor of 10. Because the base current is approximately proportional to the collector and emitter currents, they vary in the same way.

History

The bipolar point-contact transistor was invented in December 1947 at the Bell Telephone Laboratories by John Bardeen and Walter Brattain under the direction of William Shockley. The junction version known as the bipolar junction transistor, invented by Shockley in 1948, enjoyed three decades as the device of choice in the design of discrete and integrated circuits. Nowadays, the use of the BJT has declined in favour of CMOS technology in the design of digital integrated circuits. The incidental low performance BJTs inherent in CMOS ICs, however, are often utilized as bandgap voltage reference, silicon bandgap temperature sensor and to handle electrostatic discharge.

Germanium Transistors

The germanium transistor was more common in the 1950s and 1960s, and while it exhibits a lower "cut off" voltage, typically around 0.2 V, making it more suitable for some applications, it also has a greater tendency to exhibit thermal runaway.

Early Manufacturing Techniques

Various methods of manufacturing bipolar transistors were developed.

Bipolar Transistors

- Point-contact transistor – first transistor ever constructed (December 1947), a bipolar transistor, limited commercial use due to high cost and noise.
    - o Tetrode point-contact transistor – Point-contact transistor having two emitters. It became obsolete in the middle 1950s.
- Junction transistors
    - o Grown-junction transistor – first bipolar *junction* transistor made. Invented by William Shockley at Bell Labs. Invented on June 23, 1948. Patent filed on June 26, 1948.

- o Alloy-junction transistor – emitter and collector alloy beads fused to base. Developed at General Electric and RCA in 1951.
    - – Micro-alloy transistor (MAT) – high speed type of alloy junction transistor. Developed at Philco.
    - – Micro-alloy diffused transistor (MADT) – high speed type of alloy junction transistor, speedier than MAT, a diffused-base transistor. Developed at Philco.
    - – Post-alloy diffused transistor (PADT) – high speed type of alloy junction transistor, speedier than MAT, a diffused-base transistor. Developed at Philips.
- o Tetrode transistor – high speed variant of grown-junction transistor or alloy junction transistor with two connections to base.
- o Surface-barrier transistor – high speed metal barrier junction transistor. Developed at Philco in 1953.
- o Drift-field transistor – high speed bipolar junction transistor. Invented by Herbert Kroemer at the Central Bureau of Telecommunications Technology of the German Postal Service, in 1953.
- o Spacistor – circa 1957.
- o Diffusion transistor – modern type bipolar junction transistor. Prototypes developed at Bell Labs in 1954.
    - – Diffused-base transistor – first implementation of diffusion transistor.
    - – Mesa transistor – Developed at Texas Instruments in 1957.
    - – Planar transistor – the bipolar junction transistor that made mass-produced monolithic integrated circuits possible. Developed by Dr. Jean Hoerni at Fairchild in 1959.
- o Epitaxial transistor – a bipolar junction transistor made using vapor phase deposition.

Theory and Modelling

Transistors can be thought of as two diodes (P–N junctions) sharing a common region that minority carriers can move through. A PNP BJT will function like two diodes that share an N-type cathode region, and the NPN like two diodes sharing a P-type anode region.

Connecting two diodes with wires will not make a transistor, since minority carriers will not be able to get from one P–N junction to the other through the wire.

Both types of BJT function by letting a small current input to the base control an amplified output from the collector. The result is that the transistor makes a good switch that is controlled by its base input.

The BJT also makes a good amplifier, since it can multiply a weak input signal to about 100 times its original strength. Networks of transistors are used to make powerful amplifiers with many different applications. In the discussion below, focus is on the NPN bipolar transistor.

In the NPN transistor in what is called active mode, the base–emitter voltage V_{BE} and collector–base voltage V_{CB} are positive, forward biasing the emitter–base junction and reverse-biasing the collector–base junction.

In the active mode of operation, electrons are injected from the forward biased n-type emitter region into the p-type base where they diffuse as minority carriers to the reverse-biased n-type collector and are swept away by the electric field in the reverse-biased collector–base junction. For a figure describing forward and reverse bias.

Large-signal Models

In 1954 Jewell James Ebers and John L. Moll introduced their mathematical model of transistor currents:

Ebers–Moll Model

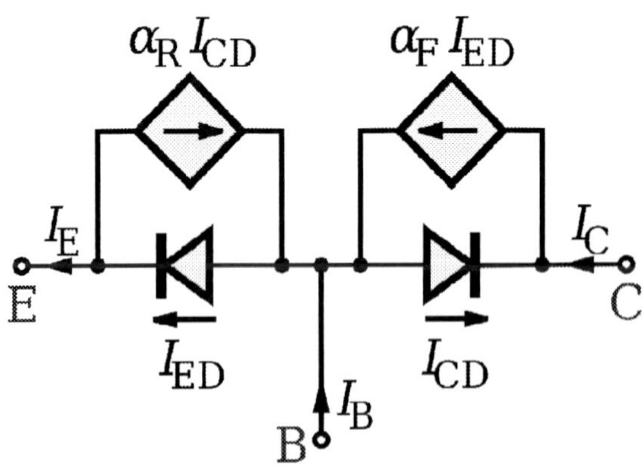

Figure: *Ebers–Moll Model for an NPN transistor. * I_B, I_C, I_E: base, collector and emitter currents * I_{CD}, I_{ED}: collector and emitter diode currents * α_F, α_R: forward and reverse common-base current gains*

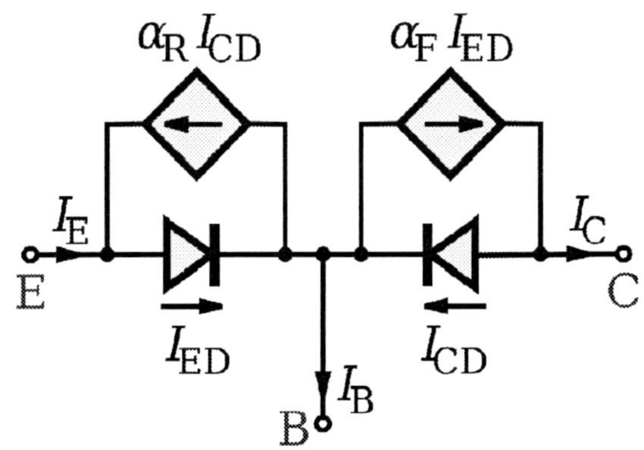

***Figure:** Ebers–Moll Model for a PNP transistor.*

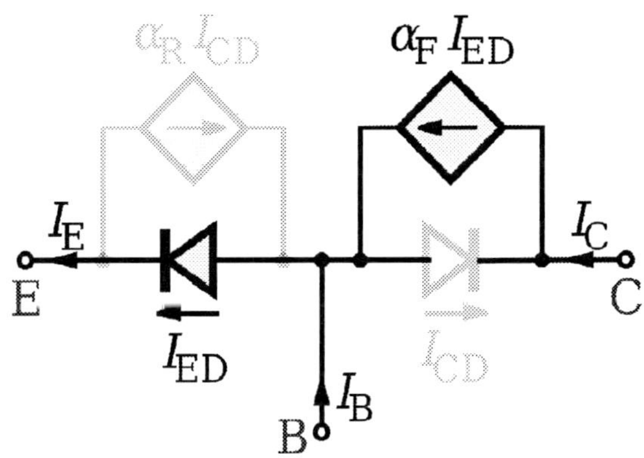

***Figure:** Approximated Ebers–Moll Model for an NPN transistor in the forward active mode. The collector diode is reverse-biased so I_{CD} is virtually zero. Most of the emitter diode current (α_F is nearly 1) is drawn from the collector, providing the amplification of the base current.*

The DC emitter and collector currents in active mode are well modelled by an approximation to the Ebers–Moll model:

$$I_E = I_{ES}\left(e^{\frac{V_{BE}}{V_T}} - 1\right)$$

$$I_C = \alpha_F I_E$$

$$I_B = (1 - \alpha_F) I_E$$

The base internal current is mainly by diffusion and

$$J_{n(\text{base})} = \frac{qD_n n_{bo}}{W} e^{\frac{V_{EB}}{V_T}}$$

where

- V_T is the thermal voltage kT/q (approximately 26 mV at 300 K≈ room temperature).
- I_E is the emitter current
- I_C is the collector current
- α_F is the common base forward short circuit current gain (0.98 to 0.998)
- α_F is the reverse saturation current of the base–emitter diode (on the order of 10^{-15} to 10^{-12} amperes)
- V_{BE} is the base–emitter voltage
- D_n is the diffusion constant for electrons in the p-type base
- W is the base width

The α and forward β parameters are as described previously. A reverse β is sometimes included in the model.

The unapproximated Ebers–Moll equations used to describe the three currents in any operating region are given below.

These equations are based on the transport model for a bipolar junction transistor.

$$i_C = I_S\left(e^{\frac{V_{BE}}{V_T}} - e^{\frac{V_{BC}}{V_T}}\right) - \frac{I_S}{\beta_R}\left(e^{\frac{V_{BC}}{V_T}} - 1\right)$$

$$i_B = \frac{I_S}{\beta_F}\left(e^{\frac{V_{BE}}{V_T}} - 1\right) + \frac{I_S}{\beta_R}\left(e^{\frac{V_{BC}}{V_T}} - 1\right)$$

$$i_E = I_S\left(e^{\frac{V_{BE}}{V_T}} - e^{\frac{V_{BC}}{V_T}}\right) + \frac{I_S}{\beta_F}\left(e^{\frac{V_{BE}}{V_T}} - 1\right)$$

where

- i_C is the collector current
- i_B is the base current
- i_E is the emitter current
- β_F is the forward common emitter current gain (20 to 500)
- β_R is the reverse common emitter current gain (0 to 20)
- I_S is the reverse saturation current (on the order of 10^{-15} to 10^{-12} amperes)

- V_T is the thermal voltage (approximately 26 mV at 300 K≈ room temperature).
- V_{BE} is the base–emitter voltage
- V_{BC} is the base–collector voltage

Base-width Modulation

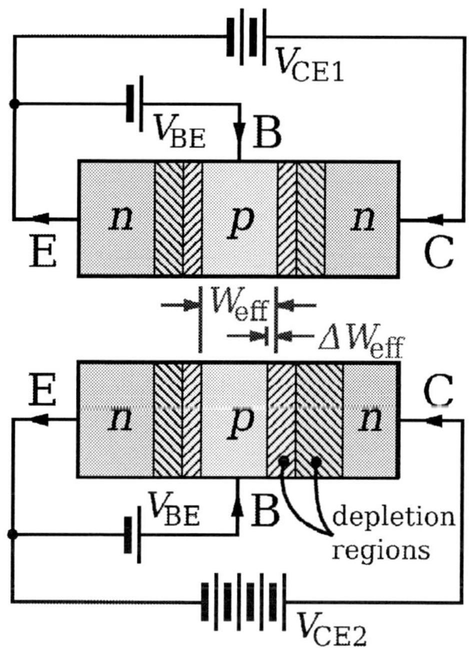

Figure: *Top: NPN base width for low collector-base reverse bias; Bottom: narrower NPN base width for large collector-base reverse bias. Hashed regions are depleted regions.*

As the collector–base voltage ($V_{CB} = V_{CE} - V_{BE}$) varies, the collector–base depletion region varies in size. An increase in the collector–base voltage, for example, causes a greater reverse bias across the collector–base junction, increasing the collector–base depletion region width, and decreasing the width of the base. This variation in base width often is called the "Early effect" after its discoverer James M. Early.

Narrowing of the base width has two consequences:

- There is a lesser chance for recombination within the "smaller" base region.
- The charge gradient is increased across the base, and consequently, the current of minority carriers injected across the emitter junction increases.

Both factors increase the collector or "output" current of the transistor in response to an increase in the collector–base voltage.

In the forward-active region, the Early effect modifies the collector current (i_C) and the forward common emitter current gain (β_F) as given by:

$$i_C = I_S e^{\frac{v_{BE}}{V_T}}\left(1+\frac{V_{CE}}{V_A}\right)$$

$$\beta_F = \beta_{F0}\left(1+\frac{V_{CB}}{V_A}\right)$$

$$r_o = \frac{V_A}{I_C}$$

where:

- V_{CE} is the collector–emitter voltage
- V_A is the Early voltage (15 V to 150 V)
- β_{F0} is forward common-emitter current gain when $V_{CB} = 0$ V
- r_o is the output impedance
- I_C is the collector current

Punchthrough

When the base–collector voltage reaches a certain (device specific) value, the base–collector depletion region boundary meets the base–emitter depletion region boundary. When in this state the transistor effectively has no base. The device thus loses all gain when in this state.

Gummel–Poon Charge-control Model

The Gummel–Poon model is a detailed charge-controlled model of BJT dynamics, which has been adopted and elaborated by others to explain transistor dynamics in greater detail than the terminal-based models typically do. This model also includes the dependence of transistor β-values upon the direct current levels in the transistor, which are assumed current-independent in the Ebers–Moll model.

Small-signal Models

The hybrid-pi model is a popular circuit model used for analyzing the small signal behaviour of bipolar junction and field effect transistors. Sometimes it is also called *Giacoletto model* because it was introduced by L.J. Giacoletto in 1969. The model can be quite accurate for low-frequency circuits and can easily be adapted for higher frequency circuits with the addition of appropriate inter-electrode capacitances and other parasitic elements.

hybrid-pi model:

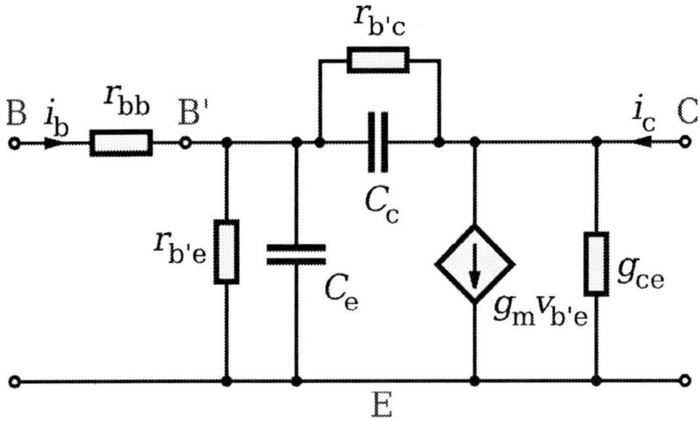

Figure: *Hybrid-pi model*

h-parameter model

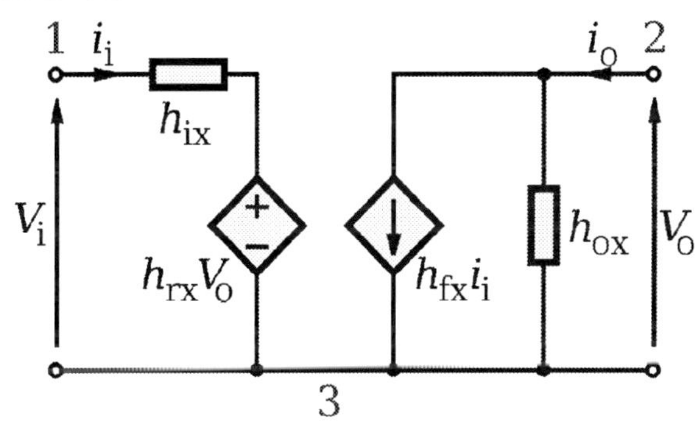

Figure: *Generalized h-parameter model of an NPN BJT.*

Replace x with e, b or c for CE, CB and CC topologies respectively.

Another model commonly used to analyze BJT circuits is the "h-parameter" model, closely related to the hybrid-pi model and the y-parameter two-port, but using input current and output voltage as independent variables, rather than input and output voltages. This two-port network is particularly suited to BJTs as it lends itself easily to the analysis of circuit behaviour, and may be used to develop further accurate models. As shown, the term *"x"* in the model represents a different BJT lead depending on the topology used. For common-emitter mode the various symbols take on the specific values as:

- x = 'e' because it is a common-emitter topology
- Terminal 1 = Base
- Terminal 2 = Collector
- Terminal 3 = Emitter
- i_i = Base current (i_b)
- i_o = Collector current (i_c)
- V_{in} = Base-to-emitter voltage (V_{BE})
- V_o = Collector-to-emitter voltage (V_{CE})

and the h-parameters are given by:

- $h_{ix} = h_{ie}$ – The input impedance of the transistor (corresponding to the base resistance r_{pi}).
- $h_{rx} = h_{re}$ – Represents the dependence of the transistor's I_B–V_{BE} curve on the value of V_{CE}. It is usually very small and is often neglected (assumed to be zero).
- $h_{fx} = h_{fe}$ – The current-gain of the transistor. This parameter is often specified as h_{FE} or the DC current-gain (β_{DC}) in datasheets.
- $h_{ox} = 1/h_{oe}$ – The output impedance of transistor. The parameter h_{oe} usually corresponds to the output admittance of the bipolar transistor and has to be inverted to convert it to an impedance.

As shown, the h-parameters have lower-case subscripts and hence signify AC conditions or analyses. For DC conditions they are specified in upper-case.

For the CE topology, an approximate h-parameter model is commonly used which further simplifies the circuit analysis. For this the h_{oe} and h_{re} parameters are neglected (that is, they are set to infinity and zero, respectively). It should also be noted that the h-parameter model as shown is suited to low-frequency, small-signal analysis. For high-frequency analyses the inter-electrode capacitances that are important at high frequencies must be added.

Etymology of h_{FE}

The 'h' refers to its being an h-parameter, a set of parameters named for their origin in a *hybrid equivalent circuit* model. 'F' is from *forward current amplification* also called the current gain. 'E' refers to the transistor operating in a *common emitter* (CE) configuration. Capital letters used in the subscript indicate that h_{FE} refers to a direct current circuit.

Applications

The BJT remains a device that excels in some applications, such as discrete circuit design, due to the very wide selection of BJT types available, and because of its high transconductance and output resistance compared to MOSFETs. The BJT is also the choice for demanding analog circuits, especially for very-high-frequency applications, such as radio-frequency circuits for wireless systems. Bipolar transistors can be combined with MOSFETs in an integrated circuit by using a BiCMOS process of wafer fabrication to create circuits that take advantage of the application strengths of both types of transistor.

Temperature Sensors

Because of the known temperature and current dependence of the forward-biased base–emitter junction voltage, the BJT can be used to measure temperature by subtracting two voltages at two different bias currents in a known ratio.

Logarithmic Converters

Because base–emitter voltage varies as the log of the base emitter and collector–emitter currents, a BJT can also be used to compute logarithms and anti-logarithms. A diode can also perform these nonlinear functions, but the transistor provides more circuit flexibility.

Vulnerabilities

Exposure of the transistor to ionizing radiation causes radiation damage. Radiation causes a buildup of 'defects' in the base region that act as recombination centres. The resulting reduction in minority carrier lifetime causes gradual loss of gain of the transistor.

Power BJTs are subject to a failure mode called secondary breakdown, in which excessive current and normal imperfections in the silicon die cause portions of the silicon inside the device to become disproportionately hotter than the others. The doped silicon has a negative temperature coefficient, meaning that it conducts more current at higher temperatures.

Thus, the hottest part of the die conducts the most current, causing its conductivity to increase, which then causes it to become progressively hotter again, until the device fails internally. The thermal runaway process associated with secondary breakdown, once triggered, occurs almost instantly and may catastrophically damage the transistor package.

If the emitter-base junction is reverse biased into avalanche or Zener mode and current flows for a short period of time, the current gain of the BJT will be permanently degraded.

Common Emitter and Common Collector Configurations

Common Emitter

In electronics, a common-emitteramplifier is one of three basic single-stage bipolar-junction-transistor (BJT) amplifier topologies, typically used as a voltage amplifier. In this circuit the base terminal of the transistor serves as the input, the collector is the output, and the emitter is *common* to both (for example, it may be tied to ground reference or a power supply rail), hence its name. The analogous field-effect transistor circuit is the common-source amplifier, and the analogous tube circuit is the common-cathode amplifier.

Emitter Degeneration

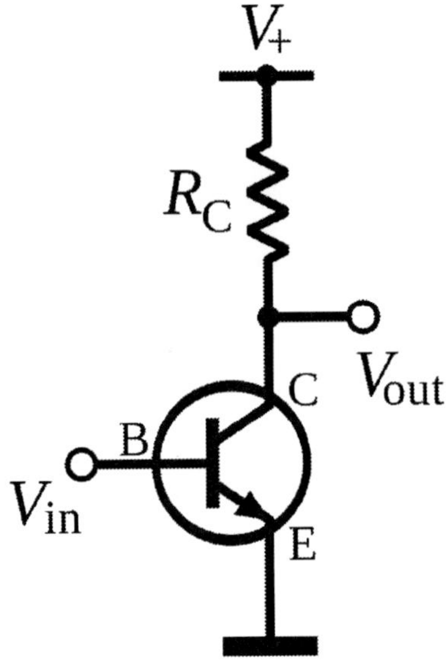

Figure: *Adding an emitter resistor decreases gain, but increases linearity and stability*

Common- emitter amplifiers give the amplifier an inverted output and can have a very high gain that may vary widely from one transistor to the next. The gain is a strong function of both temperature and bias current, and so the actual gain is somewhat unpredictable. Stability is another problem associated with such high gain circuits due to any unintentional positive feedback that may be present. Other problems associated with the circuit are the low input dynamic range imposed by the small-signal limit; there is high distortion if this limit is exceeded and the transistor ceases to behave like its small-signal model. One common way of alleviating these issues is with the use of negative feedback, which is usually implemented with *emitter degeneration*. Emitter degeneration refers to the addition of a small resistor (or any impedance) between the emitter and the common signal source (e.g., the ground reference or a power supply rail). This impedance R_E reduces the overall transconductance $G_m = g_m$ of the circuit by a factor of $g_m R_E + 1$, which makes the voltage gain

$$A_v \triangleq \frac{v_{out}}{v_{in}} = \frac{-g_m R_C}{g_m R_E + 1} \approx -\frac{R_C}{R_E} \qquad (\text{where} \quad g_m R_E \gg 1).$$

So the voltage gain depends almost exclusively on the ratio of the resistors R_C / R_E rather than the transistor's intrinsic and unpredictable characteristics. The distortion and stability characteristics of the circuit are thus improved at the expense of a reduction in gain.

Characteristics

At low frequencies and using a simplified hybrid-pi model, the following small-signal characteristics can be derived.

	Definition	*Expression (with emitter degeneration)*	*Expression (without emitter degeneration, ie., $R_E = 0$)*
Current gain	$A_i \triangleq \frac{i_{out}}{i_{in}}$	β	β
Voltage gain	$A_v \triangleq \frac{v_{out}}{v_{in}}$	$-\frac{\beta R_C}{r_\pi + (\beta+1)R_E}$	$\approx -g_m R_C$
Input impedance	$r_{in} \triangleq \frac{v_{in}}{i_{in}}$	$r_\pi + (\beta+1)R_E$	r_π
Output impedance	$r_{out} \triangleq \frac{v_{out}}{i_{out}}$	R_C	R_C

If the emitter degeneration resistor is not present, $R_E = 0\Omega$, As expected, when R_E is increased, the input impedance is increased and the voltage gain A_v is reduced.

Bandwidth

The bandwidth of the common-emitter amplifier tends to be low due to high capacitance resulting from the Miller effect. The parasitic base-collector capacitance C_{CB} appears like a larger parasitic capacitor

$C_{CB}(1 - A_v)$ (where A_v is negative) from the base to ground. This large capacitor greatly decreases the bandwidth of the amplifier as it makes the time constant of the parasitic input RC filter $r_s(1 - A_V)C_{CB}$ where r_s is the output impedance of the signal source connected to the ideal base.

The problem can be mitigated in several ways, including:

- Reduction of the voltage gain magnitude $|A_v|$ (e.g., by using emitter degeneration).
- Reduction of the output impedance r_s of the signal source connected to the base (e.g., by using an emitter follower or some other voltage follower).
- Using a cascode configuration, which inserts a low input impedance current buffer (e.g. a common base amplifier) between the transistor's collector and the load. This configuration holds the transistor's collector voltage roughly constant, thus making the base to collector gain zero and hence (ideally) removing the Miller effect.
- Using a differential amplifiertopology like an emitter follower driving a grounded-base amplifier; as long as the emitter follower is truly a common-collector amplifier, the Miller effect is removed.

The Miller effect negatively affects the performance of the common-source amplifier in the same way (and has similar solutions). When an AC signal is applied to the transistor amplifier it causes the base voltage VB to fluctuate in value at the AC signal. The positive half of the applied signal will cause an increase in the value of VB this turn will increase the base current IB and cause a corresponding increase in emitter current IE and collector current IC. As a result, the collector emitter voltage will be reduced because of the increase voltage drop across RL. The negative alternation of an AC signal will cause a decrease in IB this action then causes a corresponding decrease in IE through RL.

The output signal of a common- emitter amplifier is therefore 180 degrees out of phase with the input signal. It is also named common-emitter amplifier because the emitter of the transistor is common to both the input circuit and output circuit. The input signal is applied across the ground and the base circuit of the transistor. The output signal appears across ground and the collector of the transistor. Since the emitter is connected to the ground, it is common to signals, input and output. The common- emitter circuit is the most widely used of junction, transistor amplifiers. As compared with the common- base connection, it has higher input impedance and lower output impedance. A single power supply is easily used for biasing. In addition, higher voltage and power gains are usually obtained for common- emitter (CE) operation. Current gain in the common emitter circuit is obtained from the base and the collector circuit currents. Because a very small change in base current produces a large change in collector current, the current gain (β) is always greater than unity for the common-emitter circuit, a typical value is about 50.

Applications

Low Frequency Voltage Amplifier: A typical example of the use of a common-emitter amplifier is shown in Figure.

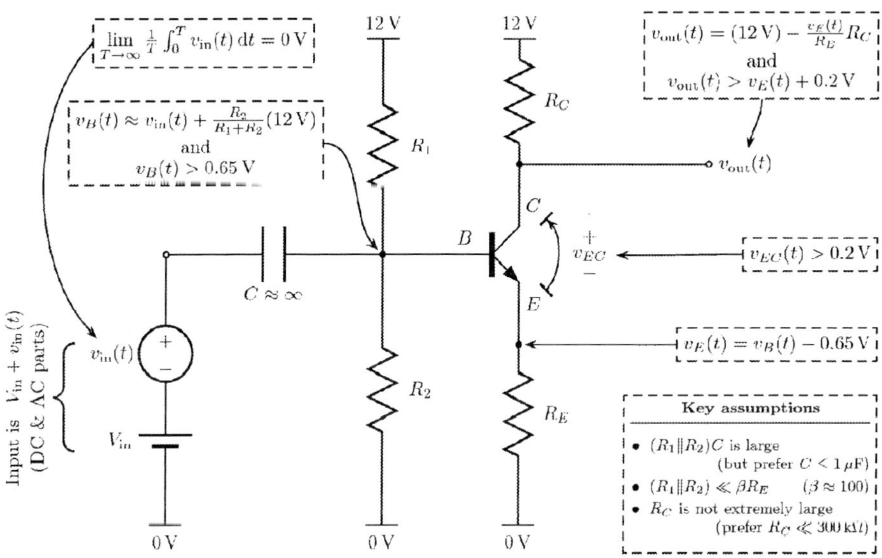

Figure: *Single-ended* npn *common-emitter amplifier with emitter degeneration. The AC-coupled circuit acts as a level-shifter amplifier. Here, the base–emitter voltage drop is assumed to be 0.65 Volts.*

The input capacitor C removes any constant component of the input, and the resistors R_1 and R_2 bias the transistor so that it will remain in active mode for the entire range of the input. The output is an inverted copy of the AC-component of the input that has been amplified by the ratio R_C/R_E and shifted by an amount determined by all four resistors. Because R_C is often large, the output impedance of this circuit can be prohibitively high. To alleviate this problem, R_C is kept as low as possible and the amplifier is followed by a voltage buffer like an emitter follower.

Radio

Common-emitter amplifiers are also used in radio frequency circuits, for example to amplify faint signals received by an antenna. In this case it is common to replace the load resistor with a tuned circuit. This may be done to limit the bandwidth to a narrow band centred around the intended operating frequency. More importantly it also allows the circuit to operate at higher frequencies as the tuned circuit can be used to resonate any inter-electrode and stray capacitances, which normally limit the frequency response. Common emitters are also commonly used as low-noise amplifiers.

Common Collector

In electronics, a common-collectoramplifier (also known as an *emitter follower* or *BJT voltage follower*) is one of three basic single-stage bipolar junction transistor (BJT) amplifier topologies, typically used as a voltage buffer. In this circuit the base terminal of the transistor serves as the input, the emitter is the output, and the collector is *common* to both (for example, it may be tied to ground reference or a power supply rail), hence its name. The analogous field-effect transistor circuit is the common-drain amplifier.

Basic Circuit

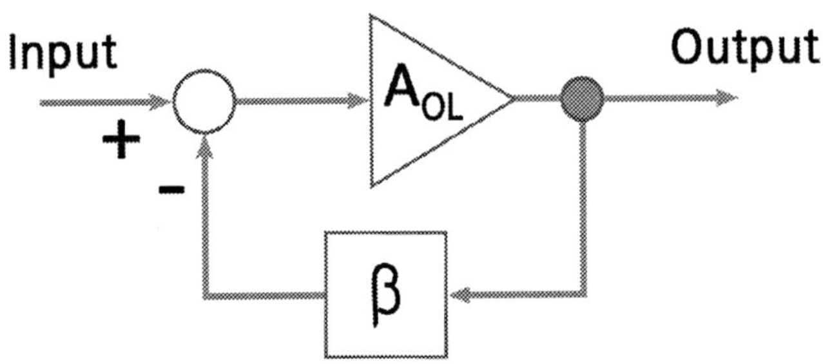

Figure: *A negative feedback amplifier*

The circuit can be explained by viewing the transistor as being under the control of negative feedback. From this viewpoint, a common-collector stage is an amplifier with full series negative feedback. In this configuration, the entire output voltage V_{OUT} is placed contrary and in series with the input voltage V_{IN}. Thus the two voltages are subtracted according to KVL (the subtractor from the function block diagram is implemented just by the input loop) and their difference $Vdiff = V_{IN} - V_{OUT}$ is applied to the base-emitter junction. The transistor monitors continuously Vdiff and adjusts its emitter voltage almost equal (less V_{BEO}) to the input voltage by passing the according collector current through the emitter resistor R_E. As a result, the output voltage follows the input voltage variations from V_{BEO} up to V_+; hence the name, emitter follower.

Intuitively, this behaviour can be also understood by realizing that the base-emitter voltage in the bipolar transistor is very insensitive to bias changes, so any change in base voltage is transmitted (to good approximation) directly to the emitter. It depends slightly on various disturbances (transistor tolerances, temperature variations, load resistance, collector resistor if it is added, etc.) since the transistor reacts to these disturbances and restores the equilibrium. It never saturates even if the input voltage reaches the positive rail.

The common collector circuit can be shown mathematically to have a voltagegain of almost unity:

$$A_v = \frac{v_{out}}{v_{in}} \approx 1$$

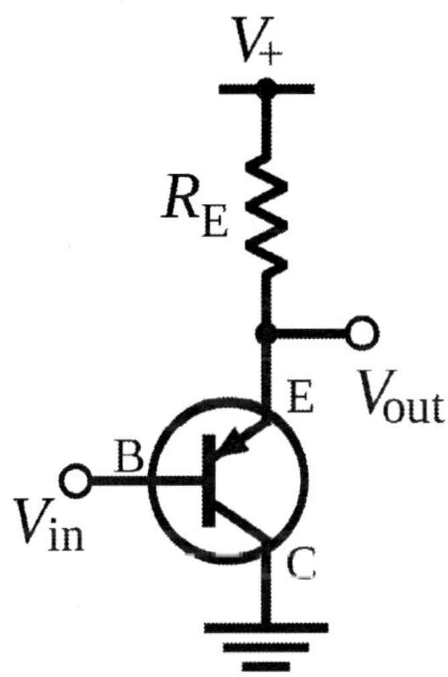

Figure: *PNP version of the emitter follower circuit, all polarities are reversed.*

A small voltage change on the input terminal will be replicated at the output (depending slightly on the transistor's gain and the value of the load resistance). This circuit is useful because it has a large input impedance, so it will not load down the previous circuit:

$$r_{in} \approx \beta_0 R_E$$

and a small output impedance, so it can drive low-resistance loads:

$$r_{out} \approx R_E \,||\, \frac{R_{source}}{\beta_0}$$

Typically, the emitter resistor is significantly larger and can be removed from the equation:

$$r_{\text{out}} \approx \frac{R_{\text{source}}}{\beta_0}$$

Applications

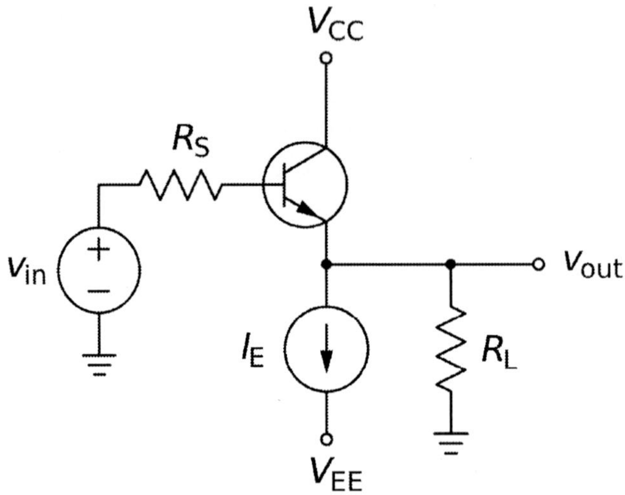

Figure: *NPN voltage follower with current source biasing suitable for integrated circuits*

The low output impedance allows a source with a large output impedance to drive a small load impedance; it functions as a voltage buffer.

In other words, the circuit has current gain (which depends largely on the h_{FE} of the transistor) instead of voltage gain. A small change to the input current results in much larger change in the output current supplied to the output load.

One aspect of buffer action is transformation of impedances. For example, the Thévenin resistance of a combination of a voltage follower driven by a voltage source with high Thévenin resistance is reduced to only the output resistance of the voltage follower (a small resistance).

That resistance reduction makes the combination a more ideal voltage source.

Conversely, a voltage follower inserted between a small load resistance and a driving stage presents a large load to the driving stage—an advantage in coupling a voltage signal to a small load. This configuration is commonly used in the output stages of class-B and class-AB amplifiers. The base circuit is modified to operate the transistor in class-B or AB mode.

In class-A mode, sometimes an active current source is used instead of R_E to improve linearity and/or efficiency.

Characteristics

At low frequencies and using a simplified hybrid-pi model, the following small-signal characteristics can be derived. (Parameter $\beta = g_m r_\pi$ and the parallel lines indicate components in parallel.)

	Definition	*Expression*	*Approximate expression*	*Conditions*
Current gain	$A_i = \frac{i_{out}}{i_{in}}$	$\beta_0 + 1$	$\approx \beta_0$	$\beta_0 \gg 1$
Voltage gain	$A_v = \frac{v_{out}}{v_{in}}$	$\frac{g_m R_E}{g_m R_E + 1}$	≈ 1	$g_m R_E \gg 1$
Input resistance	$r_{in} = \frac{v_{in}}{i_{in}}$	$r_\pi + (\beta_0 + 1) R_E$	$\approx \beta_0 R_E$	$(g_m R_E \gg 1) \wedge (\beta_0 \gg 1)$
Output resistance	$r_{out} = \frac{v_{out}}{i_{out}}$	$R_E \parallel \left(\frac{r_\pi + R_{source}}{\beta_0 + 1} \right)$	$\approx \frac{1}{g_m} + \frac{R_{source}}{\beta_0}$	$(\beta_0 \gg 1) \wedge (r_{in} \gg R_{source})$

Where R_{source} is the Thévenin equivalent source resistance.

Derivations

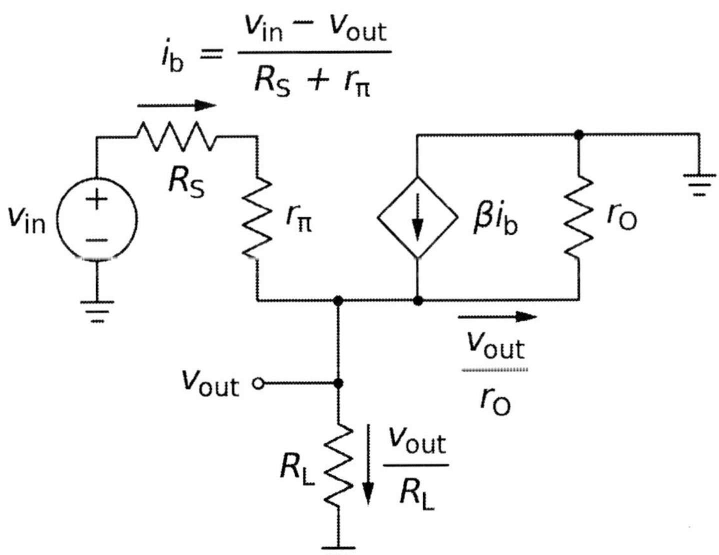

Figure: *Small-signal circuit corresponding to Figure using the hybrid-pi model for the bipolar transistor at frequencies low enough to ignore bipolar device capacitances*

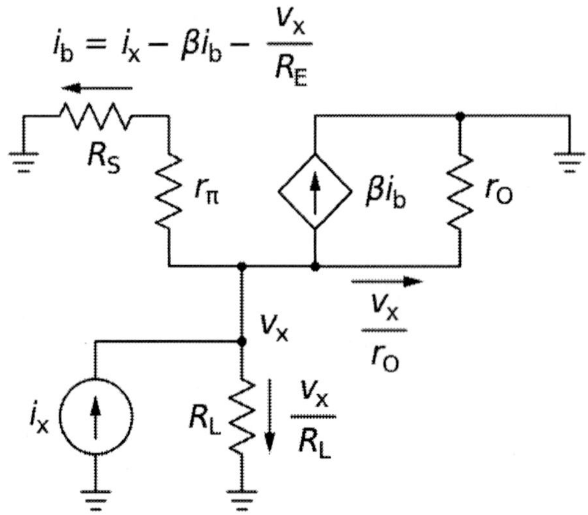

Figure: *Low-frequency small-signal circuit for bipolar voltage follower with test current at output for finding output resistance. Resistor* $R_E = R_L \| r_O$.

Figure shows a low-frequency hybrid-pi model for the circuit of Figure. Using Ohm's law various currents have been determined and these results are shown on the diagram. Applying Kirchhoff's current law at the emitter one finds:

$$(\beta+1)\frac{v_{in}-v_{out}}{R_S+r_\pi} = v_{out}\left(\frac{1}{R_L}+\frac{1}{r_O}\right).$$

Define the following resistance values:

$$\frac{1}{R_E} = \frac{1}{R_L}+\frac{1}{r_O}$$

$$R = \frac{R_S+r_\pi}{\beta+1}.$$

Then collecting terms the voltage gain is found as:

$$A_v = \frac{v_{out}}{v_{in}} = \frac{1}{1+\dfrac{R}{R_E}}.$$

From this result the gain approaches unity (as expected for a buffer amplifier) if the resistance ratio in the denominator is small. This ratio decreases with larger values of current gain β and with larger vales of R_E.

The input resistance is found as:

$$R_{\text{in}} = \frac{v_{\text{in}}}{i_{\text{b}}} = \frac{R_{\text{S}} + r_{\pi}}{1 - A_{\text{v}}}$$

$$= \left(R_{\text{S}} + r_{\pi}\right)\left(1 + \frac{R_{\text{E}}}{R}\right)$$

$$= R_{\text{S}} + r_{\pi} + (\beta + 1)R_{\text{E}} .$$

The transistor output resistance r_{O} ordinarily is large compared to the load R_{L} and therefore R_{L} dominates R_{E}. From this result, the input resistance of the amplifier is much larger than the output load resistance R_{L} for large current gain β. That is, placing the amplifier between the load and the source presents a smaller (high-resistive) load to the source than direct coupling to R_{L}, which results in less signal attenuation in the source impedance R_{S} as a consequence of voltage division.

Figure shows the small-signal circuit with the input short-circuited and a test current placed at its output. The output resistance is found using this circuit as:

$$R_{\text{out}} = \frac{v_{\text{X}}}{i_{\text{X}}} .$$

Using Ohm's law, various currents have been found, as indicated on the diagram. Collecting the terms for the base current, the base current is found as:

$$(\beta + 1)i_{\text{b}} = i_{\text{X}} - \frac{v_{\text{X}}}{R_{\text{E}}} ,$$

where R_{E} is defined above. Using this value for base current, Ohm's law provides v_{x} as:

$$v_{\text{X}} = i_{\text{b}}\left(R_{\text{S}} + r_{\pi}\right).$$

Substituting for the base current, and collecting terms,

$$R_{\text{out}} = \frac{v_{\text{X}}}{i_{\text{X}}} = R \parallel R_{\text{E}} ,$$

where $\parallel$ denotes a parallel connection and R is defined above. Because R generally is a small resistance when the current gain β is large, R dominates the output impedance which therefore also is small. A small output impedance means the series combination of the original voltage source and the voltage follower presents a Thévenin voltage source with a lower Thévenin resistance at its output node; that is, the combination of voltage source with voltage follower makes a more ideal voltage source than the original one.

Voltage Divider Bias Configuration

Bipolar transistoramplifiers must be properly biased to operate correctly. In circuits made with individual devices (discrete circuits), biasing networks consisting of resistors are commonly employed. Much more elaborate biasing arrangements are used in integrated circuits, for example, bandgap voltage references and current mirrors.

The operating point of a device, also known as *bias point*, *quiescent point*, or *Q-point*, is the point on the output characteristics that shows the DC collector–emitter voltage (V_{ce}) and the collector current (I_c) with no input signal applied. The term is normally used in connection with devices such as transistors.

Bias Circuit Requirements

Signal Requirements for Class A Amplifiers: For analog operation of a Class A amplifier, the Q-point is placed so the transistor stays in active mode (does not shift to operation in the saturation region or cut-off region) when input is applied. For digital operation, the Q-point is placed so the transistor does the contrary - switches from "on" to "off" state. Often, Q-point is established near the centre of active region of transistor characteristic to allow similar signal swings in positive and negative directions. Q-point should be stable. In particular, it should be insensitive to variations in transistor parameters (for example, should not shift if transistor is replaced by another of the same type), variations in temperature, variations in power supply voltage and so forth. The circuit must be practical: easily implemented and cost-effective.

Thermal Considerations

At constant current, the voltage across the emitter–base junction V_{BE} of a bipolar transistor *decreases* 2mV (silicon) and 1.8mV (germanium) for each 1 °C rise in temperature (reference being 25 °C). By the Ebers–Moll model, if the base–emitter voltage V_{BE} is held constant and the temperature rises, the current through the base–emitter diode I_B will increase, and thus the collector current I_C will also increase. Depending on the bias point, the power dissipated in the transistor may also increase, which will further increase its temperature and exacerbate the problem. This deleterious positive feedback results in thermal runaway. There are several approaches to mitigate bipolar transistor thermal runaway. For example,

- Negative feedback can be built into the biasing circuit so that increased collector current leads to decreased base current. Hence, the increasing collector current throttles its source.
- Heat sinks can be used that carry away extra heat and prevent the base–emitter temperature from rising.
- The transistor can be biased so that its collector is normally less than half of the power supply voltage, which implies that collector–emitter power dissipation is at its maximum value. Runaway is then impossible because increasing collector current leads to a decrease in dissipated power; this notion is known as the *half-voltage principle.*

The circuits below primarily demonstrate the use of negative feedback to prevent thermal runaway.

Types of Bias Circuit for Class A Amplifiers

The following discussion treats five common biasing circuits used with Class A bipolar transistor amplifiers:

1. Fixed bias
2. Collector-to-base bias
3. Fixed bias with emitter resistor
4. Voltage divider bias
5. Emitter bias.

Fixed Bias (Base Bias)

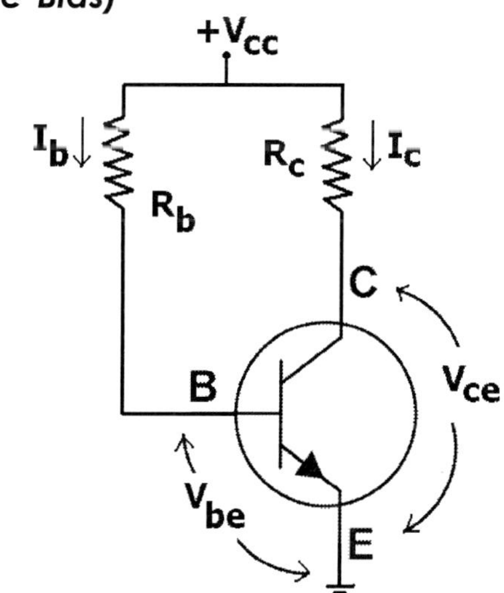

Figure: *Fixed bias (Base bias)*

This form of biasing is also called *base bias*. In the example image on the right, the single power source (for example, a battery) is used for both collector and base of a transistor, although separate batteries can also be used.

In the given circuit,

$$V_{cc} = I_B R_B + V_{be}$$

Therefore,

$$I_B = (V_{cc} - V_{be})/R_B$$

For a given transistor, V_{be} does not vary significantly during use. As V_{cc} is of fixed value, on selection of R_B, the base current I_B is fixed. Therefore this type is called *fixed bias* type of circuit.

Also for given circuit,

$$V_{cc} = I_C R_C + V_{ce}$$

Therefore,

$$V_{ce} = V_{cc} - I_C R_C$$

The common-emitter current gain of a transistor is an important parameter in circuit design, and is specified on the data sheet for a particular transistor. It is denoted as β on this page.

Because

$$I_C = \beta I_B$$

we can obtain I_C as well. In this manner, operating point given as (V_{ce}, I_C) can be set for given transistor.

Merits:

- It is simple to shift the operating point anywhere in the active region by merely changing the base resistor (R_B).
- A very small number of components are required.

Demerits:

- The collector current does not remain constant with variation in temperature or power supply voltage. Therefore the operating point is unstable.
- Changes in V_{be} will change I_B and thus cause R_E to change. This in turn will alter the gain of the stage.
- When the transistor is replaced with another one, considerable change in the value of β can be expected. Due to this change the operating point will shift.

- For small-signal transistors (e.g., not power transistors) with relatively high values of β (i.e., between 100 and 200), this configuration will be prone to thermal runaway. In particular, the stability factor, which is a measure of the change in collector current with changes in reverse saturation current, is approximately β+1. To ensure absolute stability of the amplifier, a stability factor of less than 25 is preferred, and so small-signal transistors have large stability factors.

Usage: Due to the above inherent drawbacks, fixed bias is rarely used in linear circuits (i.e., those circuits which use the transistor as a current source). Instead, it is often used in circuits where transistor is used as a switch. However, one application of fixed bias is to achieve crude automatic gain control in the transistor by feeding the base resistor from a DC signal derived from the AC output of a later stage.

Collector-to-base Bias

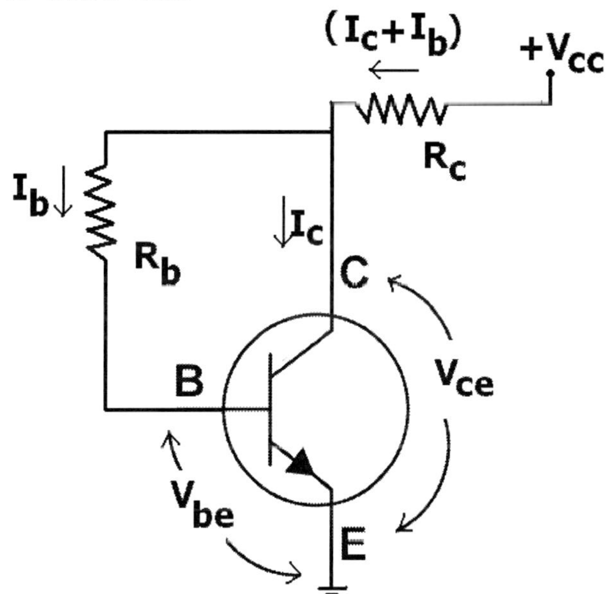

Figure: *Collector-to-base bias*

This configuration employs negative feedback to prevent thermal runaway and stabilize the operating point. In this form of biasing, the base resistor R_B is connected to the collector instead of connecting it to the DC source V_{cc}. So any thermal runaway will induce a voltage drop across the R_C resistor that will throttle the transistor's base current.

From Kirchhoff's voltage law, the voltage V_{R_b} across the base resistor R_b is

$$V_{R_b} = V_{cc} - \overbrace{(I_c + I_b)R_c}^{\text{Voltage drop across } R_c} - \overbrace{V_{be}}^{\text{Voltage at base}} .$$

By the Ebers–Moll model, $I_c = \beta I_b$, and so

$$V_{R_b} = V_{cc} - (\overbrace{\beta I_b}^{I_c} + I_b)R_c - V_{be} = V_{cc} - I_b(\beta + 1)R_c - V_{be}.$$

From Ohm's law, the base current $I_b = V_{R_b} / R_b$, and so

$$\overbrace{I_b R_b}^{V_{R_b}} = V_{cc} - I_b(\beta + 1)R_c - V_{be}.$$

Hence, the base current I_b is

$$I_b = \frac{V_{cc} - V_{be}}{R_b + (\beta + 1)R_c}$$

If V_{be} is held constant and temperature increases, then the collector current I_c increases. However, a larger I_c causes the voltage drop across resistor R_c to increase, which in turn reduces the voltage V_{R_b} across the base resistor R_b. A lower base-resistor voltage drop reduces the base current I_b, which results in less collector current I_c. Because an increase in collector current with temperature is opposed, the operating point is kept stable.

Merits:

- Circuit stabilizes the operating point against variations in temperature and β (i.e. replacement of transistor)

Demerits:

- In this circuit, to keep I_c independent of β, the following condition must be met:

$$I_c = \beta I_b = \frac{\beta(V_{cc} - V_{be})}{R_b + R_c + \beta R_c} \approx \frac{(V_{cc} - V_{be})}{R_c}$$

which is the case when

$$\beta R_c \gg R_b.$$

- As β-value is fixed (and generally unknown) for a given transistor, this relation can be satisfied either by keeping R_c fairly large or making R_b very low.

- If R_c is large, a high V_{cc} is necessary, which increases cost as well as precautions necessary while handling.
- If is low, the reverse bias of the collector–base region is small, which limits the range of collector voltage swing that leaves the transistor in active mode.
- The resistor R_b causes an AC feedback, reducing the voltage gain of the amplifier. This undesirable effect is a trade-off for greater Q-point stability.

Usage: The feedback also decreases the input impedance of the amplifier as seen from the base, which can be advantageous. Due to the gain reduction from feedback, this biasing form is used only when the trade-off for stability is warranted.

Fixed Bias with Emitter Resistor

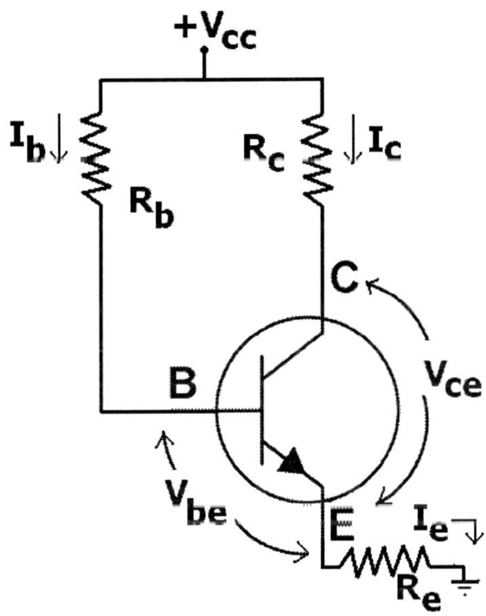

Figure: *Fixed bias with emitter resistor*

The fixed bias circuit is modified by attaching an external resistor to the emitter. This resistor introduces negative feedback that stabilizes the Q-point. From Kirchhoff's voltage law, the voltage across the base resistor is

$V_{Rb} = V_{CC} - I_e R_e - V_{be}.$

From Ohm's law, the base current is

$I_b = V_{Rb} / R_b.$

The way feedback controls the bias point is as follows. If V_{be} is held constant and temperature increases, emitter current increases. However, a larger I_e increases the emitter voltage $V_e = I_eR_e$, which in turn reduces the voltage V_{Rb} across the base resistor. A lower base-resistor voltage drop reduces the base current, which results in less collector current because $I_c = \beta\ I_B$. Collector current and emitter current are related by $I_c = \alpha\ I_e$ with $\alpha \approx 1$, so increase in emitter current with temperature is opposed, and operating point is kept stable.

Similarly, if the transistor is replaced by another, there may be a change in I_C (corresponding to change in β-value, for example). By similar process, the change is negated and operating point kept stable.

For the given circuit,

$$I_B = (V_{CC} - V_{be})/(R_B + (\beta+1)R_E).$$

Merits: The circuit has the tendency to stabilize operating point against changes in temperature and β-value.

Demerits:

- In this circuit, to keep I_C independent of β the following condition must be met:

$$I_C = \beta I_B = \frac{\beta(V_{CC} - V_{be})}{R_B + (\beta+1)R_E} \approx \frac{(V_{CC} - V_{be})}{R_E}$$

  which is approximately the case if

$$(\beta + 1)R_E >> R_B.$$

- As β-value is fixed for a given transistor, this relation can be satisfied either by keeping R_E very large, or making R_B very low.
- If R_E is of large value, high V_{CC} is necessary. This increases cost as well as precautions necessary while handling.
- If R_B is low, a separate low voltage supply should be used in the base circuit. Using two supplies of different voltages is impractical.
- In addition to the above, R_E causes ac feedback which reduces the voltage gain of the amplifier.

Usage:

The feedback also increases the input impedance of the amplifier when seen from the base, which can be advantageous. Due to the above disadvantages, this type of biasing circuit is used only with careful consideration of the trade-offs involved.

Collector-Stabilized Biasing

Voltage Divider Biasing

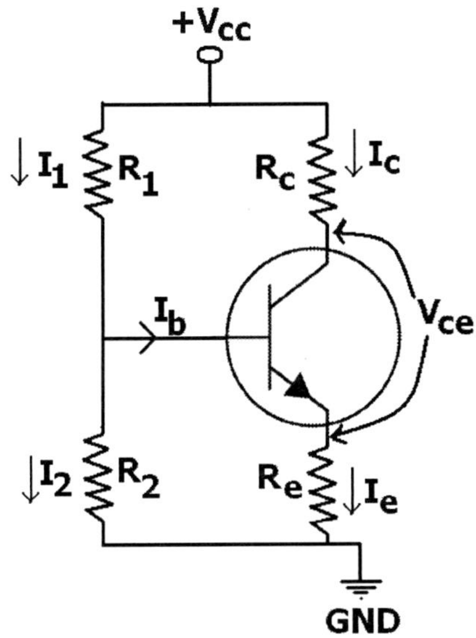

Figure: *Voltage divider bias*

The voltage divider is formed using external resistors R_1 and R_2. The voltage across R_2 forward biases the emitter junction. By proper selection of resistors R_1 and R_2, the operating point of the transistor can be made independent of β. In this circuit, the voltage divider holds the base voltage fixed independent of base current provided the divider current is large compared to the base current. However, even with a fixed base voltage, collector current varies with temperature (for example) so an emitter resistor is added to stabilize the Q-point, similar to the above circuits with emitter resistor.

In this circuit the base voltage is given by:

$$V_B = \text{voltage across } R_2 = V_{cc}\frac{R_2}{(R_1+R_2)} - I_B\frac{R_1R_2}{(R_1+R_2)}$$

$$\approx V_{cc}\frac{R_2}{(R_1+R_2)} \text{provided } I_B << I_2 = V_B / R_2 .$$

Also $V_B = V_{be} + I_E R_E$

For the given circuit,

$$I_B = \frac{\frac{V_{CC}}{1+R_1/R_2} - V_{be}}{(\beta+1)R_E + R_1 \parallel R_2}.$$

Merits:

- Unlike above circuits, only one dc supply is necessary.
- Operating point is almost independent of β variation.
- Operating point stabilized against shift in temperature.

Demerits:

- In this circuit, to keep I_C independent of β the following condition must be met:

$$I_C = \beta I_B = \beta \frac{\frac{V_{CC}}{1 + R_1 / R_2} - V_{be}}{(\beta + 1)R_E + R_1 \parallel R_2} \approx \frac{\frac{V_{CC}}{1 + R_1 / R_2} - V_{be}}{R_E},$$

which is approximately the case if

$$(\beta + 1)R_E >> R_1 \parallel R_2$$

where $R_1 \parallel R_2$ denotes the equivalent resistance of R_1 and R_2 connected in parallel.

- As β-value is fixed for a given transistor, this relation can be satisfied either by keeping R_E fairly large, or making $R_1 \parallel R_2$ very low.
- If R_E is of large value, high V_{CC} is necessary. This increases cost as well as precautions necessary while handling.
- If $R_1 \parallel R_2$ is low, either R_1 is low, or R_2 is low, or both are low. A low R_1 raises V_B closer to V_C, reducing the available swing in collector voltage, and limiting how large R_C can be made without driving the transistor out of active mode. A low R_2 lowers V_{be}, reducing the allowed collector current. Lowering both resistor values draws more current from the power supply and lowers the input resistance of the amplifier as seen from the base.
- AC as well as DC feedback is caused by R_E, which reduces the AC voltage gain of the amplifier. A method to avoid AC feedback while retaining DC feedback is discussed below.

Usage: The circuit's stability and merits make it widely used for linear circuits.

Voltage divider with AC bypass capacitor:

The standard voltage divider circuit discussed above faces a drawback - AC feedback caused by resistor R_E reduces the gain. This can be avoided by placing a capacitor (C_E) in parallel with R_E, as shown in circuit diagram.

This capacitor is usually chosen to have a low enough reactance at the signal frequencies of interest such that R_E is essentially shorted at AC, thus grounding the emitter. Feedback is therefore only present at DC to stabilize the operating point, in which case any AC advantages of feedback are lost.

This idea can also be used to shunt only a portion of R_E, thereby retaining some AC feedback.

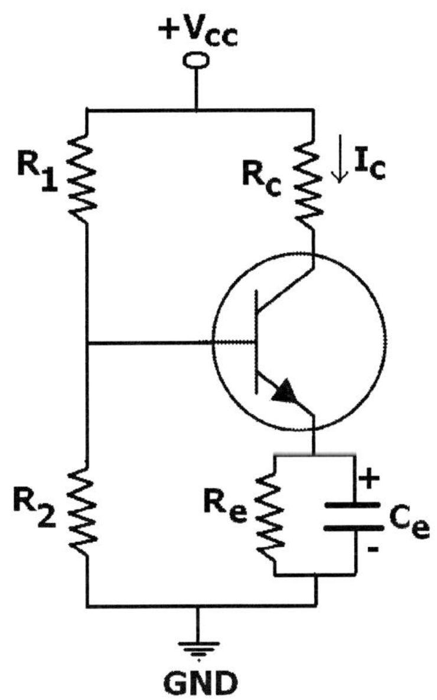

Figure: *Voltage divider with capacitor*

Emitter bias

When a split supply (dual power supply) is available, this biasing circuit is the most effective, and provides zero bias voltage at the emitter or collector for load. The negative supply V_{EE} is used to forward-bias the emitter junction through R_E. The positive supply V_{CC} is used to reverse-bias the collector junction. Only two resistors are necessary for the common collector stage and four resistors for the common emitter or common base stage.

We know that,

$V_B - V_E = V_{be}$

If R_B is small enough, base voltage will be approximately zero. Therefore emitter current is,

$I_E = (V_{EE} - V_{be})/R_E$

The operating point is independent of β if $R_E >> R_B/\beta$

Merit: Good stability of operating point similar to voltage divider bias.

Demerit: This type can only be used when a split (dual) power supply is available.

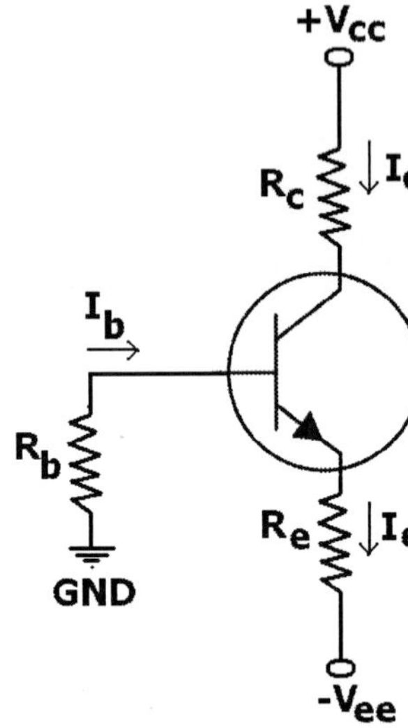

Figure: *Emitter bias*

Class B and AB amplifiers

Signal Requirements: Class B and AB amplifiers employ 2 active devices to cover the complete 360 deg of input signal flow. Each transistor is therefore biased to perform over approximately 180 deg of the input signal. Class B bias is when the collector current I_c with no signal is just conducting (about 1% of maximum possible value). Class AB bias is when the collector current I_c is about 1/4 of maximum possible value. The class AB push–pull output amplifier circuit below could be the basis for a moderate-power audio amplifier.

Q3 is a common emitter stage that provides amplification of the signal and the DC bias current through D1 and D2 to generate a bias voltage for the output devices. The output pair are arranged in Class AB push–pull, also called a complementary pair. The diodes D1 and D2 provide a small amount of constant voltage bias for the output pair, just biasing them into the conducting state so that crossover distortion is minimized. That is, the diodes push the output stage into class-AB mode (assuming that the base-emitter drop of the output transistors is reduced by heat dissipation).

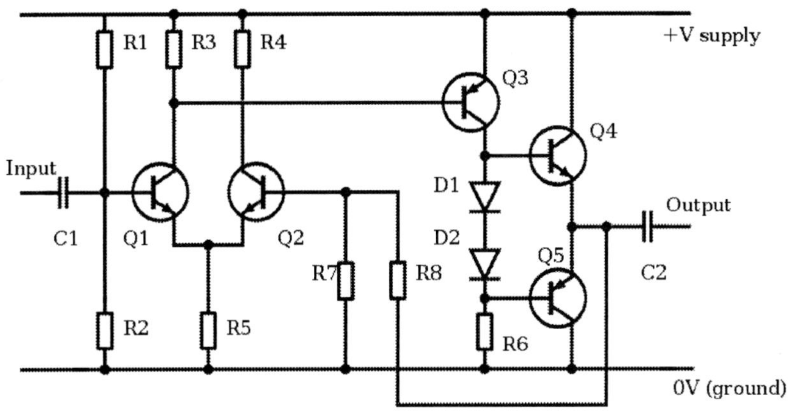

Figure: *A practical amplifier circuit*

This design automatically stabilizes its operating point, since overall feedback internally operates from DC up through the audio range and beyond. The use of fixed diode bias requires the diodes to be both electrically and thermally matched to the output transistors. If the output transistors conduct too much, they can easily overheat and destroy themselves, as the full current from the power supply is not limited at this stage.

A common solution to help stabilize the output device operating point is to include some emitter resistors, typically an ohm or so. Calculating the values of the circuit's resistors and capacitors is done based on the components employed and the intended use of the amplifier.

Field Effect Transistor (FET)

The field-effect transistor (FET) is a transistor that uses an electric field to control the shape and hence the conductivity of a channel of one type of charge carrier in a semiconductor material. FETs are unipolar transistors as they involve single-carrier-type operation. The *concept* of the FET predates the bipolar junction transistor (BJT), though it was not physically implemented until *after* BJTs due to the limitations of semiconductor materials and the relative ease of manufacturing BJTs compared to FETs at the time.

History

The field-effect transistor was first patented by Julius Edgar Lilienfeld in 1926 and by Oskar Heil in 1934, but practical semiconducting devices (the JFET) were developed only much later after the transistor effect was observed and explained by the team of William Shockley at Bell Labs in 1947. The MOSFET, which largely superseded the JFET and had a more profound effect on electronic development, was invented by DawonKahng and Martin Atalla in 1960.

Basic Information

FETs can be majority-charge-carrier devices, in which the current is carried predominantly by majority carriers, or minority-charge-carrier devices, in which the current is mainly due to a flow of minority carriers. The device consists of an active channel through which charge carriers, electrons or holes, flow from the source to the drain. Source and drain terminal conductors are connected to the semiconductor through ohmic contacts. The conductivity of the channel is a function of the potential applied across the gate and source terminals.

The FET's three terminals are:

- Source (S), through which the carriers enter the channel. Conventionally, current entering the channel at S is designated by I_S.
- Drain (D), through which the carriers leave the channel. Conventionally, current entering the channel at D is designated by I_D. Drain-to-source voltage is V_{DS}.
- Gate (G), the terminal that modulates the channel conductivity. By applying voltage to G, one can control I_D.

More About Terminals

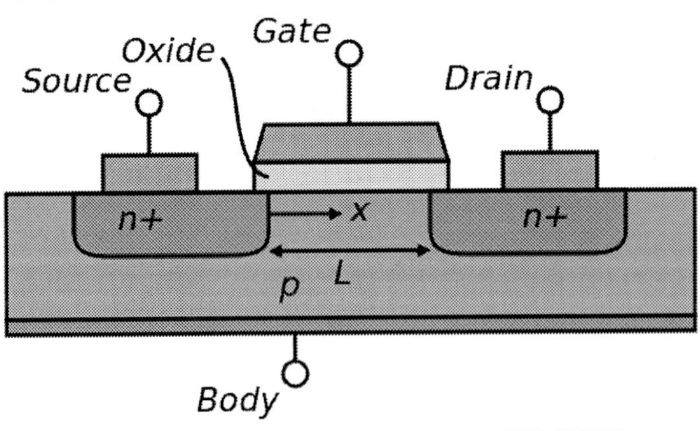

Figure: *Cross section of an n-type MOSFET*

All FETs have *source*, *drain*, and *gate* terminals that correspond roughly to the *emitter*, *collector*, and *base* of BJTs. Most FETs have a fourth terminal called the *body*, *base*, *bulk*, or *substrate*. This fourth terminal serves to bias the transistor into operation; it is rare to make non-trivial use of the body terminal in circuit designs, but its presence is important when setting up the physical layout of an integrated circuit. The size of the gate, length L in the diagram, is the distance between source and drain. The *width* is the extension of the transistor, in the direction perpendicular to the cross section in the diagram. Typically the width is much larger than the length of the gate. A gate length of 1 μm limits the upper frequency to about 5 GHz, 0.2 μm to about 30 GHz.

The names of the terminals refer to their functions. The gate terminal may be thought of as controlling the opening and closing of a physical gate. This gate permits electrons to flow through or blocks their passage by creating or eliminating a channel between the source and drain. Electrons flow from the source terminal towards the drain terminal if influenced by an applied voltage. The body simply refers to the bulk of the semiconductor in which the gate, source and drain lie. Usually the body terminal is connected to the highest or lowest voltage within the circuit, depending on the type of the FET. The body terminal and the source terminal are sometimes connected together since the source is often connected to the highest or lowest voltage within the circuit, although there are several uses of FETs which do not have such a configuration, such as transmission gates and cascode circuits.

FET Operation

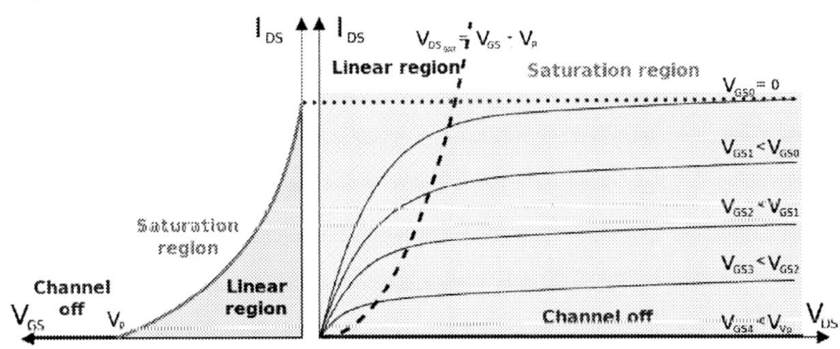

Figure: *I–V characteristics and output plot of a JFET n-channel transistor.*

The FET controls the flow of electrons (or electron holes) from the source to drain by affecting the size and shape of a "conductive channel" created and influenced by voltage (or lack of voltage) applied across the gate and source terminals. (For simplicity, this discussion assumes that the body and source are connected.) This conductive channel is the "stream" through which electrons flow from source to drain.

In an n-channel *depletion-mode* device, a negative gate-to-source voltage causes a *depletion region* to expand in width and encroach on the channel from the sides, narrowing the channel. If the depletion region expands to completely close the channel, the resistance of the channel from source to drain becomes large, and the FET is effectively turned off like a switch. This is called *pinch-off,* and the voltage at which it occurs is called the *pinch-off voltage.* Conversely, a positive gate-to-source voltage increases the channel size and allows electrons to flow easily.

In an n-channel *enhancement-mode* device, a conductive channel does not exist naturally within the transistor, and a positive gate-to-source voltage is necessary to create one. The positive voltage attracts free-floating electrons within the body towards the gate, forming a conductive channel. But first, enough electrons must be attracted near the gate to counter the dopant ions added to the body of the FET; this forms a region free of mobile carriers called a depletion region, and the voltage at which this occurs is referred to as the *threshold voltage* of the FET. Further gate-to-source voltage increase will attract even more electrons towards the gate which are able to create a conductive channel from source to drain; this process is called *inversion.*

In a p-channel *depletion-mode* device, a positive voltage from gate to body creates a depletion layer by forcing the positively charged holes away from the gate-insulator/semiconductor interface, leaving exposed a carrier-free region of immobile, negatively charged acceptor ions.

For either enhancement- or depletion-mode devices, at drain-to-source voltages much less than gate-to-source voltages, changing the gate voltage will alter the channel resistance, and drain current will be proportional to drain voltage (referenced to source voltage). In this mode the FET operates like a variable resistor and the FET is said to be operating in a *linear mode* or *ohmic mode.*

If drain-to-source voltage is increased, this creates a significant asymmetrical change in the shape of the channel due to a gradient of voltage potential from source to drain. The shape of the inversion region becomes "pinched-off" near the drain end of the channel. If drain-to-source voltage is increased further, the pinch-off point of the channel begins to move away from the drain towards the source. The FET is said to be in *saturation mode*; although some authors refer to it as *active mode*, for a better analogy with bipolar transistor operating regions. The saturation mode, or the region between ohmic and saturation, is used when amplification is needed. The in-between region is sometimes considered to be part of the ohmic or linear region, even where drain current is not approximately linear with drain voltage.

Even though the conductive channel formed by gate-to-source voltage no longer connects source to drain during saturation mode, carriers are not blocked from flowing. Considering again an n-channel enhancement-mode device, a depletion region exists in the p-type body, surrounding the conductive channel and drain and source regions. The electrons which comprise the channel are free to move out of the channel through the depletion region if attracted to the drain by drain-to-source voltage. The depletion region is free of carriers and has a resistance similar to silicon. Any increase of the drain-to-source voltage will increase the distance from drain to the pinch-off point, increasing the resistance of the depletion region in proportion to the drain-to-source voltage applied. This proportional change causes the drain-to-source current to remain relatively fixed, independent of changes to the drain-to-source voltage, quite unlike its ohmicbehaviour in the linear mode of operation. Thus, in saturation mode, the FET behaves as a constant-current source rather than as a resistor, and can effectively be used as a voltage amplifier. In this case, the gate-to-source voltage determines the level of constant current through the channel.

Composition

The FET can be constructed from a number of semiconductors, with silicon being by far the most common. Most FETs are made with conventional bulk semiconductor processing techniques, using a single crystal semiconductorwafer as the active region, or channel.

Among the more unusual body materials are amorphous silicon, polycrystalline silicon or other amorphous semiconductors in thin-film transistors or organic field-effect transistors (OFETs) that are based on organic semiconductors; often, OFET gate insulators and electrodes are made of organic materials, as well. Such FETs are manufactured using a variety of materials such as silicon carbide (SiC), gallium arsenide (GaAs), gallium nitride (GaN), and indium gallium arsenide (InGaAs). In June 2011, IBM announced that it had successfully used graphene-based FETs in an integrated circuit. These transistors are capable of about 2.23 GHz cutoff frequency, much higher than standard silicon FETs.

Types of field-effect Transistors

The channel of a FET is doped to produce either an n-type semiconductor or a p-type semiconductor. The drain and source may be doped of opposite type to the channel, in the case of depletion mode FETs, or doped of similar type to the channel as in enhancement mode FETs. Field-effect transistors are also distinguished by the method of insulation between channel and gate.

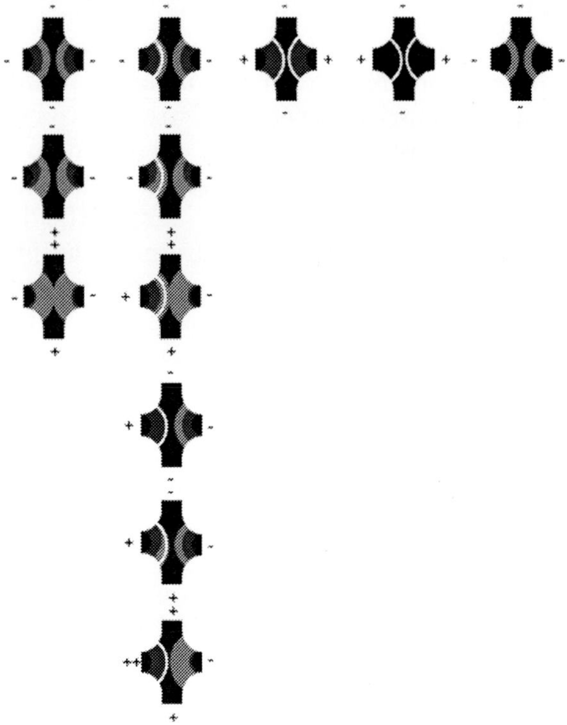

Figure: *Depletion-type FETs under typical voltages: JFET, poly-silicon MOSFET, double-gate MOSFET, metal-gate MOSFET, MESFET. depletion , electrons , holes , metal , insulator . Top=source, bottom=drain, left=gate, right=bulk. Voltages that lead to channel formation are not shown.*

Types of FETs include:

- The CNTFET (carbon nanotube field-effect transistor).
- The DEPFET is a FET formed in a fully depleted substrate and acts as a sensor, amplifier and memory node at the same time. It can be used as an image (photon) sensor.
- The DGMOSFET (dual-gate MOSFET) is a FET with two insulated gates.
- The DNAFET (DNA field-effect transistor) is a specialized FET that acts as a biosensor, by using a gate made of single-strand DNA molecules to detect matching DNA strands.

- The FREDFET (fast-reverse or fast-recovery epitaxial diode FET) is a specialized FET designed to provide a very fast recovery (turn-off) of the body diode.
- The HEMT (high-electron-mobility transistor), also called a HFET (heterostructure FET), can be made using bandgap engineering in a ternary semiconductor such as AlGaAs. The fully depleted wide-band-gap material forms the isolation between gate and body.
- The HIGFET (heterostructure insulated gate field-effect transisitor) is now used mainly in research.
- The IGBT (insulated-gate bipolar transistor) is a device for power control. It has a structure akin to a MOSFET coupled with a bipolar-like main conduction channel. These are commonly used for the 200–3000 V drain-to-source voltage range of operation. Power MOSFETs are still the device of choice for drain-to-source voltages of 1 to 200 V.
- The ISFET (ion-sensitive field-effect transistor) can be used to measure ion concentrations in a solution; when the ion concentration (such as H^+) changes, the current through the transistor will change accordingly.
- The JFET (junction field-effect transistor) uses a reverse biased p–n junction to separate the gate from the body.
- The MESFET (metal–semiconductor field-effect transistor) substitutes the p–n junction of the JFET with a Schottky barrier; and is used in GaAs and other III-V semiconductor materials.
- The MODFET (modulation-doped field-effect transistor) uses a quantum well structure formed by graded doping of the active region.
- The MOSFET (metal–oxide–semiconductor field-effect transistor) utilizes an insulator (typically SiO_2) between the gate and the body.
- The NOMFET is a nanoparticle organic memory field-effect transistor.
- The OFET (organic field-effect transistor) uses an organic semiconductor in its channel.
- The GNRFET (graphenenanoribbon field-effect transistor) uses a graphenenanoribbon for its channel.
- The VeSFET (vertical-slit field-effect transistor) is a square-shaped junctionless FET with a narrow slit connecting the

source and drain at opposite corners. Two gates occupy the other corners, and control the current through the slit.

- The TFET (tunnel field-effect transistor) is based on band-to-band tunnelling.

Advantages of FET

The main advantage of the FET is its high input resistance, on the order of 100 MΩ or more. Thus, it is a voltage-controlled device, and shows a high degree of isolation between input and output. It is a unipolar device, depending only on majority current flow. It is less noisy and is thus found in FM tuners and in low-noise amplifiers for VHF and satellite receivers. It is relatively immune to radiation. It exhibits no offset voltage at zero drain current and hence makes an excellent signal chopper. It typically has better thermal stability than a bipolar junction transistor (BJT).

Disadvantages of FET

It has a relatively low gain-bandwidth product compared to a BJT. The MOSFET has a drawback of being very susceptible to overload voltages, thus requiring special handling during installation. The fragile insulating layer of the MOSFET between the gate and channel makes it vulnerable to electrostatic damage during handling. This is not usually a problem after the device has been installed in a properly designed circuit.

Uses

The most commonly used FET is the MOSFET. The CMOS (complementary metal oxide semiconductor) process technology is the basis for modern digitalintegrated circuits. This process technology uses an arrangement where the (usually "enhancement-mode") p-channel MOSFET and n-channel MOSFET are connected in series such that when one is on, the other is off.

In FETs, electrons can flow in either direction through the channel when operated in the linear mode. The naming convention of drain terminal and source terminal is somewhat arbitrary, as the devices are typically (but not always) built symmetrically from source to drain. This makes FETs suitable for switching analog signals between paths (multiplexing). With this concept, one can construct a solid-state mixing board, for example. A common use of the FET is as an amplifier. For example, due to its large input resistance and low output resistance, it is effective as a buffer in common-drain (source follower) configuration.

CMOS Circuits

Complementary metal–oxide–semiconductor (CMOS) is a technology for constructing integrated circuits. CMOS technology is used in microprocessors, microcontrollers, static RAM, and other digital logic circuits. CMOS technology is also used for several analog circuits such as image sensors (CMOS sensor), data converters, and highly integrated transceivers for many types of communication. Frank Wanlass patented CMOS in 1963 (US patent 3,356,858).

CMOS is also sometimes referred to as complementary-symmetry metal–oxide–semiconductor (or COS-MOS). The words "complementary-symmetry" refer to the fact that the typical digital design style with CMOS uses complementary and symmetrical pairs of p-type and n-typemetal oxide semiconductor field effect transistors (MOSFETs) for logic functions.

Two important characteristics of CMOS devices are high noise immunity and low static power consumption. Since one transistor of the pair is always off, the series combination draws significant power only momentarily during switching between on and off states. Consequently, CMOS devices do not produce as much waste heat as other forms of logic, for example transistor–transistor logic (TTL) or NMOS logic, which normally have some standing current even when not changing state. CMOS also allows a high density of logic functions on a chip. It was primarily for this reason that CMOS became the most used technology to be implemented in VLSI chips.

The phrase "metal–oxide–semiconductor" is a reference to the physical structure of certain field-effect transistors, having a metal gate electrode placed on top of an oxide insulator, which in turn is on top of a semiconductor material. Aluminium was once used but now the material is polysilicon. Other metal gates have made a comeback with the advent of high-k dielectric materials in the CMOS process, as announced by IBM and Intel for the 45 nanometre node and beyond.

Technical Details

"CMOS" refers to both a particular style of digital circuitry design, and the family of processes used to implement that circuitry on integrated circuits (chips). CMOS circuitry dissipates less power than logic families with resistive loads. Since this advantage has increased and grown more important, CMOS processes and variants have come to dominate, thus the vast majority of modern integrated circuit manufacturing is on CMOS processes. As of 2010, CPUs with the best performance per watt each year have been CMOS static logic since 1976.

CMOS circuits use a combination of p-channel and n-channel metal–oxide–semiconductor field-effect transistors (MOSFETs) to implement logic gates. Although CMOS logic can be implemented with discrete devices for demonstrations, commercial CMOS products are integrated circuits composed of up to millions of transistors of both types on a rectangular piece of silicon of between 10 and 400 mm^2.

Inversion

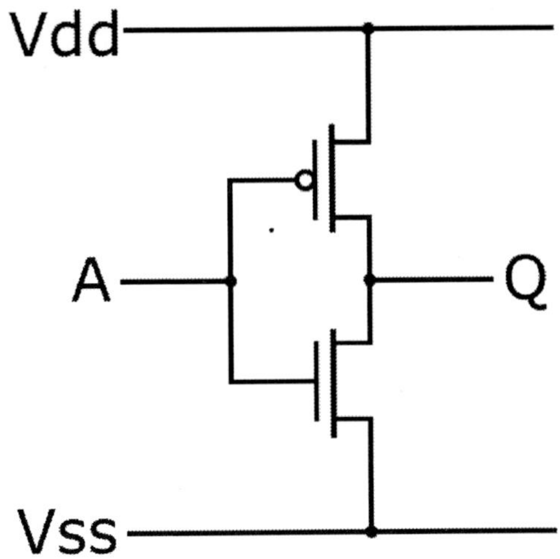

Figure: *Static CMOS Inverter*

CMOS circuits are constructed in such a way that all PMOS transistors must have either an input from the voltage source or from another PMOS transistor. Similarly, all NMOS transistors must have either an input from ground or from another NMOS transistor. The composition of a PMOS transistor creates low resistance between its source and drain contacts when a low gatevoltage is applied and high resistance when a high gate voltage is applied. On the other hand, the composition of an NMOS transistor creates high resistance between source and drain when a low gate voltage is applied and low resistance when a high gate voltage is applied. CMOS accomplishes current reduction by complementing every nMOSFET with a pMOSFET and connecting both gates and both drains together. A high voltage on the gates will cause the nMOSFET to conduct and the pMOSFET to not conduct while a low voltage on the gates causes the reverse. This arrangement greatly reduces power consumption and heat generation. However, during the switching time both MOSFETs conduct briefly as the gate voltage goes from one state to another. This induces a brief spike in power consumption and becomes a serious issue at high frequencies.

The image on the right shows what happens when an input is connected to both a PMOS transistor (top of diagram) and an NMOS transistor (bottom of diagram). When the voltage of input A is low, the NMOS transistor's channel is in a high resistance state. This limits the current that can flow from Q to ground. The PMOS transistor's channel is in a low resistance state and much more current can flow from the supply to the output. Because the resistance between the supply voltage and Q is low, the voltage drop between the supply voltage and Q due to a current drawn from Q is small. The output therefore registers a high voltage.

On the other hand, when the voltage of input A is high, the PMOS transistor is in an OFF (high resistance) state so it would limit the current flowing from the positive supply to the output, while the NMOS transistor is in an ON (low resistance) state, allowing the output to drain to ground. Because the resistance between Q and ground is low, the voltage drop due to a current drawn into Q placing Q above ground is small. This low drop results in the output registering a low voltage.

In short, the outputs of the PMOS and NMOS transistors are complementary such that when the input is low, the output is high, and when the input is high, the output is low. Because of this behaviour of input and output, the CMOS circuits' output is the inverse of the input.

The power supplies for CMOS are called V_{DD} and V_{SS}, or V_{CC} and Ground(GND) depending on the manufacturer. V_{DD} and V_{SS} are carryovers from conventional MOS circuits and stand for the drain and source supplies. These do not apply directly to CMOS since both supplies are really source supplies. V_{CC} and Ground are carryovers from TTL logic and that nomenclature has been retained with the introduction of the 54C/74C line of CMOS.

Duality

An important characteristic of a CMOS circuit is the duality that exists between its PMOS transistors and NMOS transistors. A CMOS circuit is created to allow a path always to exist from the output to either the power source or ground. To accomplish this, the set of all paths to the voltage source must be the complement of the set of all paths to ground. This can be easily accomplished by defining one in terms of the NOT of the other. Due to the De Morgan's laws based logic, the PMOS transistors in parallel have corresponding NMOS transistors in series while the PMOS transistors in series have corresponding NMOS transistors in parallel.

Logic

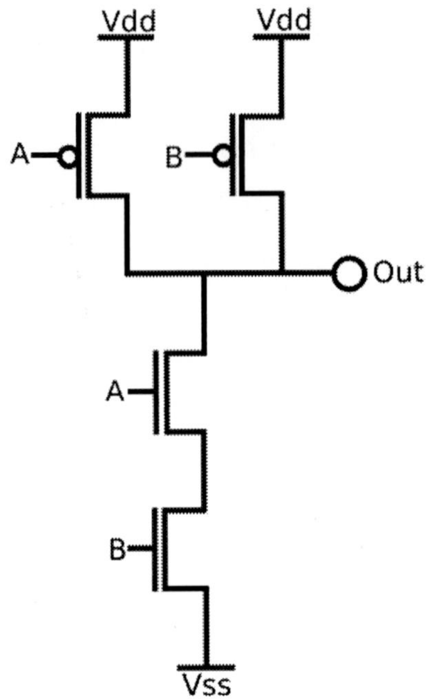

Figure: *NAND gate in CMOS logic*

More complex logic functions such as those involving ANDandOR gates require manipulating the paths between gates to represent the logic. When a path consists of two transistors in series, both transistors must have low resistance to the corresponding supply voltage, modelling an AND. When a path consists of two transistors in parallel, either one or both of the transistors must have low resistance to connect the supply voltage to the output, modelling an OR.

Shown on the right is a circuit diagram of a NAND gate in CMOS logic. If both of the A and B inputs are high, then both the NMOS transistors (bottom half of the diagram) will conduct, neither of the PMOS transistors (top half) will conduct, and a conductive path will be established between the output and V_{ss} (ground), bringing the output low. If either of the A or B inputs is low, one of the NMOS transistors will not conduct, one of the PMOS transistors will, and a conductive path will be established between the output and V_{dd} (voltage source), bringing the output high.

An advantage of CMOS over NMOS is that *both* low-to-high and high-to-low output transitions are fast since the pull-up transistors have low resistance when switched on, unlike the load resistors in NMOS logic. In addition, the output signal swings the full voltage between the low and high rails. This strong, more nearly symmetric response also makes CMOS more resistant to noise.

Example: NAND Gate in Physical Layout

This example shows a NAND logic device drawn as a physical representation as it would be manufactured. The physical layout perspective is a "bird's eye view" of a stack of layers. The circuit is constructed on a P-type substrate. The polysilicon, diffusion, and n-well are referred to as "base layers" and are actually inserted into trenches of the P-type substrate. The contacts penetrate an insulating layer between the base layers and the first layer of metal (metal1) making a connection.

The inputs to the NAND (illustrated in green colour) are in polysilicon. The CMOS transistors (devices) are formed by the intersection of the polysilicon and diffusion; N diffusion for the N device & P diffusion for the P device (illustrated in salmon and yellow colour in grespectively).

The output ("out") is connected together in metal (illustrated in cyan colouring). Connections between metal and polysilicon or diffusion are made through contacts (illustrated as black squares). The physical layout example matches the NAND logic circuit given in the previous example.

The N device is manufactured on a P-type substrate while the P device is manufactured in an N-type well (n-well). A P-type substrate "tap" is connected to V_{SS} and an N-type n-well tap is connected to V_{DD} to prevent latchup.

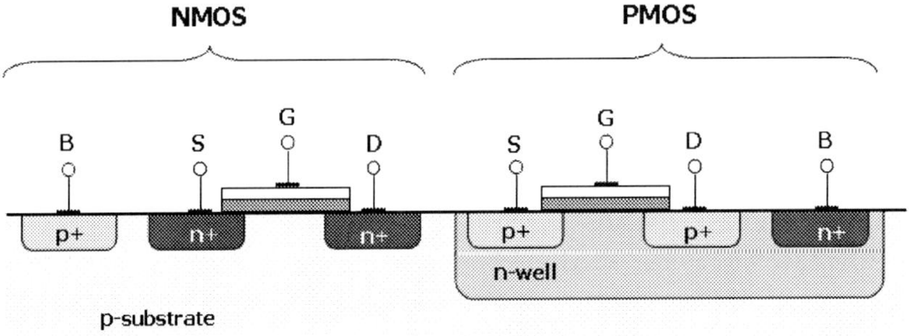

Figure: *Cross section of two transistors in a CMOS gate, in an N-well CMOS process*

Power: Switching and Leakage

CMOS logic dissipates less power than NMOS logic circuits because CMOS dissipates power only when switching ("dynamic power"). On a typical ASIC in a modern 90 nanometer process, switching the output might take 120 picoseconds, and happens once every ten nanoseconds. NMOS logic dissipates power whenever the transistor is on, because there is a current path from V_{dd} to V_{ss} through the load resistor and the n-type network.

Static CMOS gates are very power efficient because they dissipate nearly zero power when idle. Earlier, the power consumption of CMOS devices was not the major concern while designing chips. Factors like speed and area dominated the design parameters. As the CMOS technology moved below sub-micron levels the power consumption per unit area of the chip has risen tremendously.

Broadly classifying, power dissipation in CMOS circuits occurs because of two components:

- Static dissipation
- Sub threshold conduction when the transistors are off.

Both NMOS and PMOS transistors have a gate–source threshold voltage, below which the current (called *sub threshold* current) through the device drops exponentially. Historically, CMOS designs operated at supply voltages much larger than their threshold voltages (V_{dd} might have been 5 V, and V_{th} for both NMOS and PMOS might have been 700 mV). A special type of the CMOS transistor with near zero threshold voltage is the native transistor.

- Tunnelling current through gate oxide.

SiO_2 is a very good insulator, but at very small thickness levels electrons can tunnel across the very thin insulation; the probability drops off exponentially with oxide thickness. Tunnelling current becomes very important for transistors below 130 nm technology with gate oxides of 20 Å or thinner.

- Leakage current through reverse biased diodes.

Small reverse leakage currents are formed due to formation of reverse bias between diffusion regions and wells (for e.g., p-type diffusion vs. n-well), wells and substrate (for e.g., n-well vs. p-substrate). In modern process diode leakage is very small compared to sub threshold and tunnelling currents, so these may be neglected during power calculations.

- Contention current in ratioed circuit
- Dynamic Dissipation
- Charging and discharging of load capacitances.

CMOS circuits dissipate power by charging the various load capacitances (mostly gate and wire capacitance, but also drain and some source capacitances) whenever they are switched. In one complete cycle of CMOS logic, current flows from V_{DD} to the load capacitance to charge it and then flows from the charged load capacitance to ground during discharge. Therefore in one complete charge/discharge cycle, a total of $Q=C_L V_{DD}$ is thus transferred from V_{DD} to ground. Multiply by the switching frequency on the load capacitances to get the current used, and multiply by voltage again to get the characteristic switching power dissipated by a CMOS device: $P = CV^2 f$.

Since most gates do not operate/switch at every clock cycle, they are often accompanied by a factor α, called the activity factor. Now, the dynamic power dissipation may be re-written as $P = \alpha CV^2 f$.

A clock in a system has an activity factor $\alpha=1$, since it rises and falls every cycle. Most data has an activity factor of 0.1. If correct load capacitance is estimated on a node together with its activity factor, the dynamic power dissipation at that node can be calculated effectively.

- Short circuit power dissipation

Since there is a finite rise/fall time for both pMOS and nMOS, during transition, for example, from off to on, both the transistors will be on for a small period of time in which current will find a path directly from V_{DD} to ground, hence creating a short circuit current. Short circuit power dissipation increases with rise and fall time of the transistors.

An additional form of power consumption became significant in the 1990s as wires on chip became narrower and the long wires became more resistive. CMOS gates at the end of those resistive wires. During the middle of these transitions, both the NMOS and PMOS logic networks are partially conductive, and current flows directly from V_{dd} to V_{SS}. The power thus used is called *crowbar* power. Careful design which avoids weakly driven long skinny wires has ameliorated this effect, and crowbar power is nearly always substantially smaller than switching power.

To speed up designs, manufacturers have switched to constructions that have lower voltage thresholds but because of this a modern NMOS transistor with a V_{th} of 200 mV has a significant subthreshold leakage current. Designs (e.g. desktop processors) which include vast numbers of circuits which are not actively switching still consume power because of this leakage current. Leakage power is a significant portion of the total power consumed by such designs. Multi-threshold CMOS (MTCMOS), now available from foundries, is one approach to managing leakage power. With MTCMOS, high V_{th} transistors are used when switching speed is not critical, while low V_{th} transistors are used in speed sensitive paths. Further technology advances that use even thinner gate dielectrics have an additional leakage component because of current tunnelling through the extremely thin gate dielectric. Using high-k dielectrics instead of silicon dioxide that is the conventional gate dielectric allows similar device performance, but with a thicker gate insulator, thus avoiding this current. Leakage power reduction using new material and system designs is critical to sustaining scaling of CMOS.

Analog CMOS

Besides digital applications, CMOS technology is also used in analog applications. For example, there are CMOS operational amplifier ICs available in the market. Transmission gates may be used instead of signal relays. CMOS technology is also widely used for RF circuits all the way to microwave frequencies, in mixed-signal (analog+digital) applications.

Temperature Range

Conventional CMOS devices work over a range of –55 °C to +125 °C. There were theoretical indications as early as August 2008 that silicon CMOS will work down to –233 °C (40 K). Functioning temperatures near 40 K have since been achieved using overclocked AMD Phenom II processors with a combination of liquid nitrogen and liquid helium cooling.

Single electron CMOS transistors

Ultra small (L=20 nm, W=20 nm) CMOS transistors achieve the single electron limit when operated at cryogenic temperature over a range of –269 °C (4 K) to about –258 °C (15 K). The transistor displays Coulomb blockade due to progressive charging of electrons one by one. The number of electrons confined in the channel is driven by the gate voltage, starting from an occupation of zero electrons, and it can be set to 1 or many.

Chapter

3

Transistor Amplifiers and Oscillators

Transistor Amplifiers

An electronic amplifier, amplifier, or (informally) amp is an electronic device that increases the power of a signal. It does this by taking energy from a power supply and controlling the output to match the input signal shape but with a larger amplitude. In this sense, an amplifier modulates the output of the power supply.

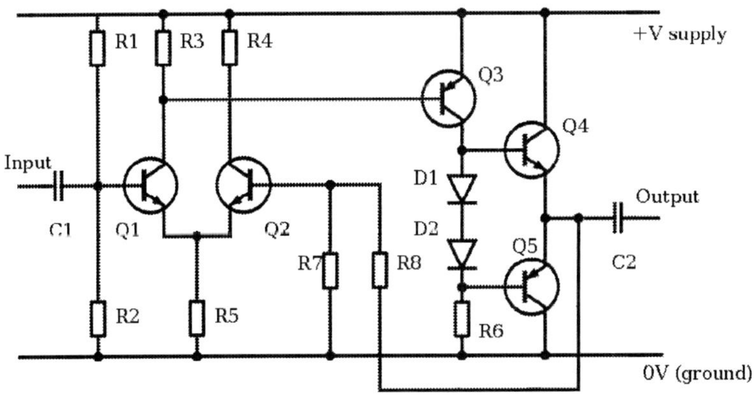

Figure: *A practical amplifier circuit*

Numerous types of electronic amplifiers are specialized to various applications. An amplifier can refer to anything from an electrical circuit that uses a single active component, to a complete system such as a packaged audio hi-fi amplifier.

Figures of Merit

Amplifier quality is characterized by a list of specifications that includes:

- Gain, the ratio between the magnitude of output and input signals
- Bandwidth, the width of the useful frequency range
- Efficiency, the ratio between the power of the output and total power consumption
- Linearity, the degree of proportionality between input and output
- Noise, a measure of undesired noise mixed into the output
- Output dynamic range, the ratio of the largest and the smallest useful output levels
- Slew rate, the maximum rate of change of the output
- Rise time, settling time, ringing and overshoot that characterize the step response
- Stability, the ability to avoid self-oscillation

Amplifier Types

Amplifiers are described according to their input and output properties. They have some kind of gain, or multiplication factor that relates the magnitude of the output signal to the input signal. The gain may be specified as the ratio of output voltage to input voltage (voltage gain), output power to input power (power gain), or some combination of current, voltage, and power. In many cases, with input and output in the same unit, gain is unitless (though often expressed in decibels). For others this is not necessarily so. For example, a transconductance amplifier has a gain with units of conductance (output current per input voltage). The power gain of an amplifier depends on the source and load impedances used as well as its voltage gain; while an RF amplifier may have its impedances optimized for power transfer, audio and instrumentation amplifiers are normally employed with amplifier input and output impedances optimized for least loading and highest quality. So an amplifier that is said to have a gain of 20 dB might have a voltage gain of ten times and an available power gain of much more than 20 dB (100 times power ratio), yet be delivering a much lower power gain if, for example, the input is a 600 ohm microphone and the output is a 47 kilohm power amplifier's input socket.

In most cases an amplifier should be linear; that is, the gain should be constant for any combination of input and output signal. If the gain is not constant, e.g., by clipping the output signal at the limits of its capabilities, the output signal is distorted. There are however cases where variable gain is useful.

There are many types of electronic amplifiers, commonly used in radio and televisiontransmitters and receivers, high-fidelity ("hi-fi") stereo equipment, microcomputers and other electronic digital equipment, and guitar and other instrument amplifiers. Critical components include active devices, such as vacuum tubes or transistors. A brief introduction to the many types of electronic amplifier follows.

Power Amplifier

The term *power amplifier* is a relative term with respect to the amount of power delivered to the load and/or sourced by the supply circuit. In general a power amplifier is designated as the last amplifier in a transmission chain (the *output stage*) and is the amplifier stage that typically requires most attention to power efficiency. Efficiency considerations lead to various classes of power amplifier based on the biasing of the output transistors or tubes.

Power Amplifiers by Application

- Audio power amplifiers
- RF power amplifier, such as for transmitter final stages.
- Servo motor controllers, where linearity is not important.
- Piezoelectric audio amplifier includes a DC-to-DC converter to generate the high voltage output required to drive piezoelectric speakers.

Power Amplifier Circuits

Power amplifier circuits include the following types:

- Vacuum tube/valve, hybrid or transistor power amplifiers
- Push-pull output or single-ended output stages

Vacuum-tube (valve) Amplifiers

According to Symons, while semiconductor amplifiers have largely displaced valve amplifiers for low power applications, valve amplifiers are much more cost effective in high power applications such as "radar, countermeasures equipment, or communications equipment" (p. 56). Many microwave amplifiers are specially designed valves, such as the klystron, gyrotron, travelling wave tube, and crossed-field amplifier, and these microwave valves provide much greater single-device power output at microwave frequencies than solid-state devices (p. 59).

Valves/tube amplifiers also have niche uses in other areas, such as

- electric guitar amplification
- in Russian military aircraft, for their EMP tolerance
- niche audio for their sound qualities (recording, and audiophile equipment)

Transistor Amplifiers

The essential role of this active element is to magnify an input signal to yield a significantly larger output signal. The amount of magnification (the "forward gain") is determined by the external circuit design as well as the active device. Many common active devices in transistor amplifiers are bipolar junction transistors (BJTs) and metal oxide semiconductor field-effect transistors (MOSFETs).

Applications are numerous, some common examples are audio amplifiers in a home stereo or PA system, RF high power generation for semiconductor equipment, to RF and Microwave applications such as radio transmitters. Transistor-based amplifier can be realized using various configurations: for example with a bipolar junction transistor we can realize common base, common collector or common emitter amplifier; using a MOSFET we can realize common gate, common source or common drain amplifier. Each configuration has different characteristic (gain, impedance...).

Operational Amplifiers (op-amps)

Figure: *An LM741 general purpose op-amp*

An operational amplifier is an amplifier circuit with very high open loop gain and differential inputs that employs external feedback to control its transfer function, or *gain*. Though the term today commonly applies to integrated circuits, the original operational amplifier design used valves.

Fully Differential Amplifiers

A fully differential amplifier is a solid state integrated circuit amplifier that uses external feedback to control its transfer function or gain. It is similar to the operational amplifier, but also has differential output pins. These are usually constructed using BJTs or FETs.

Video Amplifiers

These deal with video signals and have varying bandwidths depending on whether the video signal is for SDTV, EDTV, HDTV 720p or 1080i/p etc.. The specification of the bandwidth itself depends on what kind of filter is used—and at which point (-1 dB or -3 dB for example) the bandwidth is measured. Certain requirements for step response and overshoot are necessary for an acceptable TV image.

Oscilloscope Vertical Amplifiers

These deal with video signals that drive an oscilloscope display tube, and can have bandwidths of about 500 MHz. The specifications on step response, rise time, overshoot, and aberrations can make designing these amplifiers difficult. One of the pioneers in high bandwidth vertical amplifiers was the Tektronix company.

Distributed Amplifiers

These use transmission lines to temporally split the signal and amplify each portion separately to achieve higher bandwidth than possible from a single amplifier. The outputs of each stage are combined in the output transmission line. This type of amplifier was commonly used on oscilloscopes as the final vertical amplifier. The transmission lines were often housed inside the display tube glass envelope.

Switched Mode Amplifiers

These nonlinear amplifiers have much higher efficiencies than linear amps, and are used where the power saving justifies the extra complexity.

Negative Resistance Devices

Negative resistances can be used as amplifiers, such as the tunnel diode amplifier.

Microwave Amplifiers

Travelling wave Tube Amplifiers: Travelling wave tube amplifiers (TWTAs) are used for high power amplification at low microwave frequencies. They typically can amplify across a broad spectrum of frequencies; however, they are usually not as tunable as klystrons.

Klystrons

Klystrons are specialized linear-beam vacuum-devices, designed to provide high power, widely tunable amplification of millimetre and sub-millimetre waves. Klystrons are designed for large scale operations and despite having a narrower bandwidth than TWTAs, they have the advantage of coherently amplifying a reference signal so its output may be precisely controlled in amplitude, frequency and phase.

Musical Instrument Amplifiers

An audio power amplifier is usually used to amplify signals such as music or speech. Several factors are especially important in the selection of musical instrument amplifiers (such as guitar amplifiers) and other audio amplifiers (although the whole of the sound system – components such as microphones to loudspeakers – affect these parameters):

- Frequency response – not just the frequency range but the requirement that the signal level varies so little across the audible frequency range that the human ear notices no variation. A typical specification for audio amplifiers may be 20 Hz to 20 kHz +/- 0.5dB.
- Power output – the power level obtainable with little distortion, to obtain a sufficiently loud sound pressure level from the loudspeakers.
- Low distortion – all amplifiers and transducers distort to some extent. They cannot be perfectly linear, but aim to pass signals without affecting the harmonic content of the sound more than the human ear can tolerate. That tolerance of distortion, and indeed the possibility that some "warmth" or second harmonic distortion (Tube sound) improves the "musicality" of the sound, are subjects of great debate.

Classification of Amplifier Stages and Systems

Many alternative classifications address different aspects of amplifier designs, and they all express some particular perspective relating the design parameters to the objectives of the circuit. Amplifier design is always a compromise of numerous factors, such as cost, power consumption, real-world device imperfections, and a multitude of performance specifications. Below are several different approaches to classification:

Input and Output Variables

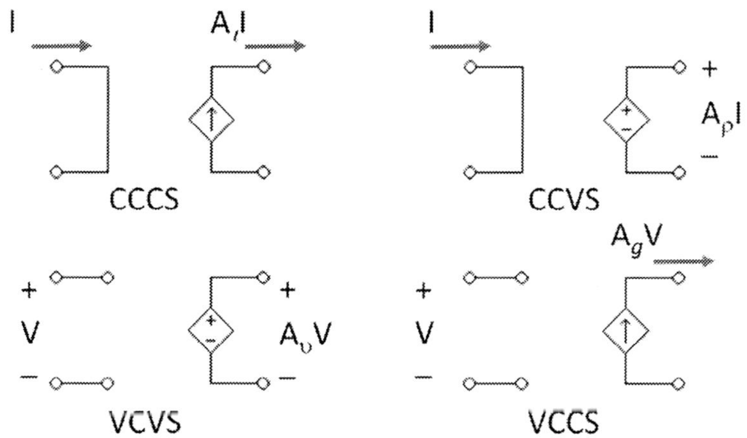

Figure: *The four types of dependent source—control variable on left, output variable on right*

Electronic amplifiers use one variable presented as either a current and voltage. Either current or voltage can be used as input and either as output, leading to four types of amplifiers. In idealized form they are represented by each of the four types of dependent source used in linear analysis, as shown in the figure, namely:

Input	***Output***	***Dependent source***	***Amplifier type***
I	I	current controlled current source CCCS	current amplifier
I	V	current controlled voltage source CCVS	transresistance amplifier
V	I	voltage controlled current source VCCS	transconductance amplifier
V	V	voltage controlled voltage source VCVS	voltage amplifier

Each type of amplifier in its ideal form has an ideal input and output resistance that is the same as that of the corresponding dependent source:

Amplifier type	***Dependent source***	***Input impedance***	***Output impedance***
Current	CCCS	0	∞
Transresistance	CCVS	0	0
Transconductance	VCCS	∞	∞
Voltage	VCVS	∞	0

In practice the ideal impedances are only approximated. For any particular circuit, a small-signal analysis is often used to find the impedance actually achieved. A small-signal AC test current I_x is applied to the input or output node, all external sources are set to AC zero, and the corresponding alternating voltage V_x across the test current source determines the impedance seen at that node as $R = V_x / I_x$.

Amplifiers designed to attach to a transmission line at input and/or output, especially RF amplifiers, do not fit into this classification approach. Rather than dealing with voltage or current individually, they ideally couple with an input and/or output impedance matched to the transmission line impedance, that is, match *ratios* of voltage to current. Many real RF amplifiers come close to this ideal. Although, for a given appropriate source and load impedance, RF amplifiers can be characterized as amplifying voltage or current, they fundamentally are amplifying power.

Common Terminal

One set of classifications for amplifiers is based on which device terminal is common to both the input and the output circuit. In the case of bipolar junction transistors, the three classes are common emitter, common base, and common collector. For field-effect transistors, the corresponding configurations are common source, common gate, and common drain; for triode vacuum devices, common cathode, *common grid*, and *common plate*. The output voltage of a common plate amplifier is the same as the input (this arrangement is used as the input presents a high impedance and does not load the signal source, although it does not amplify the voltage), i.e., the output at the cathode follows the input at the grid; consequently it was commonly called a *cathode follower*. By analogy the terms *emitter follower* and *source follower* are sometimes used.

Unilateral or Bilateral

When an amplifier has an output that exhibits no feedback to its input side, it is called 'unilateral'. The input impedance of a unilateral amplifier is independent of the load, and the output impedance is independent of the signal source impedance.

If feedback connects part of the output back to the input of the amplifier it is called a 'bilateral' amplifier. The input impedance of a bilateral amplifier is dependent upon the load, and the output impedance is dependent upon the signal source impedance.

All amplifiers are bilateral to some degree; however they may often be modelled as unilateral under operating conditions where feedback is small enough to neglect for most purposes, simplifying analysis.

Negative feedback is often applied deliberately to tailor amplifier behaviour. Some feedback, which may be positive or negative, is unavoidable and often undesirable, introduced, for example, by parasitic elements such as the inherent capacitance between input and output of a device such as a transistor and capacitative coupling due to external wiring. Excessive frequency-dependent positive feedback may cause what is intended/expected to be an amplifier to become an oscillator.

Linear unilateral and bilateral amplifiers can be represented as two-port networks.

Inverting or Non-inverting

Another way to classify amps is the phase relationship of the input signal to the output signal. An 'inverting' amplifier produces an output 180 degrees out of phase with the input signal (that is, a polarity inversion or mirror image of the input as seen on an oscilloscope). A 'non-inverting' amplifier maintains the phase of the input signal waveforms. An emitter follower is a type of non-inverting amplifier, indicating that the signal at the emitter of a transistor is following (that is, matching with unity gain but perhaps an offset) the input signal.

This description can apply to a single stage of an amplifier, or to a complete amplifier system.

Function

Other amplifiers may be classified by their function or output characteristics. These functional descriptions usually apply to complete amplifier systems or sub-systems and rarely to individual stages.

- A servo amplifier indicates an integrated feedback loop to actively control the output at some desired level. A DC servo indicates use at frequencies down to DC levels, where the rapid fluctuations of an audio or RF signal do not occur. These are often used in mechanical actuators, or devices such as DC motors that must maintain a constant speed or torque. An AC servo amp can do this for some ac motors.
- A linear amplifier responds to different frequency components independently, and does not generate harmonic distortion or Intermodulation distortion.

A nonlinear amplifier does generate distortion. For example, it may output to a lamp that must be either fully on or off based on a threshold in a continuously variable input. In other examples, a nonlinear amplifier in an analog computer can provide a special transfer function, such as logarithmic—or a following tuned circuit removes harmonics generated by a nonlinear RF amplifier.

Even the most linear amplifier has some nonlinearities, since the amplifying devices—transistors or vacuum tubes—follow nonlinear power laws such as square-laws and rely on circuitry techniques to reduce those effects.

- A wideband amplifier has a precise amplification factor over a wide frequency range, and is often used to boost signals for relay in communications systems. A narrowband amp amplifies a specific narrow range of frequencies, to the exclusion of other frequencies.
- An RF amplifier amplifies signals in the radio frequency range of the electromagnetic spectrum, and is often used to increase the sensitivity of a receiver or the output power of a transmitter.
- An audio amplifier amplifies audio frequencies. This category subdivides into small signal amplification, and power amps that are optimised to driving speakers, sometimes with multiple amps grouped together as separate or bridgeable channels to accommodate different audio reproduction requirements. Frequently used terms within audio amplifiers include:
    - Preamplifier (preamp), which may include a phono preamp with RIAA equalization, or tape head preamps with CCIR equalisation filters. They may include filters or tone control circuitry.
    - Power amplifier (normally drives loudspeakers), headphone amplifiers, and public address amplifiers.
    - Stereo amplifiers imply two channels of output (left and right), though the term simply means "solid" sound (referring to three-dimensional)—so quadraphonic stereo was used for amplifiers with four channels. 5.1 and 7.1 systems refer to Home theatre systems with 5 or 7 normal spacial channels, plus a subwoofer channel.
- Buffer amplifiers, which may include emitter followers, provide a high impedance input for a device (perhaps another amplifier, or perhaps an energy-hungry load such as lights) that would otherwise draw too much current from the source. Line drivers are a type of buffer that feeds long or interference-prone interconnect cables, possibly with differential outputs through twisted pair cables.

- A special type of amplifier is widely used in measuring instruments for signal processing, and many other uses. These are called operational amplifiers or op-amps. The "operational" name is because this type of amplifier can be used in circuits that perform mathematical algorithmic functions, or "operations" on input signals to obtain specific types of output signals. A typical modern op-amp has differential inputs (one "inverting", one "non-inverting") and one output.

An *idealised* op-amp has the following characteristics:

- Infinite input impedance (so it does not load the circuitry at its input)
- Zero output impedance
- Infinite gain
- Zero propagation delay

The performance of an op-amp with these characteristics is entirely defined by the (usually passive) components that form a negative feedback loop around it. The amplifier itself does not effect the output.

Modern op-amps are usually provided as integrated circuits, rather than constructed from discrete components. All real-world op-amps fall short of the idealised specification above—but some modern components have remarkable performance and come close in some respects.

Interstage Coupling Method

Amplifiers are sometimes classified by the coupling method of the signal at the input, output, or between stages. Different types of these include:

Resistive capacitive (RC) coupled amplifier, using a network of resistors and capacitors

By design these amplifiers cannot amplify DC signals as the capacitors block the DC component of the input signal. RC-coupled amplifiers were used very often in circuits with vacuum tubes or discrete transistors. In the days of the integrated circuit a few more transistors on a chip are much cheaper and smaller than a capacitor.

Inductive-capacitive (LC) coupled amplifier, using a network of inductors and capacitors. This kind of amplifier is most often used in selective radio-frequency circuits.

Transformer coupled amplifier, using a transformer to match impedances or to decouple parts of the circuits

Quite often LC-coupled and transformer-coupled amplifiers cannot be distinguished as a transformer is some kind of inductor.

Direct coupled amplifier, using no impedance and bias matching components. This class of amplifier was very uncommon in the vacuum tube days when the anode (output) voltage was at greater than several hundred volts and the grid (input) voltage at a few volts minus. So they were only used if the gain was specified down to DC (e.g., in an oscilloscope). In the context of modern electronics developers are encouraged to use directly coupled amplifiers whenever possible.

Frequency Range

Depending on the frequency range and other properties amplifiers are designed according to different principles.

- Frequency ranges down to DC are only used when this property is needed. DC amplification leads to specific complications that are avoided if possible; DC-blockingcapacitors are added to remove DC and sub-sonic frequencies from audio amplifiers.
- Depending on the frequency range specified different design principles must be used. Up to the MHz range only "discrete" properties need be considered; e.g., a terminal has an input impedance.
- As soon as any connection within the circuit gets longer than perhaps 1% of the wavelength of the highest specified frequency (e.g., at 100 MHz the wavelength is 3 m, so the critical connection length is approx. 3 cm) design properties radically change. For example, a specified length and width of a PCB trace can be used as a selective or impedance-matching entity.
- Above a few hundred MHz, it gets difficult to use discrete elements, especially inductors. In most cases, PCB traces of very closely defined shapes are used instead.

The frequency range handled by an amplifier might be specified in terms of bandwidth (normally implying a response that is 3 dB down when the frequency reaches the specified bandwidth), or by specifying a frequency response that is within a certain number of decibels between a lower and an upper frequency (e.g. "20 Hz to 20 kHz plus or minus 1 dB").

Power amplifier Classes

Power amplifier circuits (output stages) are classified as A, B, AB and C for analog designs, and class D and E for switching designs based on the proportion of each input cycle (conduction angle), during which an amplifying device is passing current. The image of the conduction angle is derived from amplifying a sinusoidal signal. If the device is always on, the conducting angle is 360°. If it is on for only half of each cycle, the angle is 180°. The angle of flow is closely related to the amplifier power efficiency. The various classes are introduced below, followed by a more detailed discussion under their individual headings further down.

In the illustrations below, a bipolar junction transistor is shown as the amplifying device, but the same attributes are found if with MOSFETs or vacuum tubes.

Conduction Angle Classes

Class A: 100% of the input signal is used (conduction angle È = 360°). The active element remains conducting all of the time.

Class B: 50% of the input signal is used (È = 180°); the active element carries current half of each cycle, and is turned off for the other half.

Class AB: Class AB is intermediate between class A and B, the two active elements conduct more than half of the time

Class C: Less than 50% of the input signal is used (conduction angle È < 180°).

A "Class D" amplifier uses some form of pulse-width modulation to control the output devices; the conduction angle of each device is no longer related directly to the input signal but instead varies in pulse width. These are sometimes called "digital" amplifiers because the output device is switched fully on or off, and not carrying current proportional to the signal amplitude.

Additional Classes: There are several other amplifier classes, although they are mainly variations of the previous classes. For example, class-G and class-H amplifiers are marked by variation of the supply rails (in discrete steps or in a continuous fashion, respectively) following the input signal. Wasted heat on the output devices can be reduced as excess voltage is kept to a minimum. The amplifier that is fed with these rails itself can be of any class. These kinds of amplifiers are more complex, and are mainly used for specialized applications, such as very high-power units. Also, class-E and class-F amplifiers are commonly described in literature for radio-frequency applications where efficiency of the traditional classes is important, yet several aspects deviate substantially from their ideal values. These classes use harmonic tuning of their output networks to achieve higher efficiency and can be considered a subset of class C due to their conduction-angle characteristics.

Class A

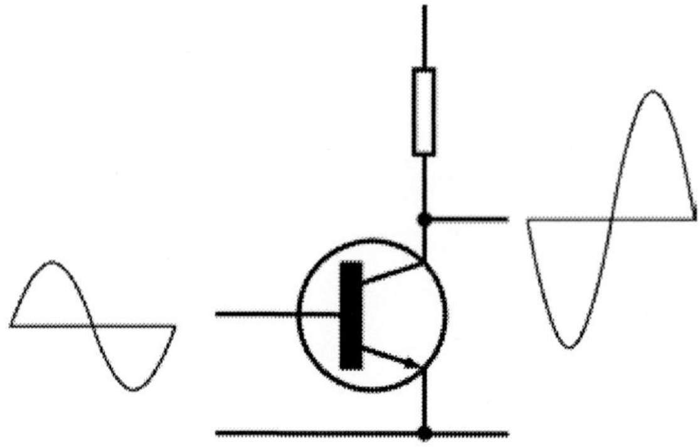

Figure: *Class-A amplifier*

Amplifying devices operating in class A conduct over the whole of the input cycle. A *class-A amplifier* is distinguished by the *output stage* being biased into class A Subclass A2 is sometimes used to refer to vacuum-tube class-A stages where the grid is allowed to be driven slightly positive on signal peaks, resulting in slightly more power than normal class A (A1; where the grid is always negative), but incurring more distortion.

Advantages of class-A Amplifiers

- Class-A designs are simpler than other classes; for example class-AB and -B designs require two devices (push–pull output) to handle both halves of the waveform; class A can use a single device single-ended.
- The amplifying element is biased so the device is always conducting to some extent, normally implying the quiescent (small-signal) collector current (for transistors; drain current for FETs or anode/plate current for vacuum tubes) is close to the most linear portion of its transconductance curve.
- Because the device is never shut off completely there is no "turn on" time, little problem with charge storage, and generally better high frequency performance and feedback loop stability (and usually fewer high-order harmonics).
- The point at which the device comes closest to being cut off is not close to zero signal, so the problem of crossover distortion associated with class-AB and -B designs is avoided.

Disadvantage of Class-A amplifiers

- They are very inefficient. A theoretical maximum of 50% is obtainable with inductive output coupling and only 25% with capacitive coupling, unless deliberate use of nonlinearities is made (such as in square-law output stages). In a power amplifier, this not only wastes power and limits battery operation, increase costs and may restrict the output devices that can be used (for example, ruling out some audio triodes to accommodate modern low-efficiency loudspeakers. Inefficiency comes not just from the fact that the device is always conducting to some extent. (That happens even with class AB, yet its efficiency can be close to class B.) It is that the standing current is roughly half the maximum output current (though this can be less with a square law output stage), and a large part of the power supply voltage develops across the output device at low signal levels (as with classes AB and B, but unlike output stages such as class D). If high output powers are needed from a class-A circuit, the power waste (and the accompanying heat) becomes significant. For every watt delivered to the load, the amplifier itself, *at best*, dissipates another watt. For large powers this means very large and expensive power supplies and heat sinking.

Class-A designs have largely been superseded by the more efficient designs for power amplifiers, though they remain popular with some hobbyists, mostly for their simplicity. There is a market for expensive high fidelity class-A amps considered a "cult item" amongst audiophiles mainly for their absence of crossover distortion and reduced odd-harmonic and high-order harmonic distortion.

Single-ended and Triode class-A amplifiers

Some aficionados who prefer class-A amplifiers also prefer the use of thermionic valve (or "tube") designs instead of transistors, especially in Single-ended triode output configurations for several claimed reasons:

- Single-ended output stages (be they tube or transistor) have an asymmetrical transfer function, meaning that even order harmonics in the created distortion tend not to be canceled (as they are in push–pull output stages); for tubes, or FETs, most of the distortion is second-order harmonics, from the square lawtransfer characteristic, which to some produces a "warmer" and more pleasant sound.

- For those who prefer low distortion figures, the use of tubes with class A (generating little odd-harmonic distortion, as mentioned above) together with symmetrical circuits (such as push–pull output stages, or balanced low-level stages) results in the cancellation of most of the even distortion harmonics, hence the removal of most of the distortion.
- Distortion is characteristic of the sound of electric guitar amplifiers.
- Historically, valve amplifiers often used a class-A power amplifier simply because valves are large and expensive; many class-A designs use only a single device.

Transistors are much cheaper, and so more elaborate designs that give greater efficiency but use more parts are still cost-effective. A classic application for a pair of class-A devices is the long-tailed pair, which is exceptionally linear, and forms the basis of many more complex circuits, including many audio amplifiers and almost all op-amps.

Class-A amplifiers are often used in output stages of high quality op-amps (although the accuracy of the bias in low cost op-amps such as the 741 may result in class A or class AB or class B, varying from device to device or with temperature). They are sometimes used as medium-power, low-efficiency, and high-cost audio power amplifiers. The power consumption is unrelated to the output power. At idle (no input), the power consumption is essentially the same as at high output volume. The result is low efficiency and high heat dissipation.

Class B

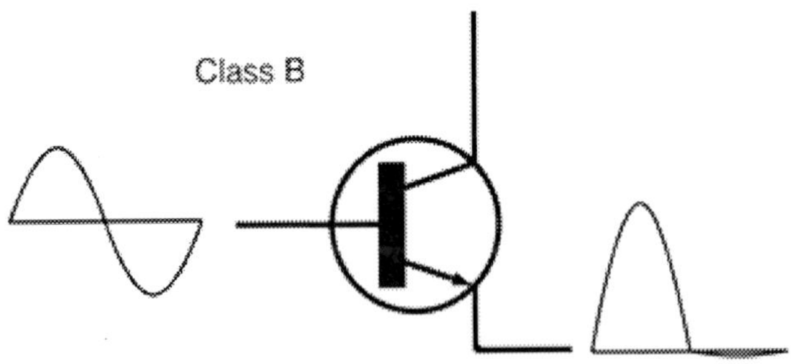

Figure: *Class-B amplifier*

Class-B amplifiers only amplify half of the input wave cycle, thus creating a large amount of distortion, but their efficiency is greatly improved and is much better than class A. Class-B amplifiers are also favoured in battery-operated devices, such as transistor radios. Class

B has a maximum theoretical efficiency of $\pi/4$. (i.e. 78.5%) This is because the amplifying element is switched off altogether half of the time, and so cannot dissipate power. A single class-B element is rarely found in practice, though it has been used for driving the loudspeaker in the early IBM Personal Computers with beeps, and it can be used in RF power amplifier where the distortion levels are less important. However, class C is more commonly used for this.

A practical circuit using class-B elements is the push–pull stage, such as the very simplified complementary pair arrangement shown below. Here, complementary or quasi-complementary devices are each used for amplifying the opposite halves of the input signal, which is then recombined at the output.

This arrangement gives excellent efficiency, but can suffer from the drawback that there is a small mismatch in the cross-over region – at the "joins" between the two halves of the signal, as one output device has to take over supplying power exactly as the other finishes. This is called crossover distortion. An improvement is to bias the devices so they are not completely off when they're not in use. This approach is called *class AB* operation.

Class AB

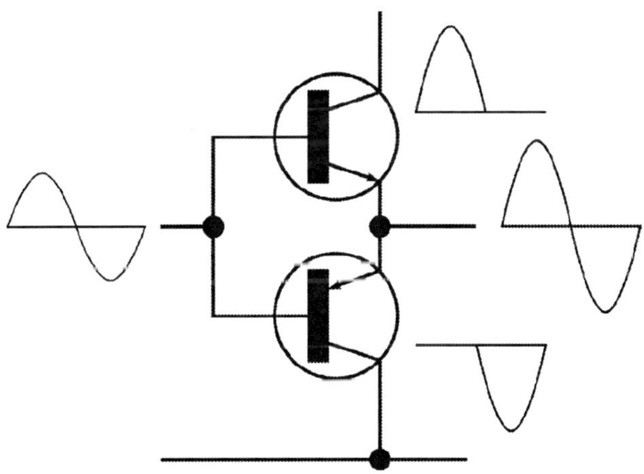

***Figure:** Class-AB push–pull amplifier*

Class AB is widely considered a good compromise for audio power amplifiers, since much of the time the music is quiet enough that the signal stays in the "class A" region, where it is amplified with good fidelity, and by definition if passing out of this region, is large enough that the distortion products typical of class B are relatively small. The crossover distortion can be reduced further by using negative feedback.

In class-AB operation, each device operates the same way as in class B over half the waveform, but also conducts a small amount on the other half. As a result, the region where both devices simultaneously are nearly off (the "dead zone") is reduced. The result is that when the waveforms from the two devices are combined, the crossover is greatly minimised or eliminated altogether. The exact choice of quiescent current, the standing current through both devices when there is no signal, makes a large difference to the level of distortion (and to the risk of thermal runaway, that may damage the devices); often the bias voltage applied to set this quiescent current has to be adjusted with the temperature of the output transistors (for example in the circuit at the beginning of the article the diodes would be mounted physically close to the output transistors, and chosen to have a matched temperature coefficient). Another approach (often used as well as thermally tracking bias voltages) is to include small value resistors in series with the emitters.

Class AB sacrifices some efficiency over class B in favour of linearity, thus is less efficient (below 78.5% for full-amplitude sinewaves in transistor amplifiers, typically; much less is common in class-AB vacuum-tube amplifiers). It is typically much more efficient than class A.

Sometimes a numeral is added for vacuum-tube stages. If the grid voltage is always negative with respect to the cathode the class is AB_1. If the grid is allowed to go slightly positive (hence drawing grid current, adding more distortion, but giving slightly higher output power) on signal peaks the class is AB_2.

Class C

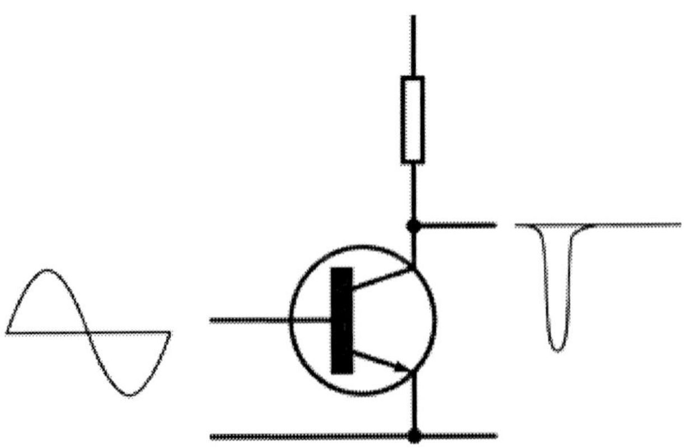

Figure: *Class-C amplifier*

Class-C amplifiers conduct less than 50% of the input signal and the distortion at the output is high, but high efficiencies (up to 90%)

are possible. The usual application for class-C amplifiers is in RF transmitters operating at a single fixed carrier frequency, where the distortion is controlled by a tuned load on the amplifier. The input signal is used to switch the active device causing pulses of current to flow through a tuned circuit forming part of the load.

The class-C amplifier has two modes of operation: tuned and untuned. The diagram shows a waveform from a simple class-C circuit without the tuned load. This is called untuned operation, and the analysis of the waveforms shows the massive distortion that appears in the signal. When the proper load (e.g., an inductive-capacitive filter plus a load resistor) is used, two things happen. The first is that the output's bias level is clamped with the average output voltage equal to the supply voltage. This is why tuned operation is sometimes called a *clamper*. This allows the waveform to be restored to its proper shape despite the amplifier having only a one-polarity supply. This is directly related to the second phenomenon: the waveform on the centre frequency becomes less distorted. The residual distortion is dependent upon the bandwidth of the tuned load, with the centre frequency seeing very little distortion, but greater attenuation the farther from the tuned frequency that the signal gets.

The tuned circuit resonates at one frequency, the fixed carrier frequency, and so the unwanted frequencies are suppressed, and the wanted full signal (sine wave) is extracted by the tuned load. The signal bandwidth of the amplifier is limited by the Q-factor of the tuned circuit but this is not a serious limitation. Any residual harmonics can be removed using a further filter.

In practical class-C amplifiers a tuned load is invariably used. In one common arrangement the resistor shown in the circuit above is replaced with a parallel-tuned circuit consisting of an inductor and capacitor in parallel, whose components are chosen to resonate the frequency of the input signal. Power can be coupled to a load by transformer action with a secondary coil wound on the inductor. The average voltage at the drain is then equal to the supply voltage, and the signal voltage appearing across the tuned circuit varies from near zero to near twice the supply voltage during the rf cycle. The input circuit is biassed so that the active element (e.g. transistor) conducts for only a fraction of the rf cycle, usually one third (120 degrees) or less.

The active element conducts only while the drain voltage is passing through its minimum. By this means, power dissipation in the active device is minimised, and efficiency increased. Ideally, the active element would pass only an instantaneous current pulse while the voltage across it is zero: it then disspates no power and 100% efficiency is achieved. However practical devices have a limit to the peak current they can pass, and the pulse must therefore be widened, to around 120 degrees, to obtain a reasonable amount of power, and the efficiency is then 60-70%.

Class D

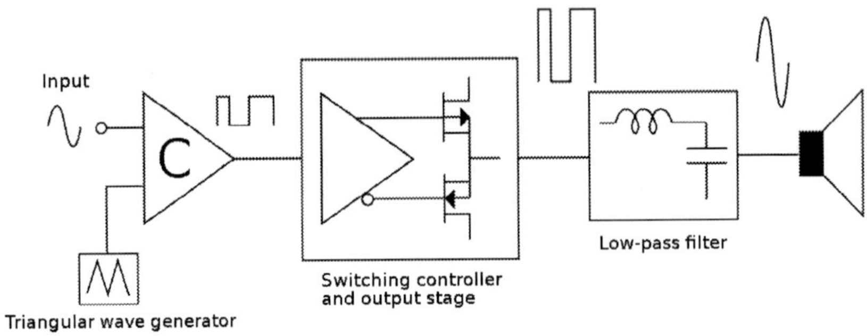

Figure: *Block diagram of a basic switching or PWM (class-D) amplifier.*

Figure: *Boss Audio class-D mono car audio amplifier with a low pass filter for powering subwoofers*

In the class-D amplifier the input signal is converted to a sequence of higher voltage output pulses. The averaged-over-time power values of these pulses are directly proportional to the instantaneous amplitude of the input signal. The frequency of the output pulses is typically ten or more times the highest frequency in the input signal to be amplified. The output pulses contain inaccurate spectral components (that is, the pulse frequency and its harmonics), which must be removed by a low-pass passive filter. The resulting filtered signal is then an amplified replica of the input.

These amplifiers use pulse width modulation, pulse density modulation (sometimes referred to as pulse frequency modulation) or a more advanced form of modulation such as Delta-sigma modulation (for example, in the Analog Devices AD1990 class-D audio power amplifier). Output stages such as those used in pulse generators are examples of class-D amplifiers. The term *class D* is usually applied to devices intended to reproduce signals with a bandwidth well below the switching frequency.

Class-D amplifiers can be controlled by either analog or digital circuits. The digital control introduces additional distortion called *quantization error* caused by its conversion of the input signal to a digital value.

The main advantage of a class-D amplifier is power efficiency. Because the output pulses have a fixed amplitude, the switching elements (usually MOSFETs, but valves (vacuum tubes) and bipolar transistors were once used) are switched either completely on or completely off, rather than operated in linear mode. A MOSFET operates with the lowest resistance when fully on and thus (excluding when fully off) has the lowest power dissipation when in that condition. Compared to an equivalent class-AB device, a class-D amplifier's lower losses permit the use of a smaller heat sink for the MOSFETs while also reducing the amount of input power required, allowing for a lower-capacity power supply design. Therefore, class-D amplifiers are typically smaller than an equivalent class-AB amplifier.

Class-D amplifiers have been widely used to control motors, but they are now also used as audio power amplifiers, with some extra circuitry to allow analogue to be converted to a much higher frequency pulse width modulated signal.

High quality class-D audio power amplifiers have now appeared on the market. These designs have been said to rival traditional AB amplifiers in terms of quality. An early use of class-D amplifiers was high-power subwoofer amplifiers in cars. Because subwoofers are generally limited to a bandwidth of no higher than 150 Hz, the switching speed for the amplifier does not have to be as high as for a full range amplifier, allowing simpler designs. Class-D amplifiers for driving subwoofers are relatively inexpensive in comparison to class-AB amplifiers.

The letter *D* used to designate this amplifier class is simply the next letter after *C*, and does not stand for *digital*. Class-D and class-E amplifiers are sometimes mistakenly described as "digital" because the output waveform superficially resembles a pulse-train of digital symbols, but a class-D amplifier merely converts an input waveform into a continuously pulse-width modulated (square wave) analog signal. (A digital waveform would be pulse-code modulated.)

Additional Classes

Class E: The class-E/F amplifier is a highly efficient switching power amplifier, typically used at such high frequencies that the switching time becomes comparable to the duty time. As said in the class-D amplifier, the transistor is connected via a serial LC circuit to the load, and connected via a large L (inductor) to the supply voltage. The supply voltage is connected to ground via a large capacitor to prevent any RF signals leaking into the supply. The class-E amplifier adds a C (capacitor) between the transistor and ground and uses a defined L_1 to connect to the supply voltage.

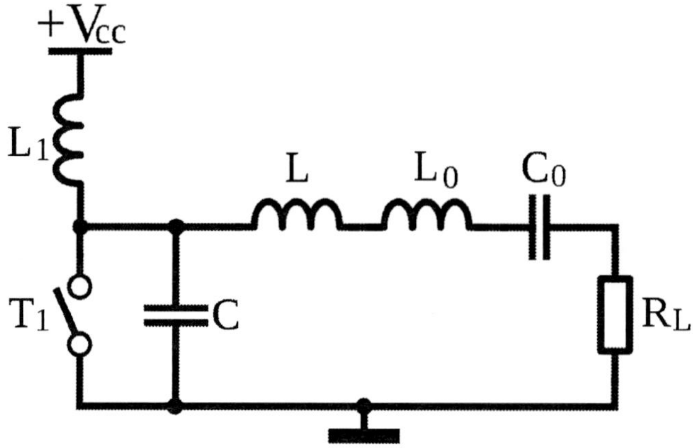

Figure: *Class-E amplifier*

The following description ignores DC, which can be added easily afterwards. The above mentioned C and L are in effect a parallel LC circuit to ground. When the transistor is on, it pushes through the serial LC circuit into the load and some current begins to flow to the parallel LC circuit to ground. Then the serial LC circuit swings back and compensates the current into the parallel LC circuit. At this point the current through the transistor is zero and it is switched off. Both LC circuits are now filled with energy in C and L_0. The whole circuit performs a damped oscillation. The damping by the load has been adjusted so that some time later the energy from the Ls is gone into the load, but the energy in both C_0 peaks at the original value to in turn restore the original voltage so that the voltage across the transistor is zero again and it can be switched on.

With load, frequency, and duty cycle (0.5) as given parameters and the constraint that the voltage is not only restored, but peaks at the original voltage, the four parameters (L, L_0, C and C_0) are determined. The class-E amplifier takes the finite on resistance into account and tries to make the current touch the bottom at zero. This means that the voltage and the current at the transistor are symmetric with respect to time. The Fourier transform allows an elegant formulation to generate the complicated LC networks and says that the first harmonic is passed into the load, all even harmonics are shorted and all higher odd harmonics are open.

Class E uses a significant amount of second-harmonic voltage. The second harmonic can be used to reduce the overlap with edges with finite sharpness. For this to work, energy on the second harmonic has to flow from the load into the transistor, and no source for this is visible in the circuit diagram. In reality, the impedance is mostly reactive and the only reason for it is that class E is a class F amplifier with a much simplified load network and thus has to deal with imperfections.

In many amateur simulations of class-E amplifiers, sharp current edges are assumed nullifying the very motivation for class E and measurements near the transit frequency of the transistors show very symmetric curves, which look much similar to class-F simulations.

The class-E amplifier was invented in 1972 by Nathan O. Sokal and Alan D. Sokal, and details were first published in 1975. Some earlier reports on this operating class have been published in Russian.

Class F

In push–pull amplifiers and in CMOS, the even harmonics of both transistors just cancel. Experiment shows that a square wave can be generated by those amplifiers. Theoretically square waves consist of odd harmonics only. In a class-D amplifier, the output filter blocks all harmonics; i.e., the harmonics see an open load. So even small currents in the harmonics suffice to generate a voltage square wave. The current is in phase with the voltage applied to the filter, but the voltage across the transistors is out of phase. Therefore, there is a minimal overlap between current through the transistors and voltage across the transistors. The sharper the edges, the lower the overlap.

While in class D, transistors and the load exist as two separate modules, class F admits imperfections like the parasitics of the transistor and tries to optimise the global system to have a high impedance at the harmonics. Of course there has to be a finite voltage across the transistor to push the current across the on-state resistance. Because the combined current through both transistors is mostly in the first harmonic, it looks like a sine. That means that in the middle of the square the maximum of current has to flow, so it may make sense to have a dip in the square or in other words to allow some overswing of the voltage square wave. A class-F load network by definition has to transmit below a cutoff frequency and reflect above.

Any frequency lying below the cutoff and having its second harmonic above the cutoff can be amplified, that is an octave bandwidth. On the other hand, an inductive-capacitive series circuit with a large inductance and a tunable capacitance may be simpler to implement. By reducing the duty cycle below 0.5, the output amplitude can be modulated. The voltage square waveform degrades, but any overheating is compensated by the lower overall power flowing. Any load mismatch behind the filter can only act on the first harmonic current waveform, clearly only a purely resistive load makes sense, then the lower the resistance, the higher the current.

Class F can be driven by sine or by a square wave, for a sine the input can be tuned by an inductor to increase gain. If class F is implemented with a single transistor, the filter is complicated to short the even harmonics. All previous designs use sharp edges to minimise the overlap.

Classes G and H

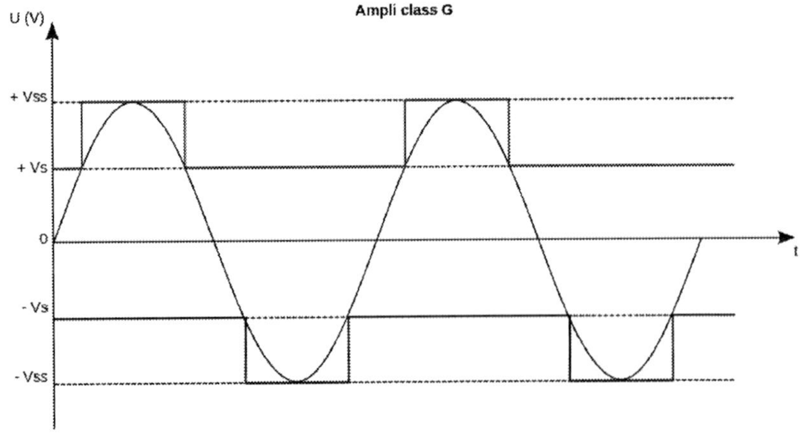

There are a variety of amplifier designs that enhance class-AB output stages with more efficient techniques to achieve greater efficiencies with low distortion. These designs are common in large audio amplifiers since the heatsinks and power transformers would be prohibitively large (and costly) without the efficiency increases. The terms "class G" and "class H" are used interchangeably to refer to different designs, varying in definition from one manufacturer or paper to another.

Class-G amplifiers (which use "rail switching" to decrease power consumption and increase efficiency) are more efficient than class-AB amplifiers. These amplifiers provide several power rails at different voltages and switch between them as the signal output approaches each level. Thus, the amplifier increases efficiency by reducing the wasted power at the output transistors. Class-G amplifiers are more efficient than class AB but less efficient when compared to class D, without the negative EMI effects of class D.

Class-H amplifiers take the idea of class G one step further creating an infinitely variable supply rail. This is done by modulating the supply rails so that the rails are only a few volts larger than the output signal at any given time. The output stage operates at its maximum efficiency all the time. Switched-mode power supplies can be used to create the tracking rails. Significant efficiency gains can be achieved but with the drawback of more complicated supply design and reduced THD performance. In common designs, a voltage drop of about 10V is maintained over the output transistors in Class H circuits. The picture above shows positive supply voltage of the output stage and the voltage at the speaker output. The boost of the supply voltage is shown for a real music signal.

The voltage signal shown is thus a larger version of the input, but has been changed in sign (inverted) by the amplification. Other arrangements of amplifying device are possible, but that given (that is, common emitter, common source or common cathode) is the easiest to understand and employ in practice. If the amplifying element is linear, the output is a faithful copy of the input, only larger and inverted. In practice, transistors are not linear, and the output only approximates the input. nonlinearity from any of several sources is the origin of distortion within an amplifier. The class of amplifier (A, B, AB or C) depends on how the amplifying device is biased. The diagrams omit the bias circuits for clarity.

Any real amplifier is an imperfect realization of an ideal amplifier. An important limitation of a real amplifier is that the output it generates is ultimately limited by the power available from the power supply. An amplifier saturates and clips the output if the input signal becomes too large for the amplifier to reproduce or exceeds operational limits for the device.

Doherty Amplifiers

The Doherty, a hybrid configuration, is receiving new attention. It was invented in 1934 by William H. Doherty for Bell Laboratories—whose sister company, Western Electric, manufactured radio transmitters. The Doherty amplifier consists of a class-B *primary* or *carrier* stages in parallel with a class-C *auxiliary* or *peak* stage. The input signal splits to drive the two amplifiers, and a combining network sums the two output signals. Phase shifting networks are used in inputs and outputs. During periods of low signal level, the class-B amplifier efficiently operates on the signal and the class-C amplifier is cutoff and consumes little power. During periods of high signal level, the class-B amplifier delivers its maximum power and the class-C amplifier delivers up to its maximum power. The efficiency of previous AM transmitter designs was proportional to modulation but, with average modulation typically around 20%, transmitters were limited to less than 50% efficiency. In Doherty's design, even with zero modulation, a transmitter could achieve at least 60% efficiency.

As a successor to Western Electric for broadcast transmitters, the Doherty concept was considerably refined by Continental Electronics Manufacturing Company of Dallas, TX. Perhaps, the ultimate refinement was the screen-grid modulation scheme invented by Joseph B. Sainton. The Sainton amplifier consists of a class-C primary or carrier stage in parallel with a class-C auxiliary or peak stage. The stages are split and combined through 90-degree phase shifting networks as in the Doherty amplifier. The unmodulated radio frequency carrier is applied to the control grids of both tubes. Carrier modulation is applied to the screen grids of both tubes. The bias point of the carrier and peak tubes is different, and is established such that the peak tube is cutoff when modulation is absent (and the amplifier is producing rated unmodulated carrier power) whereas both tubes contribute twice the rated carrier power during 100% modulation (as four times the carrier power is required to achieve 100% modulation). As both tubes operate in class C, a significant improvement in efficiency is thereby achieved in the final stage. In addition, as the tetrode carrier and peak tubes require very little drive power, a significant improvement in efficiency within the driver stage is achieved as well (317C, et al.). The released version of the Sainton amplifier employs a cathode-follower modulator, not a push–pull modulator. Previous Continental Electronics designs, by James O. Weldon and others, retained most of the characteristics of the Doherty amplifier but added screen-grid modulation of the driver (317B, et al.).

The Doherty amplifier remains in use in very-high-power AM transmitters, but for lower-power AM transmitters, vacuum-tube amplifiers in general were eclipsed in the 1980s by arrays of solid-state amplifiers, which could be switched on and off with much finer granularity in response to the requirements of the input audio. However, interest in the Doherty configuration has been revived by cellular-telephone and wireless-Internet applications where the sum of several constant envelope users creates an aggregate AM result. The main challenge of the Doherty amplifier for digital transmission modes is in aligning the two stages and getting the class-C amplifier to turn on and off very quickly.

Recently, Doherty amplifiers have found widespread use in cellular base station transmitters for GHz frequencies. Implementations for transmitters in mobile devices have also been demonstrated.

Special Classes

Various newer classes of amplifier, as defined by the technical details of their topology, have been developed on the basis of previously existing operating classes. For example, Crown's K and I-Tech Series as well as several other models use Crown's patented class I (or BCA) technology. Lab.gruppen use a form of class-D amplifier called class TD or tracked class D that tracks the waveform to more accurately amplify it without the drawbacks of traditional class-D amplifiers.

"Class S" was the name of a design published by A M Sandman in *Wireless World* (sept. 1982). It had some elements in common with a current dumping design. It comprises a class A input stage coupled to a class B output stage, with a specific feedback design. Technics used a modified design in their class AA (marketed) output stage.

"Class T" is a trademark of TriPath company, which manufactures audio amplifier ICs. This new class T is a revision of the common class-D amplifier, but with changes to ensure fidelity over the full audio spectrum, unlike traditional class-D designs. It operates at different frequencies depending on the power output, with values ranging from as low as 200 kHz to 1.2 MHz, using a proprietary modulator. Tripath ceased operations in 2007, its patents acquired by Cirrus Logic for their Mixed-Signal Audio division. Some Kenwood Recorder use class-W amplifiers.

"Class Z" is a trademark of Zetex Semiconductors (now part of Diodes Inc. of Dallas, TX) and is a direct-digital-feedback technology. Zetex-patented circuits are being used in the latest power amplifiers by NAD Electronics of Canada.

Implementation

Amplifiers are implemented using active elements of different kinds:

- The first active elements were relays. They were for example used in transcontinental telegraph lines: a weak current was used to switch the voltage of a battery to the outgoing line.
- For transmitting audio, carbon microphones were used as the active element. This was used to modulate a radio-frequency source in one of the first AM audio transmissions, by Reginald Fessenden on Dec. 24, 1906.
- In the 1960s, the transistor started to take over. These days, discrete transistors are still used in high-power amplifiers and in specialist audio devices.
- Up to the early 1970s, most amplifiers used vacuum tubes. Today, tubes are used for specialist audio applications such as guitar amplifiers and audiophile amplifiers. Many broadcast transmitters still use vacuum tubes.
- Beginning in the 1970s, more and more transistors were connected on a single chip therefore creating the integrated circuit. A large number of amplifiers commercially available today are based on integrated circuits.

For special purposes, other active elements have been used. For example, in the early days of the satellite communication, parametric amplifiers were used. The core circuit was a diode whose capacity was changed by an RF signal created locally. Under certain conditions, this RF signal provided energy that was modulated by the extremely weak satellite signal received at the earth station.

Amplifier Circuit

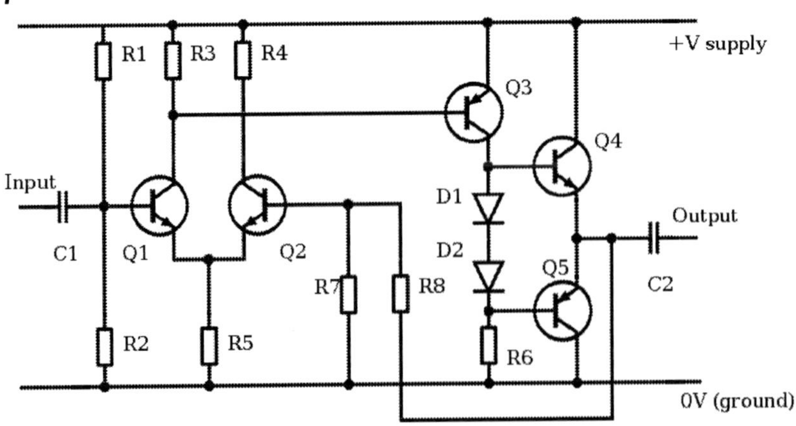

The practical amplifier circuit to the right could be the basis for a moderate-power audio amplifier. It features a typical (though substantially simplified) design as found in modern amplifiers, with a class-AB push–pull output stage, and uses some overall negative feedback. Bipolar transistors are shown, but this design would also be realizable with FETs or valves.

The input signal is coupled through capacitor C1 to the base of transistor Q1. The capacitor allows the AC signal to pass, but blocks the DC bias voltage established by resistors R1 and R2 so that any preceding circuit is not affected by it. Q1 and Q2 form a differential amplifier (an amplifier that multiplies the difference between two inputs by some constant), in an arrangement known as a long-tailed pair. This arrangement is used to conveniently allow the use of negative feedback, which is fed from the output to Q2 via R7 and R8.

The negative feedback into the difference amplifier allows the amplifier to compare the input to the actual output. The amplified signal from Q1 is directly fed to the second stage, Q3, which is a common emitter stage that provides further amplification of the signal and the DC bias for the output stages, Q4 and Q5. R6 provides the load for Q3 (A better design would probably use some form of active load here, such as a constant-current sink). So far, all of the amplifier is operating in class A. The output pair are arranged in class-AB push–pull, also called a complementary pair. They provide the majority of the current amplification (while consuming low quiescent current) and directly drive the load, connected via DC-blocking capacitor C2. The diodes D1 and D2 provide a small amount of constant voltage bias for the output pair, just biasing them into the conducting state so that crossover distortion is minimized. That is, the diodes push the output stage firmly into class-AB mode (assuming that the base-emitter drop of the output transistors is reduced by heat dissipation).

This design is simple, but a good basis for a practical design because it automatically stabilises its operating point, since feedback internally operates from DC up through the audio range and beyond. Further circuit elements would probably be found in a real design that would roll off the frequency response above the needed range to prevent the possibility of unwanted oscillation. Also, the use of fixed diode bias as shown here can cause problems if the diodes are not both electrically and thermally matched to the output transistors – if the output transistors turn on too much, they can easily overheat and destroy themselves, as the full current from the power supply is not limited at this stage.

A common solution to help stabilise the output devices is to include some emitter resistors, typically an ohm or so. Calculating the values of the circuit's resistors and capacitors is done based on the components employed and the intended use of the amp.

Notes on Implementation

Real world amplifiers are imperfect.

- One consequence is that the power supply itself may influence the output, and must itself be considered when designing the amplifier
- a power amplifier is effectively an input signal controlled power regulator - regulating the power sourced from the power supply or mains to the amplifier's load. The power output from a power amplifier cannot exceed the power input to it.
- The amplifier circuit has an "open loop" performance, that can be described by various parameters (gain, slew rate, output impedance, distortion, bandwidth, signal to noise ratio, etc.)
- Many modern amplifiers use negative feedback techniques to hold the gain at the desired value and to reduce distortion. Negative loop feedback has the intended effect of electrically damping loudspeaker motion, thereby damping the mechanical dynamic performance of the loudspeaker.
- When assessing rated amplifier power output it is useful to consider the load to be applied, the form of signal - i.e. speech or music, duration of power output needed - e.g. short-time or continuous, and dynamic range required - e.g. recorded programme or live
- In the case of high-powered audio applications requiring long cables to the load - e.g. cinemas and shipping centres - instead of using heavy gauge cables it may be more efficient to connect to the load at line output voltage with matching transformers at source and loads.
- To prevent instability and/or overheating, care is need to ensure solid state amplifiers are adequately loaded. Most have a rated minimum load impedance.
- All amplifiers generate heat through electrical losses. This heat must be dissipated via natural or forced air cooling. Heat can damage or reduce service life of electronic components. Consideration should be given to the heating effects of or upon adjacent equipment.

Different methods of supplying power result in many different methods of bias. Bias is a technique by which the active devices are set up to operate in a particular region, or by which the DC component of the output signal is set to the midpoint between the maximum voltages available from the power supply. Most amplifiers use several devices at each stage; they are typically matched in specifications except for polarity. Matched inverted polarity devices are called complementary pairs. Class-A amplifiers generally use only one device, unless the power supply is set to provide both positive and negative voltages, in which case a dual device symmetrical design may be used. Class-C amplifiers, by definition, use a single polarity supply.

Amplifiers often have multiple stages in cascade to increase gain. Each stage of these designs may be a different type of amp to suit the needs of that stage. For instance, the first stage might be a class-A stage, feeding a class-AB push–pull second stage, which then drives a class-G final output stage, taking advantage of the strengths of each type, while minimizing their weaknesses.

Coupling (Electronics)

In electronics and telecommunication, coupling is the desirable or undesirable transfer of energy from one medium, such as a metallic wire or an optical fibre, to another medium, including fortuitous transfer.

Coupling is also the transfer of electrical energy from one circuit segment to another. For example, energy is transferred from a power source to an electrical load by means of conductive coupling, which may be either resistive or hard-wire. An AC potential may be transferred from one circuit segment to another having a DC potential by use of a capacitor. Electrical energy may be transferred from one circuit segment to another segment with different impedance by use of a transformer. This is known as impedance matching. These are examples of electrostatic and electrodynamic inductive coupling.

Types of Coupling

Electrical conduction:

- hard-wire
- resistive
- natural conductor

Electromagnetic induction:

- electrodynamic — commonly called inductive coupling, also magnetic coupling

- electrostatic — commonly called capacitive coupling
- evanescent wave coupling

Electromagnetic radiation:

- radio — wireless telecommunications
- electromagnetic interference (EMI) — Sometimes called radio frequency interference (RFI), is unwanted coupling. Electromagnetic compatibility (EMC) requires techniques to avoid such unwanted coupling, such as electromagnetic shielding.
- Microwave power transmission

Other kinds of energy coupling:

- acoustic

Decoupling Capacitor

A decoupling capacitor is a capacitor used to decouple one part of an electrical network (circuit) from another. Noise caused by other circuit elements is shunted through the capacitor, reducing the effect it has on the rest of the circuit.

For example, if the voltage level for a device is fixed, changing power demands are manifested as changing current demand. The power supply must accommodate these variations in current draw with as little change as possible in the power supply voltage. When the current draw in a device changes, the power supply cannot respond to that change instantaneously. As a consequence, the voltage at the device changes for a brief period before the power supply responds. The voltage regulator adjusts the amount of current it is supplying to keep the output voltage constant but can only effectively maintain the output voltage for events at frequencies from DC to a few hundred kHz, depending on the regulator (some are effective at regulating in the low MHz). For transient events that occur at frequencies above this range, there is a time lag before the voltage regulator responds to the new current demand level.

This is where the decoupling capacitor comes in. The decoupling capacitor works as the device's local energy storage. The capacitor cannot provide DC power because it stores only a small amount of energy but this energy can respond very quickly to changing current demands. The capacitors effectively maintain power-supply voltage at frequencies from hundreds of kHz to hundreds of MHz (in the milliseconds to nanoseconds range). Decoupling capacitors are not useful for events occurring above or below this range. An alternative name is bypass capacitor as it is used to bypass the power supply or other high impedance component of a circuit.

Decoupling

One common kind of decoupling is of a powered circuit from signals in the power supply. Sometimes, for various reasons, a power supply supplies an AC signal superimposed on the DC power line. Such a signal is often undesirable in the powered circuit. A decoupling capacitor can prevent the powered circuit from seeing that signal, thus decoupling it from that aspect of the power supply circuit.

Another kind of decoupling is stopping a portion of a circuit from being affected by switching that happens in another portion. Switching in subcircuit A may cause fluctuations in the power supply or other electrical lines, but you do not want subcircuit B, which has nothing to do with that switching, to be affected. A decoupling capacitor can decouple subcircuits A and B so that B doesn't see any effects of the switching.

To decouple a subcircuit from AC signals or voltage spikes on a power supply or other line, a bypass capacitor is often used. A bypass capacitor is to shunt energy from those signals or transients past the subcircuit to be decoupled, right to the return path. For a power supply line, a bypass capacitor from the supply voltage line to the power supply return (neutral) would be used.

High frequencies and transient currents flow through a capacitor, in this case in preference to the harder path through the decoupled circuit, but DC cannot go through the capacitor, so continues on to the decoupled circuit.

Switching Subcircuits

In a switching subcircuit switching noise must be suppressed. When a load is applied to a voltage source, it draws a certain amount of current. Typical power supply lines show inherent inductance, which results in a slower response to change in current. This in turn affects the transient voltage levels, since if the load current is zero the voltage across the load is zero as well. This sudden voltage drop would be seen by other loads as well if the inductance between two loads is much lower compared to the inductance between the loads and the output capacitors of the power supply. This is only temporary; the inductor ultimately saturates (that is the magnetic field around the conductor reaches its max), the voltage drop across the inductor reaches zero, and the supply voltage comes back to normal. But even a temporary reduction in voltage can disturb adjacent subcircuits. Decoupling caps provide instantenous current jolt which helps maintain constant voltage across a subcircuit (or provide a low impedance path for the transient currents; different descriptions are used by different industries).

To decouple other subcircuits from the effect of the sudden current demand, a decoupling capacitor can be placed between the supply voltage line and its reference (ground) next to the switched load. While the load is switched out, the capacitor charges up to full power supply voltage and otherwise does nothing. When the load is applied, the capacitor initially supplies demanded current. Ideally, by the time the capacitor runs out of charge, the power supply line inductance is saturated, and the load can draw full current at normal voltage from the power supply (and the capacitor can recharge too). Note that the voltage dip is reduced but not eliminated; i.e., the decoupling is not perfect and sometimes parallel combinations of caps are used to improve response. The best way to reduce switching noise is to design a PCB as a giant capacitor by sandwiching the power and ground planes across a dielectric material.

The size of the capacitor must be reasonable, and there is a tradeoff between capacitor size and signal quality at a given frequency. If a cap is too large it would distort the signal by charging too slowly and filtering out the signal's most needed high-frequency components.

Transient Load Decoupling

Transientload decoupling as described above is needed when there is a large load that gets switched quickly. The parasitic inductance in every (decoupling) capacitor may limit the suitable capacity and influence appropriate type if switching occurs very fast.

Logic circuits tend to do sudden switching (an ideal logic circuit would switch from low voltage to high voltage instantaneously, with no middle voltage ever observable). So logic circuit boards often have a decoupling capacitor close to each logic IC connected from each power supply connection to a nearby ground. These capacitors decouple every IC from every other IC in terms of supply voltage dips.

These capacitors are often placed at each power source as well as at each analog component in order to ensure that the supplies are as steady as possible. Otherwise, an analog component with poor power supply rejection ratio (PSRR) will copy fluctuations in the power supply onto its output.

In these applications, the decoupling capacitors are often called *bypass capacitors* to indicate that they provide an alternate path for high-frequency signals that would otherwise cause the normally steady supplies to move. Those components that require quick injections of current can *bypass* the power supply by receiving the current from the nearby capacitor. Hence, the slower power supply connection is used to charge these capacitors, and the capacitors actually provide the large quantities of high-availability current.

Placement

A transient load decoupling capacitor should usually be placed as close as possible to the device requiring the decoupled signal. The goal is to minimize the amount of line inductance and series resistance between the decoupling capacitor and that device, and the longer the conductor between the capacitor and the device, the more inductance there is.

The guidelines for placing a high-speed decoupling capacitor on a multi-layer printed circuit board depend on whether the board has dedicated power distribution planes and how closely spaced those planes are.

Since capacitors differ in their high-frequency characteristics (and capacitors with good high-frequency properties are often types with small capacity, while large capacitors usually have worse high-frequency response), decoupling often involves the use of a *combination* of capacitors. For example in logic circuits, a common arrangement is ~100 nF ceramic per logic IC (multiple ones for complex IC's), combined with electrolytic or tantalum capacitor(s) up to a few hundred μF per board / board section.

Distortion

Distortion (or warping) is the alteration of the original shape (or other characteristic) of something, such as an object, image, sound or waveform. Distortion is usually unwanted, and so engineers strive to eliminate distortion, or minimize it. In some situations, however, distortion may be desirable. The important signal processing operation of heterodyning is based on nonlinear mixing of signals to cause intermodulation. Distortion is also used as a musical effect, particularly with electric guitars.

The addition of noise or other outside signals (hum, interference) is not deemed distortion, though the effects of quantization distortion are sometimes deemed noise. A quality measure that explicitly reflects both the noise and the distortion is the Signal-to-noise-and-distortion (SINAD) ratio.

Electronic Signals

In telecommunication and signal processing, a noise-free "system" can be characterised by a transfer function, such that the output $y(t)$ can be written as a function of the input x as

$$y(t) = F(x(t))$$

When the transfer function comprises only a perfect gain constant A and perfect delay$T$

$$y(t) = A{\cdot}x(t-T)$$

the output is undistorted. Distortion occurs when the transfer function F is more complicated than this. If F is a linear function, for instance a filter whose gain and/or delay varies with frequency, the signal suffers linear distortion. Linear distortion does not change the shape of a single sinuosoid, but usually changes the shape of a multi-tone signal.

This diagram shows the behaviour of a signal (made up of a square wave followed by a sine wave) as it is passed through various distorting functions.

1. The first trace (in black) shows the input. It also shows the output from a non-distorting transfer function (straight line).
2. A high-pass filter (green trace) distorts the shape of a square wave by reducing its low frequency components. This is the cause of the "droop" seen on the top of the pulses. This "pulse distortion" can be very significant when a train of pulses must pass through an AC-coupled (high-pass filtered) amplifier. As the sine wave contains only one frequency, its shape is unaltered.
3. A low-pass filter (blue trace) rounds the pulses by removing the high frequency components. All systems are low pass to some extent. Note that the phase of the sine wave is different for the lowpass and the highpass cases, due to the phase distortion of the filters.
4. A slightly non-linear transfer function (purple), this one gently compresses the peaks of the sine wave, as may be typical of a tube audio amplifier. This generates small amounts of low order harmonics.
5. A hard-clipping transfer function (red) generates high order harmonics. Parts of the transfer function are flat, which

indicates that all information about the input signal has been lost in this region.

The transfer function of an ideal amplifier, with perfect gain and delay, is only an approximation. The true behaviour of the system is usually different. Nonlinearities in the transfer function of an active device (such as vacuum tubes, transistors, and operational amplifiers) are a common source of non-linear distortion; in passive components (such as a coaxial cable or optical fibre), linear distortion can be caused by inhomogeneities, reflections, and so on in the propagation path.

Amplitude Distortion

Amplitude distortion is distortion occurring in a system, subsystem, or device when the output amplitude is not a linear function of the input amplitude under specified conditions.

Harmonic Distortion

Harmonic distortion adds overtones that are whole number multiples of a sound wave's frequencies. Nonlinearities that give rise to amplitude distortion in audio systems are most often measured in terms of the harmonics (overtones) added to a pure sinewave fed to the system. Harmonic distortion may be expressed in terms of the relative strength of individual components, in decibels, or the Root Mean Square of all harmonic components: Total harmonic distortion (THD), as a percentage. The level at which harmonic distortion becomes audible is not straightforward. Different types of distortion (like crossover distortion) are more audible than others (like soft clipping) even if the THD measurements are identical. Harmonic distortion in RF applications is rarely expressed as THD.

Frequency Response Distortion

Non-flat frequency response is a form of distortion that occurs when different frequencies are amplified by different amounts, caused by filters. For example, the non-uniform frequency response curve of AC-coupled cascade amplifier is an example of frequency distortion. In the audio case, this is mainly caused by room acoustics, poor loudspeakers and microphones, long loudspeaker cables in combination with frequency dependent loudspeaker impedance, etc.

Phase Distortion

This form of distortion mostly occurs due to the reactive component, such as capacitive reactance or inductive reactance. Here, all the components of the input signal are not amplified with the same phase shift, hence making some parts of the output signal out of phase with the rest of the output.

Group Delay Distortion

Can be found only in dispersive media. In a waveguide, propagation velocity varies with frequency. In a filter, group delay tends to peak near the cut-off frequency, resulting in pulse distortion. When analog long distance trunks were commonplace, for example in 12 channel carrier, group delay distortion had to be corrected in repeaters.

Correction of Distortion

As the system output is given by y(t) = F(x(t)), then if the inverse function F^{-1} can be found, and used intentionally to distort either the input or the output of the system, then the distortion is corrected.

An example of such correction is where LP/vinyl recordings or FM audio transmissions are deliberately pre-emphasised by a linear filter, the reproducing system applies an inverse filter to make the overall system undistorted.

Correction is not possible if the inverse does not exist—for instance if the transfer function has flat spots (the inverse would map multiple input points to a single output point). This produces an uncorrectable loss of information. Such a situation can occur when an amplifier is overdriven—causing clipping or slew rate distortion when, for a moment, the amplifier characteristics alone and not the input signal determine the output.

Teletypewriter or Modem Signalling

In binary signalling such as FSK, distortion is the shifting of the significant instants of the signal pulses from their proper positions relative to the beginning of the start pulse. The magnitude of the distortion is expressed in percent of an ideal unit pulse length. This is sometimes called 'bias' distortion.

Telegraphic distortion is a similar older problem, distorting the ratio between "mark" and "space" intervals.

AC Equivalent Circuit

In electrical engineering and science, an equivalent circuit refers to a theoretical circuit that retains all of the electrical characteristics of a given circuit. Often, an equivalent circuit is sought that simplifies calculation, and more broadly, that is a simplest form of a more complex circuit in order to aid analysis. In its most common form, an equivalent circuit is made up of linear, passive elements. However, more complex equivalent circuits are used that approximate the nonlinear behaviour of the original circuit as well. These more complex circuits often are called macromodels of the original circuit. An example of a macromodel is the Boyle circuit for the 741 operational amplifier.

Equivalent circuits can also be used to electrically describe and model either a) continuous materials or biological systems in which current does not actually flow in defined circuits, or, b) distributed reactances, such as found in electrical lines or windings, that do not represent actual discrete components. For example, a cell membrane can be modelled as a capacitance (i.e. the lipid bilayer) in parallel with resistance-DC voltage source combinations (i.e. ion channels powered by an ion gradient across the membrane).

Examples

Thévenin and Norton Equivalents: One of linear circuit theory's most surprising properties relates to the ability to treat any two-terminal circuit no matter how complex as behaving as only a source and an impedance, which have either of two simple equivalent circuit forms:

- Thévenin equivalent - Any linear two-terminal circuit can be replaced by a single voltage source and a series impedance.
- Norton equivalent - Any linear two-terminal circuit can be replaced by a current source and a parallel impedance.

However, the single impedance can be of arbitrary complexity and may be irreducible to a simpler form.

DC and AC Equivalent Circuits

In linear circuits, due to the superposition principle, the output of a circuit is equal to the sum of the output due to its DC sources alone, and the output from its AC sources alone. Therefore, the DC and AC response of a circuit is often analyzed independently, using separate DC and AC equivalent circuits which have the same response as the original circuit to DC and AC currents respectively. The composite response is calculated by adding the DC and AC responses:

- A DC equivalent of a circuit can be constructed by replacing all capacitances with open circuits, inductances with short circuits, and reducing AC sources to zero (replacing AC voltage sources by short circuits and AC current sources by open circuits.)

- An AC equivalent circuit can be constructed by reducing all DC sources to zero (replacing DC voltage sources with short circuits and DC current sources with open circuits)

This technique is often extended to small-signal nonlinear circuits like tube and transistor circuits, by linearizing the circuit about the DC bias point Q-point, using an AC equivalent circuit made by calculating the equivalent *small signal* AC resistance of the nonlinear components at the bias point.

Two-port Networks

Linear four-terminal circuits in which a signal is applied to one pair of terminals and an output is taken from another, are often modelled as two-port networks. These can be represented by simple equivalent circuits of impedances and dependent sources. To be analyzed as a two port network the currents applied to the circuit must satisfy the *port condition*: the current entering one terminal of the port must be equal to the current leaving the other terminal. By linearizing a nonlinear circuit about its operating point, such a two-port representation can be made for transistors.

Delta and Wye Circuits

In three phase power circuits, three phase sources and loads can be connected in two different ways, called a "delta" connection and a "wye" connection. In analyzing circuits, sometimes it simplifies the analysis to convert between equivalent wye and delta circuits. This can be done with the wye-delta transform.

Feedback Amplifiers

A negative feedback amplifier (or more commonly simply a feedback amplifier) is an amplifier which combines a fraction of the output with the input so that a negative feedback opposes the original signal. The applied negative feedback improves performance (gain stability, linearity, frequency response, step response) and reduces sensitivity to parameter variations due to manufacturing or environment. Because of these advantages, negative feedback is used in this way in many amplifiers and control systems.

A negative feedback amplifier is a system of three elements: an *amplifier* with gainA_{OL}, a *feedback network* which senses the output signal and possibly transforms it in some way (for example by attenuating or filtering it), and a summing circuit acting as a *subtractor* (the circle in the figure) which combines the input and the attenuated output.

Overview

Fundamentally, all electronic devices used to provide power gain (e.g. vacuum tubes, bipolar transistors, MOS transistors) are nonlinear. Negative feedback allows gain to be traded for higher linearity (reducing distortion), amongst other things. If not designed correctly, amplifiers with negative feedback can become unstable, resulting in unwanted behaviour such as oscillation. The Nyquist stability criterion developed by Harry Nyquist of Bell Laboratories can be used to study the stability of feedback amplifiers.

Feedback amplifiers share these properties:

Pros:

- Can increase or decrease input impedance (depending on type of feedback)
- Can increase or decrease output impedance (depending on type of feedback)
- Reduces distortion (increases linearity)
- Increases the bandwidth
- Desensitizes gain to component variations
- Can control step response of amplifier

Cons:

- May lead to instability if not designed carefully
- The gain of the amplifier decreases
- The input and output impedances of the amplifier with feedback (the closed-loop amplifier) become sensitive to the gain of the amplifier without feedback (the open-loop amplifier); that exposes these impedances to variations in the open loop gain, for example, due to parameter variations or due to nonlinearity of the open-loop gain

History

The negative feedback amplifier was invented by Harold Stephen Black while a passenger on the Lackawanna Ferry (from Hoboken Terminal to Manhattan) on his way to work at Bell Laboratories (historically located in Manhattan instead of New Jersey in 1927) on August 2, 1927 (US patent 2,102,671, issued in 1937). Black had been toiling at reducing distortion in repeater amplifiers used for telephone transmission. Black submitted his invention to the U. S. Patent Office on August 8, 1928, and it took more than nine years for the patent to be issued. Black later wrote: “One reason for the delay was that the concept was so contrary to established beliefs that the Patent Office initially did not believe it would work.”

Classical Feedback

Gain Reduction: Below, the voltage gain of the amplifier with feedback, the closed-loop gainA_{fb}, is derived in terms of the gain of the amplifier without feedback, the open-loop gainA_{OL} and the feedback factor β, which governs how much of the output signal is applied to the input. The open-loop gain A_{OL} in general may be a function of both frequency and voltage; the feedback parameter β is determined by the feedback network that is connected around the amplifier. For an operational amplifier two resistors forming a voltage divider may be used for the feedback network to set β between 0 and 1. This network may be modified using reactive elements like capacitors or inductors to (a) give frequency-dependent closed-loop gain as in equalization/ tone-control circuits or (b) construct oscillators. The gain of the amplifier with feedback is derived below in the case of a voltage amplifier with voltage feedback.

Without feedback, the input voltage V'_{in} is applied directly to the amplifier input. The according output voltage is

$$V_{out} = A_{OL} \cdot V'_{in}$$

Suppose now that an attenuating feedback loop applies a fraction $\beta . V_{out}$ of the output to one of the subtractor inputs so that it subtracts from the circuit input voltage V_{in} applied to the other subtractor input. The result of subtraction applied to the amplifier input is

$$V'_{in} = V_{in} - \beta \cdot V_{out}$$

Substituting for V'_{in} in the first expression,

$$V_{out} = A_{OL}(V_{in} - \beta \cdot V_{out})$$

Rearranging

$$V_{out}(1 + \beta \cdot A_{OL}) = V_{in} \cdot A_{OL}$$

Then the gain of the amplifier with feedback, called the closed-loop gain, A_{fb} is given by,

$$A_{fb} = \frac{V_{out}}{V_{in}} = \frac{A_{OL}}{1 + \beta \cdot A_{OL}}$$

If $A_{OL} >> 1$, then $A_{fb} \approx 1 / \beta$ and the effective amplification (or closed-loop gain) A_{fb} is set by the feedback constant β, and hence set by the feedback network, usually a simple reproducible network, thus making linearizing and stabilizing the amplification characteristics straightforward. Note also that if there are conditions where βA_{OL} = "1, the amplifier has infinite amplification – it has become an oscillator, and the system is unstable. The stability characteristics of the gain feedback product βA_{OL} are often displayed and investigated on a Nyquist plot (a polar plot of the gain/phase shift as a parametric function of frequency). A simpler, but less general technique, uses Bode plots.

The combination $L = \beta A_{OL}$ appears commonly in feedback analysis and is called the loop gain. The combination $(1 + \beta A_{OL})$ also appears commonly and is variously named as the desensitivity factor or the improvement factor.

Bandwidth Extension

Feedback can be used to extend the bandwidth of an amplifier at the cost of lowering the amplifier gain. Figure shows such a comparison. The figure is understood as follows. Without feedback the so-called open-loop gain in this example has a single time constant frequency response given by

$$A_{OL}(f) = \frac{A_0}{1 + jf / f_C},$$

wheref_C is the cutoff or corner frequency of the amplifier: in this example $f_C = 10^4$ Hz and the gain at zero frequency $A_0 = 10^5$ V/V. The figure shows the gain is flat out to the corner frequency and then drops. When feedback is present the so-called closed-loop gain, as shown in the formula of the previous section, becomes,

$$A_{fb}(f) = \frac{A_{OL}}{1 + \beta A_{OL}}$$

$$= \frac{A_0 / (1 + jf / f_C)}{1 + \beta A_0 / (1 + jf / f_C)}$$

$$= \frac{A_0}{1 + jf / f_C + \beta A_0}$$

$$= \frac{A_0}{(1 + \beta A_0)\left(1 + j\frac{f}{(1 + \beta A_0) f_C}\right)}.$$

The last expression shows the feedback amplifier still has a single time constant behaviour, but the corner frequency is now increased by the improvement factor $(1 + \beta A_0)$, and the gain at zero frequency has dropped by exactly the same factor. This behaviour is called the gain-bandwidth tradeoff. In Figure, ($1 + \beta A_0$) = 10^3, so $A_{fb}(0) = 10^5 / 10^3 = 100$ V/V, and f_C increases to $10^4 \times 10^3 = 10^7$ Hz.

Multiple Poles

When the open-loop gain has several poles, rather than the single pole of the above example, feedback can result in complex poles (real and imaginary parts). In a two-pole case, the result is peaking in the frequency response of the feedback amplifier near its corner frequency, and ringing and overshoot in its step response. In the case of more than two poles, the feedback amplifier can become unstable, and oscillate.

Asymptotic Gain Model

In the above analysis the feedback network is unilateral. However, real feedback networks often exhibit feed forward as well, that is, they feed a small portion of the input to the output, degrading performance of the feedback amplifier. A more general way to model negative feedback amplifiers including this effect is with the asymptotic gain model.

Feedback and Amplifier Type

Amplifiers use current or voltage as input and output, so four types of amplifier are possible. Any of these four choices may be the open-loop amplifier used to construct the feedback amplifier. The objective for the feedback amplifier also may be any one of the four types of amplifier, not necessarily the same type as the open-loop amplifier. For example, an op amp (voltage amplifier) can be arranged to make a current amplifier instead. The conversion from one type to another is implemented using different feedback connections, usually referred to as series or shunt (parallel) connections.

Feedback amplifier type	*Input connection*	*Output connection*	*Ideal feedback*	*Two-port feedback*
Current	Shunt	Series	CCCS	g-parameter
Transresistance	Shunt	Shunt	CCVS	y-parameter
Transconductance	Series	Series	VCCS	z-parameter
Voltage	Series	Shunt	VCVS	h-parameter

The feedback can be implemented using a two-port network. There are four types of two-port network, and the selection depends upon the type of feedback. For example, for a current feedback amplifier, current at the output is sampled and combined with current at the input. Therefore, the feedback ideally is performed using an (output) current-controlled current source (CCCS), and its imperfect realization using a two-port network also must incorporate a CCCS, that is, the appropriate choice for feedback network is a g-parameter two-port.

Two-port Analysis of Feedback

One approach to feedback is the use of return ratio. Here an alternative method used in most textbooks is presented by means of an example treated in the article on asymptotic gain model.

Figure shows a two-transistor amplifier with a feedback resistor R_f. The aim is to analyze this circuit to find three items: the gain, the output impedance looking into the amplifier from the load, and the input impedance looking into the amplifier from the source.

Replacement of the Feedback Network with a Two-port

The first step is replacement of the feedback network by a two-port. Just what components go into the two-port?

On the input side of the two-port we have R_f. If the voltage at the right side of R_f changes, it changes the current in R_f that is subtracted from the current entering the base of the input transistor. That is, the input side of the two-port is a dependent current source controlled by the voltage at the top of resistor R_2.

One might say the second stage of the amplifier is just a voltage follower, transmitting the voltage at the collector of the input transistor to the top of R_2. That is, the monitored output signal is really the voltage at the collector of the input transistor. That view is legitimate, but then the voltage follower stage becomes part of the feedback network. That makes analysis of feedback more complicated.

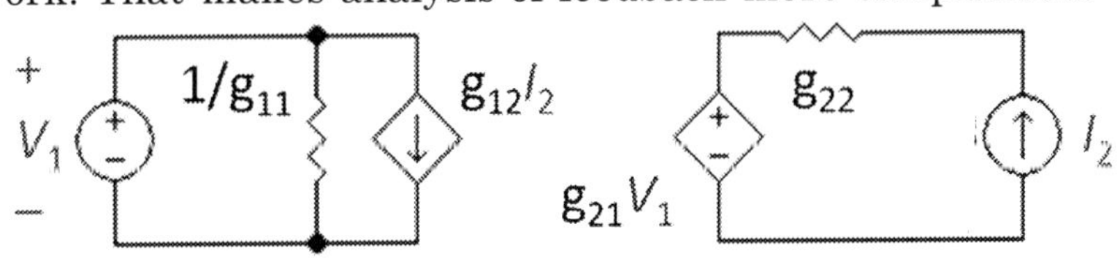

***Figure:** The g-parameter feedback network*

An alternative view is that the voltage at the top of R_2 is set by the emitter current of the output transistor. That view leads to an entirely passive feedback network made up of R_2 and R_f. The variable controlling the feedback is the emitter current, so the feedback is a current-controlled current source (CCCS). We search through the four available two-port networks and find the only one with a CCCS is the g-parameter two-port, shown in Figure. The next task is to select the g-parameters so that the two-port of Figure is electrically equivalent to the L-section made up of R_2 and R_f. That selection is an algebraic procedure made most simply by looking at two individual cases: the case with $V_1 = 0$, which makes the VCVS on the right side of the two-port a short-circuit; and the case with $I_2 = 0$. which makes the CCCS on the left side an open circuit.

The algebra in these two cases is simple, much easier than solving for all variables at once. The choice of g-parameters that make the two-port and the L-section behave the same way are shown in the table below.

g_{11}	g_{12}	g_{21}	g_{22}
$\dfrac{1}{R_f + R_2}$	$-\dfrac{R_2}{R_2 + R_f}$	$\dfrac{R_2}{R_2 + R_f}$	$R_2 // R_f$

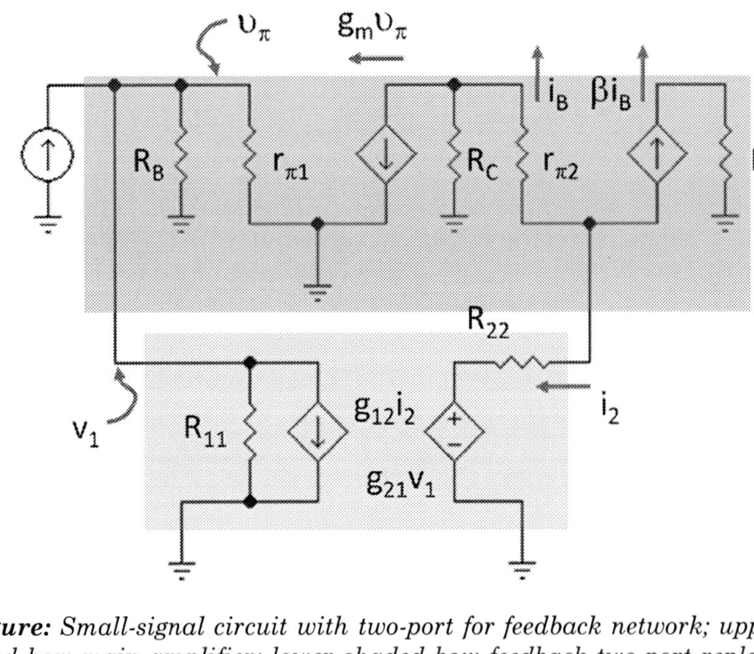

Figure: *Small-signal circuit with two-port for feedback network; upper shaded box: main amplifier; lower shaded box: feedback two-port replacing the L-section made up of* R_f *and* R_2.

Small-signal Circuit

The next step is to draw the small-signal schematic for the amplifier with the two-port in place using the hybrid-pi model for the transistors. Figure shows the schematic with notation $R_3 = R_{C2} // R_L$ and $R_{11} = 1 / g_{11}$, $R_{22} = g_{22}$.

Loaded Open-loop Gain

Figure indicates the output node, but not the choice of output variable. A useful choice is the short-circuit current output of the amplifier (leading to the short-circuit current gain). Because this variable leads simply to any of the other choices (for example, load voltage or load current), the short-circuit current gain is found below.

First the loaded open-loop gain is found. The feedback is turned off by setting $g_{12} = g_{21} = 0$. The idea is to find how much the amplifier gain is changed because of the resistors in the feedback network by themselves, with the feedback turned off. This calculation is pretty easy because R_{11}, R_B, *and* $r_{\pi 1}$ all are in parallel and $v_1 = v_\pi$. Let $R_1 = R_{11} \,//\, R_B \,//\, r_{\pi 1}$. In addition, $i_2 = -(\beta+1)\, i_B$. The result for the open-loop current gain A_{OL} is:

$$A_{OL} = \frac{\beta i_B}{i_S} = g_m R_C \left(\frac{\beta}{\beta+1}\right)\left(\frac{R_1}{R_{22} + \dfrac{r_{\pi 2} + R_C}{\beta+1}}\right).$$

Gain with Feedback

In the classical approach to feedback, the feedforward represented by the VCVS (that is, $g_{21}\, v_1$) is neglected. That makes the circuit of Figure resemble the block diagram of Figure, and the gain with feedback is then:

$$A_{FB} = \frac{A_{OL}}{1 + \beta_{FB} A_{OL}}$$

$$A_{FB} = \frac{A_{OL}}{1 + \dfrac{R_2}{R_2 + R_f} A_{OL}},$$

where the feedback factor $\beta_{FB} = -g_{12}$. Notation β_{FB} is introduced for the feedback factor to distinguish it from the transistor β.

Input and Output Resistances

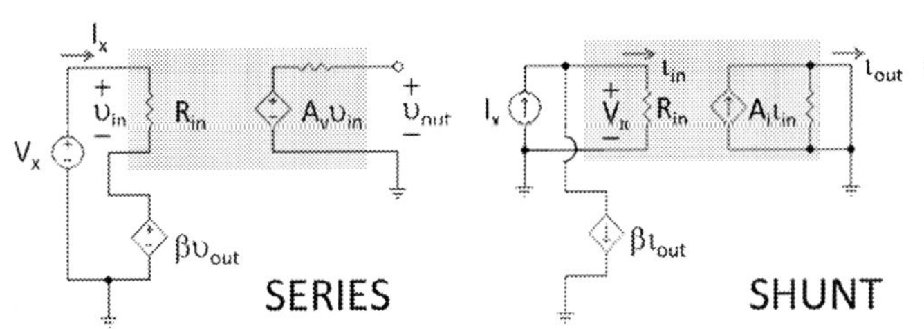

Figure: *Circuit set-up for finding feedback amplifier input resistance*

First, a digression on how two-port theory approaches resistance determination, and then its application to the amplifier at hand.

Background on Resistance Determination

Figure shows an equivalent circuit for finding the input resistance of a feedback voltage amplifier (left) and for a feedback current amplifier (right). These arrangements are typical Miller theorem applications.

In the case of the voltage amplifier, the output voltage βV_{out} of the feedback network is applied in series and with an opposite polarity to the input voltage V_x travelling over the loop (but in respect to ground, the polarities are the same).

As a result, the effective voltage across and the current through the amplifier input resistance R_{in} decrease so that the circuit input resistance increases (one might say that R_{in} apparently increases). Its new value can be calculated by applying Miller theorem (for voltages) or the basic circuit laws. Thus Kirchhoff's voltage law provides:

$$V_x = I_x R_{in} + \beta v_{out} ,$$

where $v_{out} = A_v v_{in} = A_v I_x R_{in}$. Substituting this result in the above equation and solving for the input resistance of the feedback amplifier, the result is:

$$R_{in}(fb) = \frac{V_x}{I_x} = (1 + \beta A_v) R_{in} .$$

The general conclusion to be drawn from this example and a similar example for the output resistance case is:

A series feedback connection at the input (output) increases the input (output) resistance by a factor (1 + β A_{OL}), where A_{OL} = open loop gain.

On the other hand, for the current amplifier, the output current βI_{out} of the feedback network is applied in parallel and with an opposite direction to the input current I_x.

As a result, the total current flowing through the circuit input (not only through the input resistance R_{in}) increases and the voltage across it decreases so that the circuit input resistance decreases (R_{in} apparently decreases). Its new value can be calculated by applying the dual Miller theorem (for currents) or the basic Kirchhoff's laws:

$$I_x = \frac{V_{in}}{R_{in}} + \beta i_{out} .$$

where $i_{out} = A_i i_{in} = A_i V_x / R_{in}$. Substituting this result in the above equation and solving for the input resistance of the feedback amplifier, the result is:

$$R_{in}(fb) = \frac{V_x}{I_x} = \frac{R_{in}}{(1+\beta A_i)}.$$

The general conclusion to be drawn from this example and a similar example for the output resistance case is:

A parallel feedback connection at the input (output) decreases the input (output) resistance by a factor (1 + β A_{OL}), where A_{OL} = open loop gain.

These conclusions can be generalized to treat cases with arbitrary Norton or Thévenin drives, arbitrary loads, and general two-port feedback networks. However, the results do depend upon the main amplifier having a representation as a two-port – that is, the results depend on the *same* current entering and leaving the input terminals, and likewise, the same current that leaves one output terminal must enter the other output terminal.

A broader conclusion to be drawn, independent of the quantitative details, is that feedback can be used to increase or to decrease the input and output impedances.

Application to the Example Amplifier

These resistance results now are applied to the amplifier of Figures. The *improvement factor* that reduces the gain, namely (1 + $\beta_{FB} A_{OL}$), directly decides the effect of feedback upon the input and output resistances of the amplifier. In the case of a shunt connection, the input impedance is reduced by this factor; and in the case of series connection, the impedance is multiplied by this factor. However, the impedance that is modified by feedback is the impedance of the amplifier in Figure with the feedback turned off, and does include the modifications to impedance caused by the resistors of the feedback network.

Therefore, the input impedance seen by the source with feedback turned off is $R_{in} = R_1 = R_{11} // R_B // r_{\pi 1}$, and with the feedback turned on (but no feedforward)

$$R_{in} = \frac{R_1}{1+\beta_{FB} A_{OL}},$$

where *division* is used because the input connection is *shunt*: the feedback two-port is in parallel with the signal source at the input side of the amplifier. A reminder: A_{OL} is the *loaded* open loop gain found above, as modified by the resistors of the feedback network.

The impedance seen by the load needs further discussion. The load in Figure is connected to the collector of the output transistor, and therefore is separated from the body of the amplifier by the infinite impedance of the output current source. Therefore, feedback has no effect on the output impedance, which remains simply R_{C2}. If instead we wanted to find the impedance presented at the *emitter* of the output transistor (instead of its collector), which is series connected to the feedback network, feedback would increase this resistance by the improvement factor ($1 + \beta_{FB} A_{OL}$).

Load Voltage and Load Current

The gain derived above is the current gain at the collector of the output transistor. To relate this gain to the gain when voltage is the output of the amplifier, notice that the output voltage at the load R_L is related to the collector current by Ohm's law as $v_L = i_C (R_{C2} // R_L)$. Consequently, the transresistance gain v_L / i_S is found by multiplying the current gain by $R_{C2} // R_L$:

$$\frac{v_L}{i_S} = A_{FB}(R_{C2} // R_L).$$

Similarly, if the output of the amplifier is taken to be the current in the load resistor R_L, current division determines the load current, and the gain is then:

$$\frac{i_L}{i_S} = A_{FB}\frac{R_{C2}}{R_{C2} + R_L}.$$

Is the Main Amplifier Block a Two Port?

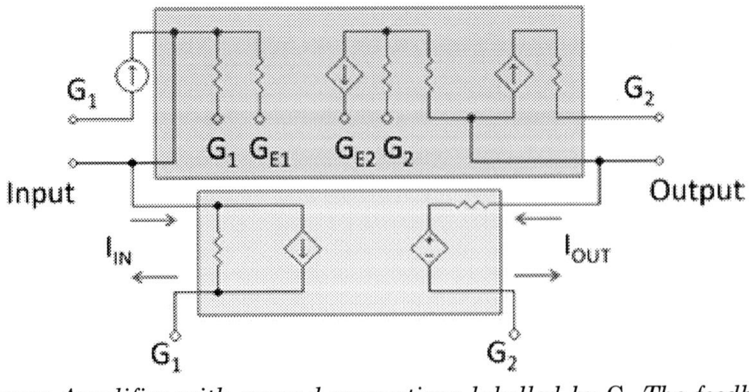

***Figure:** Amplifier with ground connections labelled by G. The feedback network satisfies the port conditions.*

Some complications follow, intended for the attentive reader.

Figure shows the small-signal schematic with the main amplifier and the feedback two-port in shaded boxes. The two-port satisfies the port conditions: at the input port, I_{in} enters and leaves the port, and likewise at the output, I_{out} enters and leaves. The main amplifier is shown in the upper shaded box. The ground connections are labelled.

Figure shows the interesting fact that the main amplifier does not satisfy the port conditions at its input and output unless the ground connections are chosen to make that happen. For example, on the input side, the current entering the main amplifier is I_S. This current is divided three ways: to the feedback network, to the bias resistor R_B and to the base resistance of the input transistor r_π. To satisfy the port condition for the main amplifier, all three components must be returned to the input side of the main amplifier, which means all the ground leads labelled G_1 must be connected, as well as emitter lead G_{E1}. Likewise, on the output side, all ground connections G_2 must be connected and also ground connection G_{E2}. Then, at the bottom of the schematic, underneath the feedback two-port and outside the amplifier blocks, G_1 is connected to G_2. That forces the ground currents to divide between the input and output sides as planned. Notice that this connection arrangement *splits the emitter* of the input transistor into a base-side and a collector-side – a physically impossible thing to do, but electrically the circuit sees all the ground connections as one node, so this fiction is permitted.

Of course, the way the ground leads are connected makes no difference to the amplifier (they are all one node), but it makes a difference to the port conditions. That is a weakness of this approach: the port conditions are needed to justify the method, but the circuit really is unaffected by how currents are traded among ground connections.

However, if there is no possible arrangement of ground conditions that will lead to the port conditions, the circuit might not behave the same way. The improvement factors ($1 + \beta_{FB} A_{OL}$) for determining input and output impedance might not work. This situation is awkward, because a failure to make a two-port may reflect a real problem (it just is not possible), or reflect a lack of imagination (for example, just did not think of splitting the emitter node in two). As a consequence, when the port conditions are in doubt, at least two approaches are possible to establish whether improvement factors are accurate: either simulate an example using Spice and compare results with use of an improvement factor, or calculate the impedance using a test source and compare results.

A more radical choice is to drop the two-port approach altogether, and use return ratios. That choice might be advisable if small-signal device models are complex, or are not available (for example, the devices are known only numerically, perhaps from measurement or from SPICE simulations).

Oscillators

An electronic oscillator is an electronic circuit that produces a repetitive, oscillating electronic signal, often a sine wave or a square wave. Oscillators convert direct current (DC) from a power supply to an alternating current signal. They are widely used in many electronic devices. Common examples of signals generated by oscillators include signals broadcast by radio and television transmitters, clock signals that regulate computers and quartz clocks, and the sounds produced by electronic beepers and video games.

Oscillators are often characterized by the frequency of their output signal:

- An audio oscillator produces frequencies in the audio range, about 16 Hz to 20 kHz.
- An RF oscillator produces signals in the radio frequency (RF) range of about 100 kHz to 100 GHz.
- A low-frequency oscillator (LFO) is an electronic oscillator that generates a frequency below ≈20 Hz. This term is typically used in the field of audio synthesizers, to distinguish it from an audio frequency oscillator.

Oscillators designed to produce a high-power AC output from a DC supply are usually called inverters. There are two main types of electronic oscillator: the linear or harmonic oscillator and the nonlinear or relaxation oscillator.

Linear Oscillator

The harmonic, or *linear*, oscillator produces a sinusoidal output. There are two types:

Feedback Oscillator: The most common form of linear oscillator is an electronic amplifier such as a transistor or op amp connected in a feedback loop with its output fed back into its input through a frequency selective electronic filter to provide positive feedback. When the power supply to the amplifier is first switched on, electronic noise in the circuit provides a signal to get oscillations started. The noise travels around the loop and is amplified and filtered until very quickly it becomes a sine wave at a single frequency.

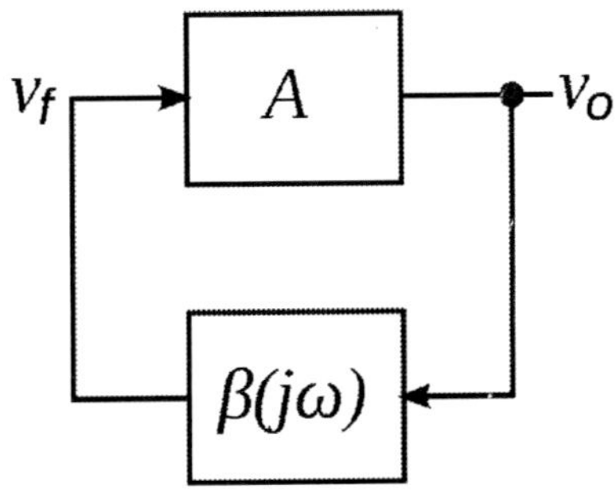

Figure: *Block diagram of a feedback linear oscillator; an amplifier* A *with its output* v_o *fed back into its input* v_f *through a filter,* β(jω).

Feedback oscillator circuits can be classified according to the type of frequency selective filter they use in the feedback loop:

- In an *RC oscillator* circuit, the filter is a network of resistors and capacitors. RC oscillators are mostly used to generate lower frequencies, for example in the audio range. Common types of RC oscillator circuits are the phase shift oscillator and the Wien bridge oscillator.
- In an *LC oscillator* circuit, the filter is a tuned circuit (often called a *tank circuit*) consisting of an inductor (L) and capacitor (C) connected together. Charge flows back and forth between the capacitor's plates through the inductor, so the tuned circuit can store electrical energy oscillating at its resonant frequency. There are small losses in the tank circuit, but the amplifier compensates for those losses and supplies the power for the output signal. LC oscillators are often used at radio frequencies, when a tunable frequency source is necessary, such as in signal generators, tunable radio transmitters and the local oscillators in radio receivers. Typical LC oscillator circuits are the Hartley, Colpitts and Clapp circuits.
- In a *crystal oscillator* circuit the filter is a piezoelectric crystal (commonly a quartz crystal). The crystal mechanically vibrates as a resonator, and its frequency of vibration determines the oscillation frequency. Crystals have very high Q-factor and

also better temperature stability than tuned circuits, so crystal oscillators have much better frequency stability than LC or RC oscillators. They are used to stabilize the frequency of most radio transmitters, and to generate the clock signal in computers and quartz clocks. Crystal oscillators often use the same circuits as LC oscillators, with the crystal replacing the tuned circuit; the Pierce oscillator circuit is commonly used. Quartz crystals are generally limited to frequencies of 30 MHz or below. Surface acoustic wave (SAW) devices are another kind of piezoelectric resonator used in crystal oscillators, which can achieve much higher frequencies. They are used in specialized applications which require a high frequency reference, for example, in cellular telephones.

Negative Resistance Oscillator: In addition to the feedback oscillators described above, which use two-port amplifying active elements such as transistors and op amps, linear oscillators can also be built using one-port (two terminal) devices with negative resistance, such as magnetron tubes, tunnel diodes and Gunn diodes. Negative resistance oscillators are often used at high frequencies in the microwave range and above, since at these frequencies feedback oscillators perform poorly due to excessive phase shift in the feedback path.

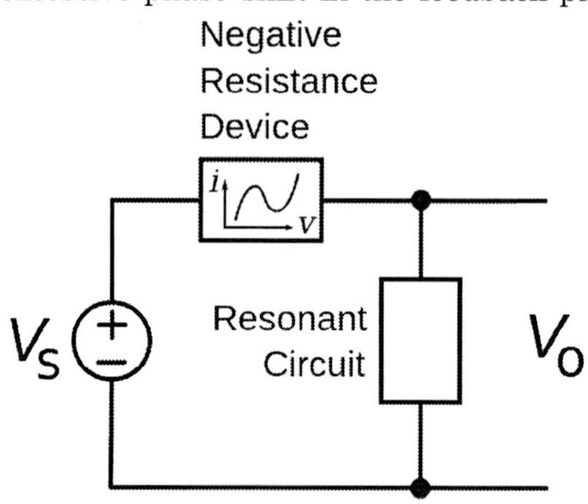

Figure: *Typical block diagram of a negative resistance oscillator. In some types the negative resistance device is connected in parallel with the resonant circuit.*

In negative resistance oscillators, a resonant circuit, such as an LC circuit, crystal, or cavity resonator, is connected across a device with negative differential resistance, and a DC bias voltage is applied to supply energy. A resonant circuit by itself is almost an oscillator;

it can store energy in the form of electronic oscillations if excited, but because it has some internal resistance or other losses the oscillations are damped and decline to zero. The negative resistance of the active device cancels the (positive) internal loss resistance in the resonator, in effect creating a resonator with no damping, which generates spontaneous continuous oscillations at its resonant frequency.

The negative resistance oscillator model is not limited to one-port devices like diodes; feedback oscillator circuits with two-port amplifying devices such as transistors and tubes also have negative resistance. At high frequencies, transistors and FETs do not need a feedback loop, but with certain loads applied to one port can become unstable at the other port and show negative resistance due to internal feedback, causing them to oscillate. So high frequency oscillators in general are designed using negative resistance techniques.

Active devices used in oscillators and approximate maximum frequencies

Device	***Frequency***
Triode vacuum tube	1 GHz
Bipolar transistor (BJT)	20 GHz
Heterojunction Bipolar Transistor (HBT)	50 GHz
Metal Semiconductor Field Effect Transistor (MESFET)	100 GHz
High Electron Mobility Transistor (HEMT)	200 GHz
Gunn diode, fundamental mode	100 GHz
Gunn diode, harmonic mode	200 GHz
IMPATT diode	300 GHz
Klystron tube	200 GHz
Magnetron tube	100 GHz
Gyrotron tube	300 GHz

Circuits

These are some of the many linear oscillator circuits:

- Armstrong oscillator
- Hartley oscillator
- Colpitts oscillator
- Clapp oscillator
- Delay line oscillator
- Pierce oscillator (crystal)

- Phase-shift oscillator
- Wien bridge oscillator
- Cross-coupled LC oscillator
- Vackαø oscillator
- Opto-electronic oscillator.
- Tri-tet oscillator
- Robinson oscillator

Relaxation Oscillator

A nonlinear or relaxation oscillator produces a non-sinusoidal output, such as a square, sawtooth or triangle wave. It contains an energy-storing element (a capacitor or, more rarely, an inductor) and a nonlinear switching circuit (a latch, Schmitt trigger, or negative resistance element) that periodically charges and discharges the energy stored in the storage element thus causing abrupt changes in the output waveform.

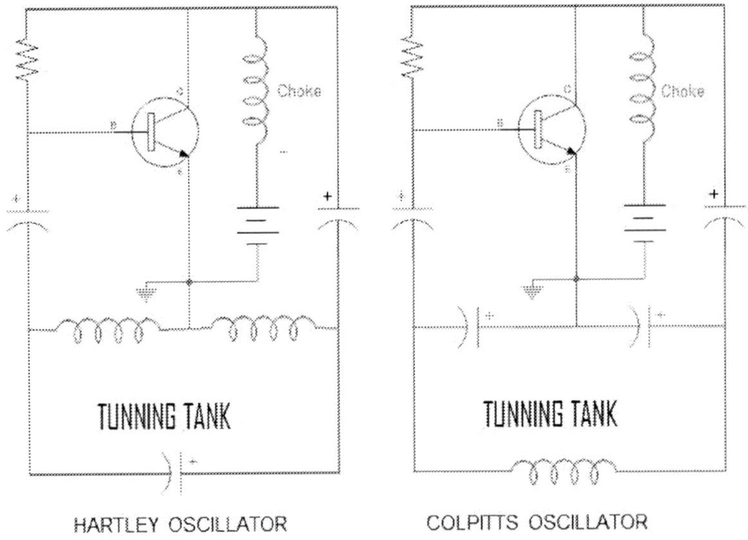

Figure: *Comparison between Hartley and Colpitts oscillators*

Square-wave relaxation oscillators are used to provide the clock signal for sequential logic circuits such as timers and counters, although crystal oscillators are often preferred for their greater stability. Triangle wave or sawtooth oscillators are used in the timebase circuits that generate the horizontal deflection signals for cathode ray tubes in analogue oscilloscopes and television sets. In function generators, this triangle wave may then be further shaped into a close approximation of a sine wave. Ring oscillators are built of a ring of active delay stages. Generally the ring has an odd number of inverting stages, so that there is no single stable state for the internal ring voltages. Instead, a single transition propagates endlessly around the ring.

Types of relaxation oscillator circuits include:

- multivibrator
- ring oscillator
- delay line oscillator
- Royer oscillator
- rotary travelling wave oscillator.

Voltage-controlled Oscillator (VCO)

An oscillator can be designed so that the oscillation frequency can be varied over some range by an input voltage or current. These voltage controlled oscillators are widely used in phase-locked loops, in which the oscillator's frequency can be locked to the frequency of another oscillator. These are ubiquitous in modern communications circuits, used in filters, modulators, demodulators, and forming the basis of frequency synthesizer circuits which are used to tune radios and televisions.

Radio frequency VCOs are usually made by adding a varactor diode to the tuned circuit or resonator in an oscillator circuit. Changing the DC voltage across the varacter changes its capacitance, which changes the resonant frequency of the tuned circuit. Voltage controlled relaxation oscillators can be constructed by charging and discharging the energy storage capacitor with a voltage controlled current source. Increasing the input voltage increases the rate of charging the capacitor, decreasing the time between switching events.

History

One of the first electronic oscillators was an oscillating arc built by Elihu Thomson in 1892. Thomson's oscillator placed an LC tuned circuit in parallel with the arc, used metal electrodes, and included a magnetic blowout. The arc oscillator was rediscovered and popularized by William Duddell in 1900. Electric arcs were used to provide illumination in the 19th century, but the arc current was unstable and they often produced hissing, humming or howling sounds. Duddell, a student at London Technical College, investigated this effect. He attached an LC circuit to the electrodes of an arc lamp, and the negative resistance of the arc excited audio frequency oscillations in the tuned circuit at its resonant frequency. Some of the energy was radiated as sound waves by the arc, producing a musical tone. To demonstrate his oscillator before the London Institute of Electrical Engineers, Duddell wired a series of tuned circuits to the arc and played a tune, "*God Save The Queen*". Duddell wasn't able to generate frequencies above the audio range with his "singing arc", but in 1902 Danish physicists ValdemarPoulsen and P. O. Pederson were able to increase the frequency produced into the radio range, inventing the Poulsen arcradio transmitter, the first continuous wave radio transmitter, which was used through the 1920s.

The vacuum tube feedback oscillator was invented around 1912, when it was discovered that feedback ("regeneration") in the recently-invented audionvacuum tube could produce oscillations. At least six researchers independently made this discovery and can be said to have some role in the invention. In the summer of 1912, Edwin Armstrong observed oscillations in audionradio receiver circuits and went on to use positive feedback in his invention of the regenerative receiver. German Alexander Meissner independently discovered positive feedback and invented oscillators in March 1913. Irving Langmuir at General Electric observed feedback in 1913. Fritz Lowenstein may have preceded the others with a crude oscillator in late 1911. In Britain, H. J. Round patented amplifying and oscillating circuits in 1913. In August 1912, Lee De Forest, the inventor of the audion, had also observed oscillations in his amplifiers, but he didn't understand its significance and tried to eliminate it until he read Armstrong's patents in 1914, which he promptly challenged. Armstrong and De Forest fought a protracted legal battle over the rights to the "regenerative" oscillator circuit which has been called "the most complicated patent litigation in the history of radio". De Forest ultimately won before the Supreme Court in 1934 on technical grounds, but most sources regard Armstrong's claim as the stronger one.

The first and most widely-used relaxation oscillator circuit, the astablemultivibrator, was invented in 1917 by French engineers Henri Abraham and Eugene Bloch. They called their cross-coupled, dual vacuum tube circuit a *multivibrateur* because the square-wave signal it produced was rich in harmonics, compared to the sinusoidal signal of other vacuum tube oscillators.

Vacuum tube feedback oscillators became the basis of radio transmission by 1920. However the triode vacuum tube oscillator performed poorly above 300 MHz because of interelectrode capacitance.

To reach higher frequencies, new "transit time" (velocity modulation) vacuum tubes were developed, in which electrons travelled in "bunches" through the tube. The first of these was the Barkhausen-Kurz oscillator (1920), the first tube to produce power in the UHF range. The most important and widely used were the klystron (R. and S. Varian, 1937) and the cavity magnetron (J. Randall and H. Boot, 1940).

Mathematical conditions for feedback oscillations, now called the Barkhausen criterion, were derived by Heinrich Georg Barkhausen in 1921. The first *stable* mathematical model of an electronic oscillator, the Van der Pol oscillator, was derived by Balthasar van der Pol in 1927. He showed that the stability of the oscillations (limit cycles) in actual oscillators was due to the nonlinearity of the amplifying device. Further advances in mathematical analysis of oscillation were made by Hendrik Wade Bode and Harry Nyquist in the 1930s. In 1969 K. Kurokawa derived necessary and sufficient conditions for oscillation in negative resistance circuits, which form the basis of modern microwave oscillator design.

RC Phase Shift

A phase-shift oscillator is a linearelectronic oscillator circuit that produces a sine wave output. It consists of an inverting amplifier element such as a transistor or op amp with its output fed back to its input through a phase-shift network consisting of resistors and capacitors. The feedback network 'shifts' the phase of the amplifier output by 180 degrees at the oscillation frequency to give positive feedback. Phase-shift oscillators are often used at audio frequency as audio oscillators.

The filter produces a phase shift that increases with frequency. It must have a maximum phase shift of more than 180 degrees at high frequencies so the phase shift at the desired oscillation frequency can be 180 degrees. The most common phase-shift network cascades three identical resistor-capacitor stages that produce a phase shift of zero at low frequencies and 270° at high frequencies.

Implementation

One of the simplest implementations for this type of oscillator uses an operational amplifier (op-amp), three capacitors and four resistors, as shown in the diagram.

The mathematics for calculating the oscillation frequency and oscillation criterion for this circuit are surprisingly complex, due to each RC stage loading the previous ones. The calculations are greatly simplified by setting all the resistors (except the negative feedbackresistor) and all the capacitors to the same values. In the diagram, if R_1=R_2=R_3=R, and C_1=C_2=C_3=C, then:

$$f_{\text{oscillation}} = \frac{1}{2\pi RC\sqrt{6}}$$

and the oscillation criterion is:

$$R_{\text{fb}} = 29 \cdot R$$

Op-amp implementation:

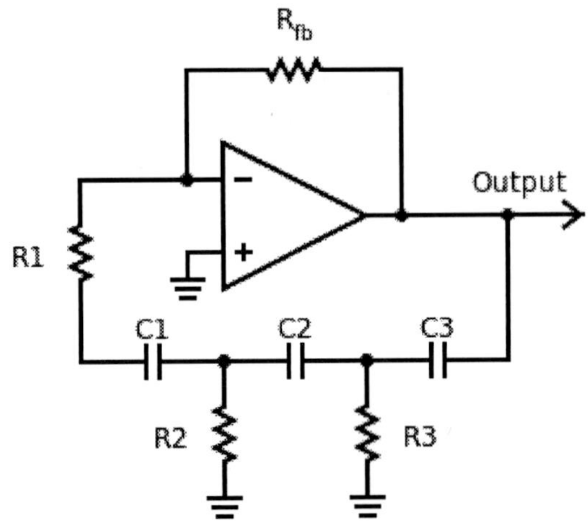

Figure: *A simple example of a phase-shift oscillator*

Without the simplification of all the resistors and capacitors having the same values, the calculations become more complex:

$$f_{\text{oscillation}} = \frac{1}{2\pi\sqrt{R_2R_3(C_1C_2 + C_1C_3 + C_2C_3) + R_1R_3(C_1C_2 + C_1C_3) + R_1R_2C_1C_2}}$$

Oscillation criterion:

$$R_{\text{fb}} = 2(R_1 + R_2 + R_3) + \frac{2R_1R_3}{R_2} + \frac{C_2R_2 + C_2R_3 + C_3R_3}{C_1}$$

$$+ \frac{2C_1R_1 + C_1R_2 + C_3R_3}{C_2} + \frac{2C_1R_1 + 2C_2R_1 + C_1R_2 + C_2R_2 + C_2R_3}{C_3}$$

$$+ \frac{C_1R_1^2 + C_3R_1R_3}{C_2R_2} + \frac{C_2R_1R_3 + C_1R_1^2}{C_3R_2} + \frac{C_1R_1^2 + C_1R_1R_2 + C_2R_1R_2}{C_3R_3}$$

In a purely ideal circuit, some sort of non-zero initial condition needs to be injected into the circuit in order to start the oscillation.

In practice, the noise present (predominantly thermal noise from the resistors) is sufficient to provide an initial signal to start the oscillation.

One potential problem with the single op-amp circuit is the high gain required to maintain the oscillation. If it is assumed that each RC segment does not effect the other, a gain of about 8 to 10 will be sufficient to enable oscillation.

As mentioned previously, each RC section loads the next section, and a larger gain (about 27 to 30) is required to keep the circuit in oscillation. An improved version of this circuit can be made by putting an op-amp buffer between each RC stage. (This also simplifies the calculations.) Òhe voltage gain of the inverting channel is always unity.

When the oscillation frequency is high enough to be near the amplifier's cutoff frequency, the amplifier will contribute significant phase shift itself, which will add to the phase shift of the feedback network. Therefore the circuit will oscillate at a frequency at which the phase shift of the feedback filter is less than 180 degrees.

FET implementation

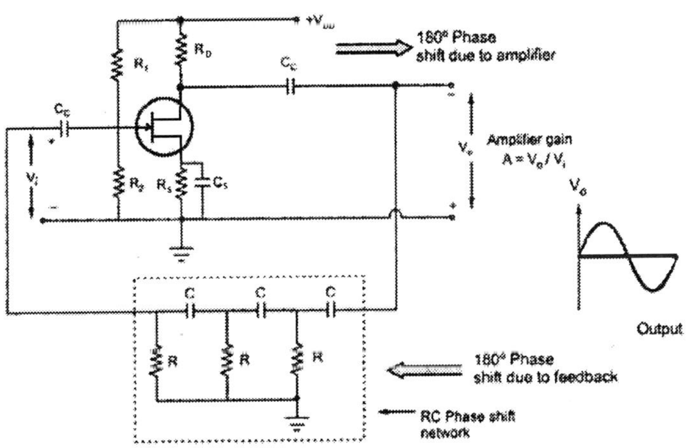

Figure: Circuit Diagram for RC-Phase Shift Oscillator using JFET

The circuit implements the RC phase shift oscillator with FET. Note that the topology used for positive feedback is voltage series feedback.

BJT Implementation

The circuit implements the RC phase shift oscillator with BJT. Note that the topology used for positive feedback is voltage shunt feedback.

Wien Bridge

A Wien bridge oscillator is a type of electronic oscillator that generates sine waves. It can generate a large range of frequencies. The oscillator is based on a bridge circuit originally developed by Max Wien in 1891. The bridge comprises four resistors and two capacitors. The oscillator can also be viewed as a positive gain amplifier combined with a bandpass filter that provides positive feedback.

The modern circuit is derived from William Hewlett's 1939 Stanford University master's degree thesis. Hewlett figured out how to make the oscillator with a stable output amplitude and low distortion. Hewlett, along with David Packard, co-founded Hewlett-Packard, and Hewlett-Packard's first product was the HP200A, a precision Wien bridge oscillator.

The frequency of oscillation is given by:

$$f = \frac{1}{2\pi RC}$$

Background

Problems with a Conventional Oscillator: The conventional oscillator circuit is designed so that it will start oscillating ("start up") and that its amplitude will be controlled.

For a linear circuit to oscillate, it must meet the Barkhausen conditions: its loop gain must be one and the phase around the loop must be an integer multiple of 360 degrees. The linear oscillator theory doesn't address how the oscillator starts up or how the amplitude is determined. The linear oscillator can support any amplitude.

In practice, the loop gain is initially larger than unity. Random noise is present in all circuits, and some of that noise will be near the desired frequency. A loop gain greater than one allows the amplitude of frequency to increase exponentially each time around the loop. With a loop gain greater than one, the oscillator will start.

Ideally, the loop gain needs to be just a little bigger than one, but in practice, it is often significantly greater than one. A larger loop gain makes the oscillator start quickly. A large loop gain also compensates for gain variations with temperature and the desired frequency of a tunable oscillator. For the oscillator to start, the loop gain must be greater than one under all possible conditions.

A loop gain greater than one has a down side. In theory, the oscillator amplitude will increase without limit. In practice, the amplitude will increase until the output runs into some limiting factor such as the power supply voltage (the amplifier output runs into the supply rails) or the amplifier output current limits. The limiting reduces the effective gain of the amplifier. (The effect is called gain compression.) In a stable oscillator, the average loop gain will be one.

Although the limiting action stabilizes the output voltage, it has two significant effects: it introduces harmonic distortion and it affects the frequency stability of the oscillator. The amount of distortion is related to the extra loop gain used for startup. If there's a lot of extra loop gain at small amplitudes, then the gain must decrease more at higher instantaneous amplitudes. That means more distortion.

The amount of distortion is also related to final amplitude of the oscillation. Although an amplifier's gain is ideally linear, in practice it is nonlinear. The nonlinear transcribing function can be viewed as a Taylor series. For small amplitudes, the higher order terms have little effect. For larger amplitudes, the nonlinearity is pronounced. Consequently, for low distortion, the oscillator's output amplitude should be a small fraction of the amplifier's dynamic range.

Bridge Oscillator

Meacham proposed a bridge oscillator to address those problems. Instead of using limiting to set an average gain of 1 around the loop, Meacham proposed a circuit that would set the loop gain to one while the amplifier was still in its linear region. As a result, the distortion would be reduced and the frequency stability would be improved. Meacham designed a quartz crystal oscillator based on a Wheatstone bridge that was a significant improvement over earlier designs.

At the oscillator frequency, Meacham's design was a linear circuit with constant gain. Consequently, there was no distortion of the sine wave. (In practice, an amplifier is not perfectly linear, so there is some distortion, but that distortion is much less than the gain compression approach.)

LC versus RC Oscillator

Hewlett's oscillator: High-gain differential amplifier with positive feedback. The Wien bridge oscillator can be considered as a combination of a differential amplifier and a Wien bridge, connected in a positive feedback loop between the op-amp output and differential input. At the oscillating frequency, the bridge is almost balanced and has very small transfer ratio. The loop gain is a product of the very high op-amp gain and the very low bridge ratio.

Conventional RC Oscillator

Low-gain single-ended amplifier with positive feedback. Rf, Rb and the op-amp compose a non-inverting amplifier with small gain of 1 + Rf/Rb ≈ 3. R1, R2, C1, C2 compose a bandpass filter. The band pass filter is connected to provide positive feedback at the frequency of oscillation. In the ideal situation, R1 = R2 = R, C1 = C2 = C and Rf/Rb = 2. Rbself heats and reduces the amplifier gain until the point is reached that there is just enough gain to sustain sinusoidal oscillation without over driving the amplifier.

Wien Bridge

Bridge circuits were a common way of measuring component values by comparing them to known values. Often an unknown component would be put in one arm of a bridge, and then the bridge would be nulled by adjusting the other arms or changing the frequency of the voltage source. The Wien bridge is one of many common bridges. Wien's bridge is used for precison measurement of capacitance in terms of resistance and frequency. It was also used to measure audio frequencies.

The Wien bridge does not require equal values of R or C. At some frequency, the reactance of the series R_1–C_1 arm will be an exact multiple of the shunt R_2–C_2 arm. If the two R_3 and R_4 arms are adjusted to the same ratio, then the bridge is balanced.

The bridge is balanced when:

$$\omega^2 = \frac{1}{R_1 R_2 C_1 C_2} \text{ and } \frac{C_1}{C_2} = \frac{R_4}{R_3} - \frac{R_2}{R_1}$$

where ω is the radian frequency. (R_3 refers to R_b in the figure at top and R_4 refers to R_f.)

The equations simplify if one chooses $R_1 = R_2$ and $C_1 = C_2$; the result is $R_4 = 2\ R_3$.

In practice, the values of R and C will never be exactly equal, but the equations above show that for fixed values in the *1* and *2* arms, the bridge will balance at some ω and some ratio of R_4/R_3.

Analysis

Analyzed From Loop Gain: According to Schilling, the loop gain of the Wien bridge oscillator is given by

$$T = \left(\frac{R_1/(1+sC_1R_1)}{R_1/(1+sC_1R_1) + R_2 + 1/(sC_2)} - \frac{R_b}{R_b + R_f}\right)A_0$$

where A_0 is the frequency-dependent gain of the op-amp. (Note, the component names in Schilling have been replaced with the component names in the figure.)

Schilling further says that the condition of oscillation is $T = 1$. Which, assuming $R_1 = R_2 = R$ and $C_1 = C_2 = C$ is satisfied by

$$\omega = \frac{1}{RC} \to f = \frac{1}{2\pi RC}$$

and

$$\frac{R_f}{R_b} = \frac{2A_0 + 3}{A_0 - 3} \text{ with } \lim_{A_0 \to \infty} \frac{R_f}{R_b} = 2$$

Another analysis, with particular reference to frequency stability and selectivity, will be found in Strauss (1970, p. 671) and Hamilton (2003, p. 449).

Frequency Determining Network

$$H(s) - \frac{R_1/(1+sC_1R_1)}{R_1/(1+sC_1R_1)+R_2+1/(sC_2)}$$

$$H(s) = \frac{sC_2R_1}{(1+sC_1R_1)(sC_2R_1/(1+sC_1R_1)+sC_2R_2+1)}$$

$$H(s) = \frac{sC_2R_1}{sC_2R_1+(1+sC_1R_1)(sC_2R_2+1)}$$

$$H(s) = \frac{sC_2R_1}{C_1C_2R_1R_2s^2+(C_2R_1+C_2R_2+C_1R_1)s+1}$$

Let R=R1=R2 and C−C1=C2

$$H(s) = \frac{sCR}{C^2R^2s^2+3CRs+1}$$

Normalize to *CR*=1.

$$H(s) = \frac{s}{s^2+3s+1}$$

Thus the frequency determining network has a zero at 0 and poles at -1.5±(sqrt(5)/2): -2.6180 and -0.38197. The resulting root locus traces the unit circle.

When the gain is 1, the two real poles meet at -1 and split into a complex pair. At gain 3, the poles cross the imaginary axis. At gain 5, the poles meet on the real axis and split into two real poles.

Amplitude Stabilization

The key to the Wien bridge oscillator's low distortion oscillation is an amplitude stabilization method that does not use clipping. The idea of using a lamp in a bridge configuration for amplitude stabilization was published by Meacham in 1938. The amplitude of electronic oscillators tends to increase until clipping or other gain limitation is reached. This leads to high harmonic distortion, which is often undesirable.

Hewlett used an incandescent bulb as a power detector, low pass filter and gain control element in the oscillator feedback path to control the output amplitude. The resistance of the light bulb filament increases as its temperature increases. The temperature of the filament depends on the power dissipated in the filament and some other factors. If the oscillator's period (an inverse of its frequency) is significantly shorter than the thermal time constant of the filament, then the temperature of the filament will be substantially constant over a cycle. The filament resistance will then determine the amplitude of the output signal. If the amplitude increases, the filament heats up and its resistance increases. The circuit is designed so that a larger filament resistance reduces loop gain, which in turn will reduce the output amplitude. The result is a negative feedback system that stabilizes the output amplitude to a constant value. With this form of amplitude control, the oscillator operates as a near ideal linear system and provides a very low distortion output signal. Oscillators that use limiting for amplitude control often have significant harmonic distortion. At low frequencies, as the time period of the Wien bridge oscillator approaches the thermal time constant of the incandescent bulb, the circuit operation becomes more nonlinear, and the output distortion rises significantly.

Light bulbs have their disadvantages when used as gain control elements in Wien bridge oscillators, most notably a very high sensitivity to vibration due to the bulb's microphonic nature amplitude modulating the oscillator output, a limitation in high frequency response due to the inductive nature of the coiled filament, and current requirements that exceed the capability of many op amps. Modern Wien bridge oscillators have used other nonlinear elements, such as diodes, thermistors, field effect transistors, or photocells for amplitude stabilization in place of light bulbs. Distortion as low as 0.0003% (3 ppm) can be achieved with modern components unavailable to Hewlett.

Wien bridge oscillators that use thermistors also exhibit "amplitude bounce" when the oscillator frequency is changed. This is due to the low damping factor and long time constant of the crude control loop, and disturbances cause the output amplitude to exhibit a decaying sinusoidal response. This can be used as a rough figure of merit, as the greater the amplitude bounce after a disLC Oscillator Introduction

High Frequency LC

Oscillators are used in many electronic circuits and systems providing the central "clock" signal that controls the sequential operation of the entire system. *Oscillators* convert a DC input (the supply voltage) into an AC output (the waveform), which can have a wide range of different wave shapes and frequencies that can be either complicated in nature or simple sine waves depending upon the application.

Oscillators are also used in many pieces of test equipment producing either sinusoidal sine waves, square, sawtooth or triangular shaped waveforms or just a train of pulses of a variable or constant width. LC Oscillators are commonly used in radio-frequency circuits because of their good phase noise characteristics and their ease of implementation. An Oscillator is basically an *Amplifier* with "Positive Feedback", or regenerative feedback (in-phase) and one of the many problems in electronic circuit design is stopping amplifiers from oscillating while trying to get oscillators to oscillate.

Oscillators work because they overcome the losses of their feedback resonator circuit either in the form of a *capacitor*, *inductor* or both in the same circuit by applying DC energy at the required frequency into this resonator circuit. In other words, an oscillator is a an amplifier which uses positive feedback that generates an output frequency without the use of an input signal. It is self sustaining.

Then an oscillator has a small signal feedback amplifier with an open-loop gain equal too or slightly greater than one for oscillations to start but to continue oscillations the average loop gain must return to unity. In addition to these reactive components, an amplifying device such as an *Operational Amplifier* or *Bipolar Transistor* is required. Unlike an amplifier there is no external AC input required to cause the Oscillator to work as the DC supply energy is converted by the oscillator into AC energy at the required frequency.

Chapter

4

Operational Amplifiers and Applications

Op-Amp

An operational amplifier (op-amp) is a DC-coupled high-gain electronic voltage amplifier with a differential input and, usually, a single-ended output. In this configuration, an op-amp produces an output potential (relative to circuit ground) that is typically hundreds of thousands of times larger than the potential difference between its input terminals.

Operational amplifiers had their origins in analog computers, where they were used to do mathematical operations in many linear, non-linear and frequency-dependent circuits. Characteristics of a circuit using an op-amp are set by external components with little dependence on temperature changes or manufacturing variations in the op-amp itself, which makes op-amps popular building blocks for circuit design.

Op-amps are among the most widely used electronic devices today, being used in a vast array of consumer, industrial, and scientific devices. Many standard IC op-amps cost only a few cents in moderate production volume; however some integrated or hybrid operational amplifiers with special performance specifications may cost over $100

US in small quantities. Op-amps may be packaged as components, or used as elements of more complex integrated circuits.

The op-amp is one type of differential amplifier. Other types of differential amplifier include the fully differential amplifier (similar to the op-amp, but with two outputs), the instrumentation amplifier (usually built from three op-amps), the isolation amplifier (similar to the instrumentation amplifier, but with tolerance to common-mode voltages that would destroy an ordinary op-amp), and negative feedback amplifier (usually built from one or more op-amps and a resistive feedback network).

Circuit Notation

The circuit symbol for an op-amp is shown to the right, where:

- V_+: non-inverting input
- V_-: inverting input
- V_{out}: output
- V_{S+}: positive power supply
- V_{S-}: negative power supply

The power supply pins (V_{S+} and V_{S-}) can be labelled in different ways. Often these pins are left out of the diagram for clarity, and the power configuration is described or assumed from the circuit.

Operation

The amplifier's differential inputs consist of a V_+ input and a V_- input, and ideally the op-amp amplifies only the difference in voltage between the two, which is called the *differential input voltage*. The output voltage of the op-amp is given by the equation:

$$V_{out} = A_{OL}(V_+ - V_-)$$

where V_+ is the voltage at the non-inverting terminal, V_- is the voltage at the inverting terminal and A_{OL} is the open-loop gain of the amplifier (the term "open-loop" refers to the absence of a feedback loop from the output to the input).

The magnitude of A_{OL} is typically very large—100,000 or more for integrated circuit op-amps—and therefore even a quite small difference between V_+ and V_- drives the amplifier output nearly to the supply voltage. Situations in which the output voltage is equal to or greater than the supply voltage are referred to as *saturation* of the amplifier. The magnitude of A_{OL} is not well controlled by the manufacturing process, and so it is impractical to use an operational amplifier as a stand-alone differential amplifier. Without negative feedback, and perhaps with positive feedback for regeneration, an op-amp acts as a comparator. If the inverting input is held at ground (0 V) directly or by a resistor, and the input voltage V_{in} applied to the non-inverting input is positive, the output will be maximum positive; if V_{in} is negative, the output will be maximum negative. Since there is no feedback from the output to either input, this is an *open loop* circuit acting as a comparator. The circuit's gain is just the A_{OL} of the op-amp.

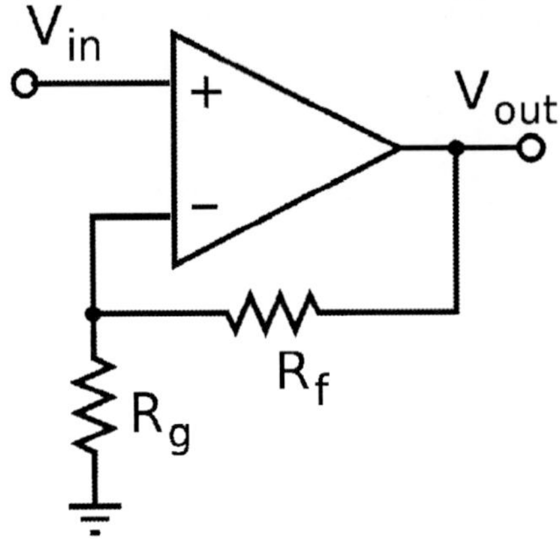

Figure: *An op-amp with negative feedback (a non-inverting amplifier)*

If predictable operation is desired, negative feedback is used, by applying a portion of the output voltage to the inverting input. The *closed loop* feedback greatly reduces the gain of the amplifier. When negative feedback is used, the circuit's overall gain and response becomes determined mostly by the feedback network rather than by the op-amp itself. If the feedback network is made of components with relatively constant, stable values, the variability of the op-amp's open loop response does not seriously affect the circuit's performance. The response of the op-amp circuit with its input, output and feedback circuits to an input is characterized mathematically by a transfer function. Designing an op-amp circuit to have a desired transfer function is in the realm of electrical engineering. The transfer functions are important in most applications of op-amps, such as in analog computers. High input impedance at the input terminals and low output impedance at the output terminal(s) are particularly useful features of an op-amp.

For example, in a non-inverting amplifier adding a negative feedback via the voltage divider R_f, R_g reduces the gain. Equilibrium will be established when V_{out} is just sufficient to reach around and "pull" the inverting input to the same voltage as V_{in}. The voltage gain of the entire circuit is determined by $1 + R_f/R_g$. As a simple example, if $V_{in} = 1$ V and $R_f = R_g$, V_{out} will be 2 V, the amount required to keep V_- at 1 V. Because of the feedback provided by R_f, R_g this is a *closed loop* circuit. Its overall gain V_{out} / V_{in} is called the *closed-loop gain*A_{CL}. Because the feedback is negative, in this case A_{CL} is less than the A_{OL} of the op-amp.

Another way of looking at it is to make two relatively valid assumptions.

One, that when an op-amp is being operated in linear (not saturated) mode, the difference in voltage between the non-inverting (+) pin and the inverting (–) pin is so small as to be considered negligible.

The second assumption is that the input impedance at both (+) and (–) pins is extremely high (at least several megohms with modern op-amps).

Thus, when the circuit to the right is operated as a non-inverting linear amplifier, V_{in} will appear at the (+) and (–) pins and create a current i through R_g equal to V_{in}/R_g. Since Kirchhoff's current law states that the same current must leave a node as enter it, and since the impedance into the (–) pin is near infinity, we can assume the overwhelming majority of the same current i travels through R_f, creating an output voltage equal to $V_{in} + i \times R_f$. By combining terms, we can easily determine the gain of this particular type of circuit.

$$i = \frac{V_{in}}{R_g}$$

$$V_{out} = V_{in} + i \times R_f = V_{in} + \left(\frac{V_{in}}{R_g} \times R_f\right) = V_{in} + \frac{V_{in} \times R_f}{R_g} = V_{in}\left(1 + \frac{R_f}{R_g}\right)$$

$$G = \frac{V_{out}}{V_{in}} = 1 + \frac{R_f}{R_g}$$

Op-amp Characteristics

Ideal op-amps:

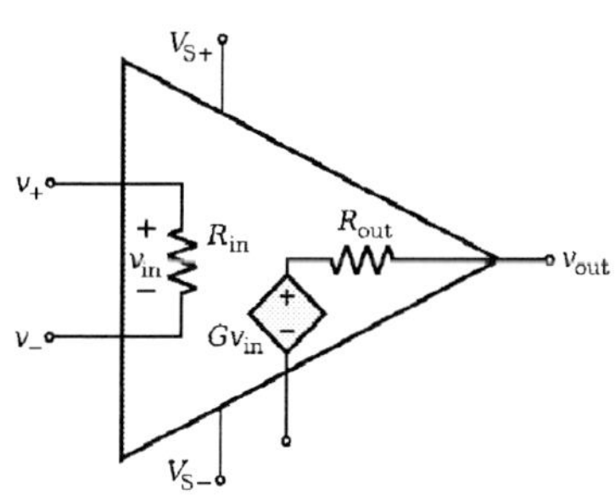

***Figure:** An equivalent circuit of an operational amplifier that models some resistive non-ideal parameters.*

An ideal op-amp is usually considered to have the following properties:

- Infinite open-loop gain
- Infinite voltage range available at the output
- Infinite bandwidth with zero phase shift and infinite slew rate
- Infinite input impedance and so zero input current and zero input offset voltage
- Zero output impedance
- Zero noise
- Infinite Common-mode rejection ratio (CMRR)
- Infinite Power supply rejection ratio.

These ideals can be summarized by the two "golden rules":

I. The output attempts to do whatever is necessary to make the voltage difference between the inputs zero.

II. The inputs draw no current.

The first rule only applies in the usual case where the op-amp is used in a closed-loop design (negative feedback, where there is a signal path of some sort feeding back from the output to the inverting input). These rules are commonly used as a good first approximation for analyzing or designing op-amp circuits.

None of these ideals can be perfectly realized. A real op-amp may be modelled with non-infinite or non-zero parameters using equivalent resistors and capacitors in the op-amp model. The designer can then include these effects into the overall performance of the final circuit. Some parameters may turn out to have negligible effect on the final design while others represent actual limitations of the final performance that must be evaluated.

Real op-amps

Real op-amps differ from the ideal model in various aspects.

DC Imperfections

Real operational amplifiers suffer from several non-ideal effects:

Finite Gain: Open-loop gain is infinite in the ideal operational amplifier but finite in real operational amplifiers. Typical devices exhibit open-loop DC gain ranging from 100,000 to over 1 million. So long as the loop gain (i.e., the product of open-loop and feedback gains) is very large, the circuit gain will be determined entirely by the amount of negative feedback (i.e., it will be independent of open-loop gain). In cases where closed-loop gain must be very high, the feedback gain will be very low, and the low feedback gain causes low loop gain; in these cases, the operational amplifier will cease to behave ideally.

Finite input impedances: The *differential input impedance* of the operational amplifier is defined as the impedance *between* its two inputs; the *common-mode input impedance* is the impedance from each input to ground. MOSFET-input operational amplifiers often have protection circuits that effectively short circuit any input differences greater than a small threshold, so the input impedance can appear to be very low in some tests. However, as long as these operational amplifiers are used in a typical high-gain negative feedback application, these protection circuits will be inactive. The input bias and leakage currents described below are a more important design parameter for typical operational amplifier applications.

Non-zero Output Impedance: Low output impedance is important for low-impedance loads; for these loads, the voltage drop across the output impedance of the amplifier will be significant. Hence, the output impedance of the amplifier limits the maximum power that can be provided. In configurations with a voltage-sensing negative feedback, the output impedance of the amplifier is effectively lowered; thus, in linear applications, op-amps usually exhibit a very low output impedance indeed. Negative feedback can not, however, reduce the limitations that R_{load} in conjunction with R_{out} place on the maximum and minimum possible output voltages; it can only reduce output errors *within* that range.

Low-impedance outputs typically require high quiescent (i.e., idle) current in the output stage and will dissipate more power, so low-power designs may purposely sacrifice low output impedance.

Input Current: Due to biasing requirements or leakage, a small amount of current (typically ~10 nanoamperes for bipolar op-amps, tens of picoamperes for JFET input stages, and only a few pA for MOSFET input stages) flows into the inputs. When large resistors or sources with high output impedances are used in the circuit, these small currents can produce large unmodelled voltage drops. If the input currents are matched, *and* the impedance looking *out* of *both* inputs are matched, then the voltages produced at each input will be equal. Because the operational amplifier operates on the *difference* between its inputs, these matched voltages will have no effect (unless the operational amplifier has poor CMRR, which is described below).

It is more common for the input currents (or the impedances looking out of each input) to be slightly mismatched, and so a small *offset voltage* (different from the input offset voltage below) can be produced. This offset voltage can create offsets or drifting in the operational amplifier. It can often be nulled externally; however, many operational amplifiers include *offset null* or *balance* pins and some procedure for using them to remove this offset. Some operational amplifiers attempt to nullify this offset automatically.

Input Offset Voltage: This voltage, which is what is required across the op-amp's input terminals to drive the output voltage to zero, is related to the mismatches in input bias current. In the perfect amplifier, there would be no input offset voltage. However, it exists in actual op-amps because of imperfections in the differential amplifier that constitutes the input stage of the vast majority of these devices. Input offset voltage creates two problems: First, due to the amplifier's high voltage gain, it virtually assures that the amplifier output will go into saturation if it is operated without negative feedback, even when the input terminals are wired together. Second, in a closed loop, negative feedback configuration, the input offset voltage is amplified along with the signal and this may pose a problem if high precision DC amplification is required or if the input signal is very small.

Common-mode Gain: A perfect operational amplifier amplifies only the voltage difference between its two inputs, completely rejecting all voltages that are common to both. However, the differential input stage of an operational amplifier is never perfect, leading to the amplification of these identical voltages to some degree. The standard measure of this defect is called the common-mode rejection ratio (denoted CMRR). Minimization of common mode gain is usually important in non-inverting amplifiers (described below) that operate at high amplification.

Output Sink Current: The output sink current is maximum current allowed to sink into the output stage. Some manufacturers show the output voltage vs. the output sink current plot, which gives an idea of the output voltage when it is sinking current from another source into the output pin.

Temperature Effects: All parameters change with temperature. Temperature drift of the input offset voltage is especially important.

Power-supply Rejection: The output of a perfect operational amplifier will be completely independent from ripples that arrive on its power supply inputs. Every real operational amplifier has a specified power supply rejection ratio (PSRR) that reflects how well the op-amp can reject changes in its supply voltage. Copious use of bypass capacitors can improve the PSRR of many devices, including the operational amplifier.

Drift: Real op-amp parameters are subject to slow change over time and with changes in temperature, input conditions, etc.

Noise: Amplifiers generate random voltage at the output even when there is no signal applied. This can be due to thermal noise and flicker noise of the devices. For applications with high gain or high bandwidth, noise becomes a very important consideration.

AC imperfections

The op-amp gain calculated at DC does not apply at higher frequencies. Thus, for high-speed operation, more sophisticated considerations must be used in an op-amp circuit design.

Finite Bandwidth: All amplifiers have finite bandwidth. To a first approximation, the op-amp has the frequency response of an integrator with gain. That is, the gain of a typical op-amp is inversely proportional to frequency and is characterized by its gain–bandwidth product (GBWP). For example, an op-amp with a GBWP of 1 MHz would have a gain of 5 at 200 kHz, and a gain of 1 at 1 MHz. This dynamic response coupled with the very high DC gain of the op-amp gives it the characteristics of a first-order low-pass filter with very high DC gain and low cutoff frequency given by the GBWP divided by the DC gain.

The finite bandwidth of an op-amp can be the source of several problems, including:

- Stability. Associated with the bandwidth limitation is a phase difference between the input signal and the amplifier output that can lead to oscillation in some feedback circuits. For example, a sinusoidal output signal meant to interfere destructively with an input signal of the same frequency will interfere constructively if delayed by 180 degrees forming positive feedback. In these cases, the feedback circuit can be stabilized by means of frequency compensation, which increases the gain or phase margin of the open-loop circuit. The circuit designer can implement this compensation externally with a separate circuit component. Alternatively, the compensation can be implemented within the operational amplifier with the addition of a dominant pole that sufficiently attenuates the high-frequency gain of the operational amplifier. The location of this pole may be fixed internally by the manufacturer or configured by the circuit designer using methods specific to the op-amp. In general, dominant-pole frequency compensation reduces the bandwidth of the op-amp even further. When the desired closed-loop gain is high, op-amp frequency compensation is often not needed because the requisite open-loop gain is sufficiently low; consequently, applications with high closed-loop gain can make use of op-amps with higher bandwidths.

- Noise, Distortion, and Other Effects. Reduced bandwidth also results in lower amounts of feedback at higher frequencies, producing higher distortion, noise, and output impedance and also reduced output phase linearity as the frequency increases.

Typical low-cost, general-purpose op-amps exhibit a GBWP of a few megahertz. Speciality and high-speed op-amps exist that can achieve a GBWP of hundreds of megahertz. For very high-frequency circuits, a current-feedback operational amplifier is often used.

Input Capacitance: Most important for high frequency operation because it further reduces the open-loop bandwidth of the amplifier.

Non-linear Imperfections

Saturation: output voltage is limited to a minimum and maximum value close to the power supply voltages. Saturation occurs when the output of the amplifier reaches this value and is usually due to:

- In the case of an op-amp using a bipolar power supply, a voltage gain that produces an output that is more positive or more negative than that maximum or minimum; or
- In the case of an op-amp using a single supply voltage, either a voltage gain that produces an output that is more positive than that maximum, or a signal so close to ground that the amplifier's gain is not sufficient to raise it above the lower threshold.

Slewing: the amplifier's output voltage reaches its maximum rate of change. Measured as the slew rate, it is usually specified in volts per microsecond. When slewing occurs, further increases in the input signal have no effect on the rate of change of the output. Slewing is usually caused by internal capacitances in the amplifier, especially those used to implement its frequency compensation.

Non-linear Input-output Relationship: The output voltage may not be accurately proportional to the difference between the input voltages. It is commonly called distortion when the input signal is a waveform. This effect will be very small in a practical circuit if substantial negative feedback is used.

Phase Reversal: In some integrated op-amps, when the published common mode voltage is violated (e.g. by one of the inputs being driven to one of the supply voltages), the output may slew to the opposite polarity from what is expected in normal operation. Under such conditions, negative feedback becomes positive, likely causing the circuit to "lock up" in that state.

Power Considerations

Limited Output Current: The output current must be finite. In practice, most op-amps are designed to limit the output current so as not to exceed a specified level – around 25 mA for a type 741 IC op-amp – thus protecting the op-amp and associated circuitry from damage. Modern designs are electronically more rugged than earlier implementations and some can sustain direct short circuits on their outputs without damage.

Limited Dissipated Power: The output current flows through the op-amp's internal output impedance, dissipating heat. If the op-amp dissipates too much power, then its temperature will increase above some safe limit. The op-amp may enter thermal shutdown, or it may be destroyed.

Modern integrated FET or MOSFET op-amps approximate more closely the ideal op-amp than bipolar ICs when it comes to input impedance and input bias currents. Bipolars are generally better when it comes to input *voltage* offset, and often have lower noise. Generally, at room temperature, with a fairly large signal, and limited bandwidth, FET and MOSFET op-amps now offer better performance.

Internal Circuitry of 741 type op-amp

Though designs vary between products and manufacturers, all op-amps have basically the same internal structure, which consists of three stages:

1. Differential amplifier — provides low noise amplification, high input impedance, usually a differential output.
2. Voltage amplifier — provides high voltage gain, a single-pole frequency roll-off, usually single-ended output.
3. Output amplifier — provides high current driving capability, low output impedance, current limiting and short circuit protection circuitry.

IC op-amps as implemented in practice are moderately complex integrated circuits. A typical example is the ubiquitous 741 op-amp designed by Dave Fullagar in Fairchild Semiconductor after the remarkable Bob Widlar LM301. Thus the basic architecture of the 741 is identical to that of the 301.

Input Stage

The input stage is a composed differential amplifier with a complex biasing circuit and a current mirror active load.

Differential Amplifier

It is implemented by two cascaded stages satisfying the conflicting requirements. The first stage consists of the NPN-based input emitter followers Q1 and Q2 that provide high input impedance. The next is the PNP-based common base pair Q3 and Q4 that eliminates the undesired Miller effect, shifts the voltage level downwards and provides a sufficient voltage gain to drive the next class A amplifier. The PNP transistors also help to increase the reverse V_{BE} rating (the base-emitter junctions of the NPN transistors Q1 and Q2 break down at around 7 V but the PNP transistors Q3 and Q4 have breakdown voltages around 50 V).

Biasing Circuit

The classical emitter-coupled differential stage is biased from the side of the emitters by connecting a constant current source to them. The series negative feedback (the emitter degeneration) makes the transistors act as voltage stabilizers; it forces them to adjust their V_{BE} voltages so that to pass the current through their collector-emitter junctions. As a result, *the quiescent current is β-independent.*

Here, the Q3/Q4 emitters are already used as inputs. Their collectors are separated and cannot be used as inputs for the quiescent current source since they behave as current sources. So, the quiescent current can be set only from the side of the bases by connecting a constant current source to them. To make it not depend on β, a negative but parallel feedback is used. For this purpose, the total quiescent current is mirrored by Q8-Q9 current mirror and the negative feedback is taken from the Q9 collector. Now it makes the transistors Q1-Q4 adjust their V_{BE} voltages so that to pass the desired quiescent current. The effect is the same as at the classical emitter-coupled pair — *the quiescent current is β-independent.* It is interesting fact that "to the extent that all PNP βs match, this clever circuit generates just the right β-dependent base current to produce a β-independent collector current". The biasing base currents are usually provided only by the negative power supply; they should come from the ground and enter the bases. But to ensure maximum high input impedances, the biasing loops are not internally closed between the base and ground; it is expected they will be closed externally by the input sources. So, the sources have to be galvanic (DC) to ensure paths for the biasing currents and low resistive enough (tens or hundreds kilohms) to not create significant voltage drops across them. Otherwise, additional DC elements should be connected between the bases and the ground (or the positive power supply).

The quiescent current is set by the 39 kΩ resistor that is common for the two current mirrors Q12-Q13 and Q10-Q11. The current determined by this resistor acts also as a reference for the other bias currents used in the chip. The Widlar current mirror built by Q10, Q11, and the 5 kΩ resistor produces a very small fraction of I_{ref} at the Q10 collector. This small constant current through Q10's collector supplies the base currents for Q3 and Q4 as well as the Q9 collector current. The Q8/Q9 current mirror tries to make Q9 collector current the same as the Q3 and Q4 collector currents and succeeds with the help of the negative feedback. The Q9 collector voltage changes until the ratio between the Q3/Q4 base and collector currents becomes equal to β. Thus Q3 and Q4's combined base currents (which are of the same order as the overall chip's input currents) are a small fraction of the already small Q10 current.

Thus the quiescent current is set by Q10-Q11 current mirror without using a current-sensing negative feedback. The voltage-sensing negative feedback only helps this process by stabilizing Q9 collector (Q3/Q4 base) voltage. The feedback loop also isolates the rest of the circuit from common-mode signals by making the base voltage of Q3/Q4 follow tightly $2V_{BE}$ below the higher of the two input voltages.

Current Mirror Active Load

The differential amplifier formed by Q1–Q4 drives an active load implemented as an improved current mirror (Q5–Q7) whose role is to convert the differential current input signal to a single ended voltage signal without the intrinsic 50% losses and to greatly increase the gain. This is achieved by copying the input signal from the left to the right side where the magnitudes of the two input signals add (Widlar used the same trick in μA702 and μA709). For this purpose, the input of the current mirror (Q5 collector) is connected to the left output (Q3 collector) and the output of the current mirror (Q6 collector) is connected to the right output of the differential amplifier (Q4 collector). Q7 increases the accuracy of the current mirror by decreasing the amount of signal current required from Q3 to drive the bases of Q5 and Q6.

Operation

Differential mode: The input voltage sources are connected through two "diode" strings, each of them consisting of two connected in series base-emitter junctions (Q1-Q3 and Q2-Q4), to the common point of Q3/Q4 bases. So, if the input voltages change slightly in opposite directions, Q3/Q4 bases stay at relatively constant voltage and the common base current does not change as well; it only vigorously steers between Q3/Q4 bases and makes the common quiescent current distribute between Q3/Q4 collectors in the same proportion. The current mirror inverts Q3 collector current and tries to pass it through Q4. In the middle point between Q4 and Q6, the signal currents (current changes) of Q3 and Q4 are subtracted. In this case (differential input signal), they are equal and opposite. Thus, the difference is twice the individual signal currents ($\Delta I - (-\Delta I) = 2\Delta I$) and the differential to single ended conversion is completed without gain losses. The open circuit signal voltage appearing at this point is given by the product of the subtracted signal currents and the total circuit impedance (the paralleled collector resistances of Q4 and Q6). Since the collectors of Q4 and Q6 appear as high differential resistances to the signal current (Q4 and Q6 behave as current sources), the open circuit voltage gain of this stage is very high.

More intuitively, the transistor Q6 can be considered as a duplicate of Q3 and the combination of Q4 and Q6 can be thought as of a varying voltage divider composed of two voltage-controlled resistors. For differential input signals, they vigorously change their instant resistances in opposite directions but the total resistance stays constant (like a potentiometer with quickly moving slider). As a result, the current stays constant as well but the voltage at the middle point changes vigorously. As the two resistance changes are equal and opposite, the effective voltage change is twice the individual change.

The base current at the inputs is not zero and the effective differential input impedance of a 741 is about 2 MΩ. The "offset null" pins may be used to place external resistors in parallel with the two 1 kΩ resistors (typically in the form of the two ends of a potentiometer) to adjust the balancing of the Q5/Q6 current mirror and thus indirectly control the output of the op-amp when zero signal is applied between the inputs.

Common Mode

If the input voltages change in the same direction, the negative feedback makes Q3/Q4 base voltage follow (with $2V_{BE}$ below) the input voltage variations. Now the output part (Q10) of Q10-Q11 current mirror keeps up the common current through Q9/Q8 constant in spite of varying voltage. Q3/Q4 collector currents and accordingly, the output voltage in the middle point between Q4 and Q6, remain unchanged.

Class A gain Stage

The section outlined in magenta is the class A gain stage. The top-right current mirror Q12/Q13 supplies this stage by a constant current load, via the collector of Q13, that is largely independent of the output voltage. The stage consists of the two NPN transistors Q15/Q19 connected in a Darlington configuration and uses the output side of a current mirror as its collector (dynamic) load to achieve high gain. The transistor Q22 prevents this stage from saturating by diverting the excessive Q15 base current (it acts as a Baker clamp).

The 30 pF capacitor provides frequency selective negative feedback around the class A gain stage as a means of frequency compensation to stabilise the amplifier in feedback configurations. This technique is called Miller compensation and functions in a similar manner to an op-amp integrator circuit. It is also known as 'dominant pole compensation' because it introduces a dominant pole (one which masks the effects of other poles) into the open loop frequency response. This pole can be as low as 10 Hz in a 741 amplifier and it introduces a "3 dB loss into the open loop response at this frequency. This internal compensation is provided to achieve unconditional stability of the amplifier in negative feedback configurations where the feedback network is non-reactive and the closed loop gain is unity or higher. Hence, the use of the operational amplifier is simplified because no external compensation is required for unity gain stability; amplifiers without this internal compensation such as the 748 may require external compensation or closed-loop gains significantly higher than unity.

Output Bias Circuitry

The green outlined section (based on Q16) is a voltage level shifter named *rubber diode, transistor Zener* or V_{BE} *multiplier*. In the circuit as shown, Q16 provides a constant voltage drop across its collector-emitter junction regardless of the current through it (it acts as a voltage stabilizer). This is achieved by introducing a negative feedback between Q16 collector and its base, i.e. by connecting a voltage divider with ratio β = 7.5 kΩ / (4.5 kΩ + 7.5 kΩ) = 0.625 composed by the two resistors. If the base current to the transistor is assumed to be zero, the negative feedback forces the transistor to increase its collector-emitter voltage up to 1 V until its base-emitter voltage reaches 0.625 V (a typical value for a BJT in the active region). This serves to bias the two output transistors slightly into conduction reducing crossover distortion (in some discrete component amplifiers, this function is usually achieved with a string of two silicon diodes).

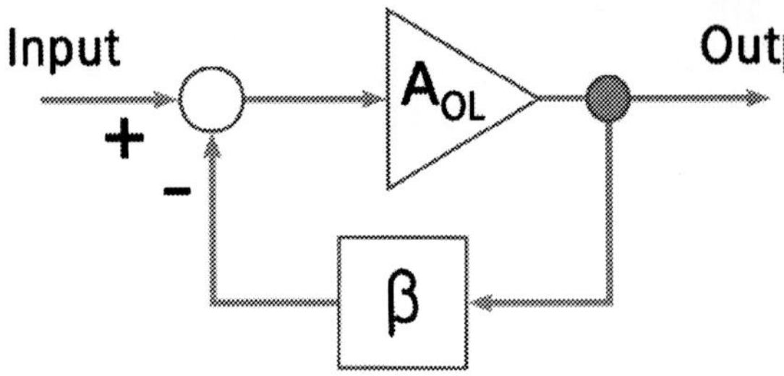

Figure: *The circuit presented as a negative feedback amplifier with constant input voltage*

The circuit can be presented as a negative feedback voltage amplifier with constant input voltage of 0.625 V and a feedback ratio of β = 0.625 (a gain of 1/β = 1.6). The same circuit but with β = 1 is used in the input current-setting part of the classical BJT current mirror.

Output Stage

The output stage (outlined in cyan) is a Class AB push-pull emitter follower (Q14, Q20) amplifier with the bias set by the V_{BE} multiplier voltage source Q16 and its base resistors. This stage is effectively driven by the collectors of Q13 and Q19. Variations in the bias with temperature, or between parts with the same type number, are common so crossover distortion and quiescent current may be subject to significant variation. The output range of the amplifier is about one volt less than the supply voltage, owing in part to V_{BE} of the output transistors Q14 and Q20.

The 25 Ω resistor in the output stage acts as a current sense to provide the output current-limiting function which limits the current in the emitter follower Q14 to about 25 mA for the 741. Current limiting for the negative output is done by sensing the voltage across Q19's emitter resistor and using this to reduce the drive into Q15's base. Later versions of this amplifier schematic may show a slightly different method of output current limiting. The output resistance is not zero, as it would be in an ideal op-amp, but with negative feedback it approaches zero at low frequencies.

Some Considerations

Note: while the 741 was historically used in audio and other sensitive equipment, such use is now rare because of the improved noise performance of more modern op-amps. Apart from generating noticeable hiss, 741s and other older op-amps may have poor common-mode rejection ratios and so will often introduce cable-borne mains hum and other common-mode interference, such as switch 'clicks', into sensitive equipment.

The "741" has come to often mean a generic op-amp IC (such as μA741, LM301, 558, LM324, TBA221 — or a more modern replacement such as the TL071). The description of the 741 output stage is qualitatively similar for many other designs (that may have quite different input stages), except:

- Some devices (μA748, LM301, LM308) are not internally compensated (require an external capacitor from output to some point within the operational amplifier, if used in low closed-loop gain applications).
- Some modern devices have "rail-to-rail output" capability, meaning that the output can range from within a few millivolts of the positive supply voltage to within a few millivolts of the negative supply voltage.

Classification

Op-amps may be classified by their construction:

- discrete (built from individual transistors or tubes/valves)
- IC (fabricated in an Integrated circuit) — most common
- hybrid

IC op-amps may be classified in many ways, including:

- Military, Industrial, or Commercial grade (for example: the LM301 is the commercial grade version of the LM101, the LM201 is the industrial version). This may define operating temperature ranges and other environmental or quality factors.
- Classification by package type may also affect environmental hardiness, as well as manufacturing options; DIP, and other through-hole packages are tending to be replaced by surface-mount devices.

- Classification by internal compensation: op-amps may suffer from high frequency instability in some negative feedback circuits unless a small compensation capacitor modifies the phase and frequency responses. Op-amps with a built-in capacitor are termed "*compensated*", or perhaps compensated for closed-loop gains down to (say) 5. All others are considered uncompensated.
- Single, dual and quad versions of many commercial op-amp IC are available, meaning 1, 2 or 4 operational amplifiers are included in the same package.
- Rail-to-rail input (and/or output) op-amps can work with input (and/or output) signals very close to the power supply rails.
- CMOS op-amps (such as the CA3140E) provide extremely high input resistances, higher than JFET-input op-amps, which are normally higher than bipolar-input op-amps.
- other varieties of op-amp include programmable op-amps (simply meaning the quiescent current, gain, bandwidth and so on can be adjusted slightly by an external resistor).
- manufacturers often tabulate their op-amps according to purpose, such as low-noise pre-amplifiers, wide bandwidth amplifiers, and so on.

Applications

Use in Electronics System Design: The use of op-amps as circuit blocks is much easier and clearer than specifying all their individual circuit elements (transistors, resistors, etc.), whether the amplifiers used are integrated or discrete. In the first approximation op-amps can be used as if they were ideal differential gain blocks; at a later stage limits can be placed on the acceptable range of parameters for each op-amp.

Circuit design follows the same lines for all electronic circuits. A specification is drawn up governing what the circuit is required to do, with allowable limits. For example, the gain may be required to be 100 times, with a tolerance of 5% but drift of less than 1% in a specified temperature range; the input impedance not less than one megohm; etc.

A basic circuit is designed, often with the help of circuit modelling (on a computer). Specific commercially available op-amps and other components are then chosen that meet the design criteria within the specified tolerances at acceptable cost. If not all criteria can be met, the specification may need to be modified.

A prototype is then built and tested; changes to meet or improve the specification, alter functionality, or reduce the cost, may be made.

Applications Without Using any Feedback

That is, the op-amp is being used as a voltage comparator. Note that a device designed primarily as a comparator may be better if, for instance, speed is important or a wide range of input voltages may be found, since such devices can quickly recover from full on or full off ("saturated") states.

A *voltage level detector* can be obtained if a reference voltage V_{ref} is applied to one of the op-amp's inputs. This means that the op-amp is set up as a comparator to detect a positive voltage. If the voltage to be sensed, E_i, is applied to op amp's (+) input, the result is a noninverting positive-level detector: when E_i is above V_{ref}, V_O equals $+V_{sat}$; when E_i is below V_{ref}, V_O equals $-V_{sat}$. If E_i is applied to the inverting input, the circuit is an inverting positive-level detector: When E_i is above V_{ref}, V_O equals $-V_{sat}$.

A *zero voltage level detector* ($E_i = 0$) can convert, for example, the output of a sine-wave from a function generator into a variable-frequency square wave. If E_i is a sine wave, triangular wave, or wave of any other shape that is symmetrical around zero, the zero-crossing detector's output will be square. Zero-crossing detection may also be useful in triggering TRIACs at the best time to reduce mains interference and current spikes.

Positive Feedback Applications

Another typical configuration of op-amps is with positive feedback, which takes a fraction of the output signal back to the non-inverting input. An important application of it is the comparator with hysteresis, the Schmitt trigger. Some circuits may use *Positive* feedback and *Negative* feedback around the same amplifier, for example Triangle waveoscillators and active filters.

Because of the wide slew-range and lack of positive feedback, the response of all the open-loop level detectors described above will be relatively slow. External overall positive feedback may be applied but (unlike internal positive feedback that may be applied within the latter stages of a purpose-designed comparator) this markedly affects the accuracy of the zero-crossing detection point. Using a general-purpose op-amp, for example, the frequency of E_i for the sine to square wave converter should probably be below 100 Hz.

Negative Feedback Applications

Non-inverting amplifier:

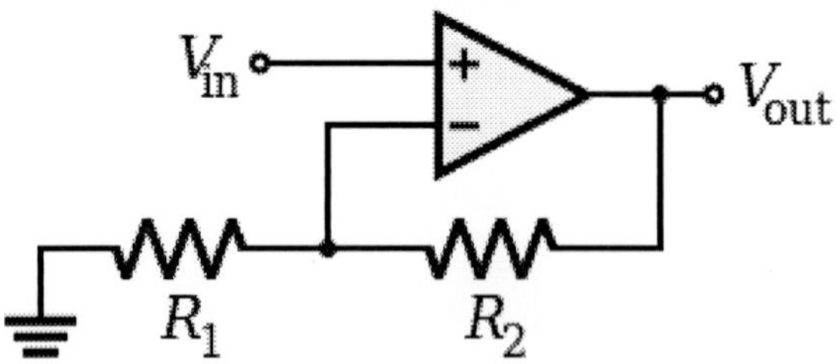

Figure: *An op-amp connected in the non-inverting amplifier configuration*

In a non-inverting amplifier, the output voltage changes in the same direction as the input voltage.

The gain equation for the op-amp is:

$$V_{out} = A_{OL}(V_+ - V_-)$$

However, in this circuit V_- is a function of V_{out} because of the negative feedback through the R_1R_2 network. R_1 and R_2 form a voltage divider, and as V_- is a high-impedance input, it does not load it appreciably. Consequently:

$$V_- = \beta \cdot V_{out}$$

where

$$\beta = \frac{R_1}{R_1 + R_2}$$

Substituting this into the gain equation, we obtain:

$$V_{out} = A_{OL}(V_{in} - \beta \cdot V_{out})$$

Solving for V_{out}:

$$V_{out} = V_{in}\left(\frac{1}{\beta + 1/A_{OL}}\right)$$

If A_{OL} is very large, this simplifies to

$$V_{out} \approx \frac{V_{in}}{\beta} = \frac{V_{in}}{\frac{R_1}{R_1 + R_2}} = V_{in}\left(1 + \frac{R_2}{R_1}\right).$$

Note that the non-inverting input of the operational amplifier will need a path for DC to ground; if the signal source might not give this, or if that source requires a given load impedance, the circuit will require another resistor — from input to ground. In either case, the ideal value for the feedback resistors (to give minimum offset voltage) will be such that the two resistances in parallel roughly equal the resistance to ground at the non-inverting input pin.

Inverting Amplifier

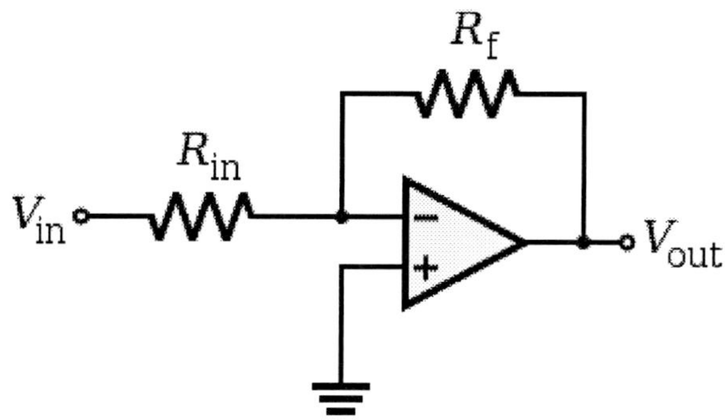

Figure: *An op-amp connected in the inverting amplifier configuration*

In an inverting amplifier, the output voltage changes in an opposite direction to the input voltage.

As with the non-inverting amplifier, we start with the gain equation of the op-amp:

$$V_{out} = A_{OL}(V_+ - V_-)$$

This time, V_- is a function of both V_{out} and V_{in} due to the voltage divider formed by R_f and R_{in}. Again, the op-amp input does not apply an appreciable load, so:

$$V_- = \frac{1}{R_f + R_{in}}\left(R_f V_{in} + R_{in} V_{out}\right)$$

Substituting this into the gain equation and solving for V_{out}:

$$V_{out} = -V_{in} \cdot \frac{A_{OL} R_f}{R_f + R_{in} + A_{OL} R_{in}}$$

If A_{OL} is very large, this simplifies to

$$V_{out} \approx -V_{in} \frac{R_f}{R_{in}}$$

A resistor is often inserted between the non-inverting input and ground (so both inputs "see" similar resistances), reducing the input offset voltage due to different voltage drops due to bias current, and may reduce distortion in some op-amps.

A DC-blockingcapacitor may be inserted in series with the input resistor when a frequency response down to DC is not needed and any DC voltage on the input is unwanted. That is, the capacitive component of the input impedance inserts a DC zero and a low-frequency pole that gives the circuit a bandpass or high-pass characteristic.

The potentials at the operational amplifier inputs remain virtually constant (near ground) in the inverting configuration. The constant operating potential typically results in distortion levels that are lower than those attainable with the non-inverting topology.

Other Applications

- audio- and video-frequency pre-amplifiers and buffers
- differential amplifiers
- differentiators and integrators
- filters
- precision rectifiers
- precision peak detectors
- voltage and current regulators
- analog calculators
- analog-to-digital converters
- digital-to-analog converters
- Voltage clamping
- oscillators and waveform generators

Most single, dual and quad op-amps available have a standardized pin-out which permits one type to be substituted for another without wiring changes. A specific op-amp may be chosen for its open loop gain, bandwidth, noise performance, input impedance, power consumption, or a compromise between any of these factors.

Historical Timeline

1941: A vacuum tube op-amp. An op-amp, defined as a general-purpose, DC-coupled, high gain, inverting feedback amplifier, is first found in U.S. Patent 2,401,779 “Summing Amplifier” filed by Karl D. Swartzel Jr. of Bell Labs in 1941. This design used three vacuum tubes to achieve a gain of 90 dB and operated on voltage rails of ±350 V. It had a single inverting input rather than differential inverting and non-inverting inputs, as are common in today’s op-amps. Throughout World War II, Swartzel’s design proved its value by being liberally used in the M9 artillery director designed at Bell Labs. This artillery director worked with the SCR584 radar system to achieve extraordinary hit rates (near 90%) that would not have been possible otherwise.

Figure: *GAP/R's K2-W: a vacuum-tube op-amp (1953)*

1947: An op-amp with an explicit non-inverting input. In 1947, the operational amplifier was first formally defined and named in a paper by Professor John R. Ragazzini of Columbia University. In this same paper a footnote mentioned an op-amp design by a student that would turn out to be quite significant. This op-amp, designed by Loebe Julie, was superior in a variety of ways. It had two major innovations. Its input stage used a long-tailed triode pair with loads matched to reduce drift in the output and, far more importantly, it was the first op-amp design to have two inputs (one inverting, the other non-inverting). The differential input made a whole range of new functionality possible, but it would not be used for a long time due to the rise of the chopper-stabilized amplifier.

1949: A chopper-stabilized op-amp. In 1949, Edwin A. Goldberg designed a chopper-stabilized op-amp. This set-up uses a normal op-amp with an additional AC amplifier that goes alongside the op-amp. The chopper gets an AC signal from DC by switching between the DC voltage and ground at a fast rate (60 Hz or 400 Hz). This signal is then amplified, rectified, filtered and fed into the op-amp's non-inverting input. This vastly improved the gain of the op-amp while significantly reducing the output drift and DC offset. Unfortunately, any design that used a chopper couldn't use their non-inverting input for any other purpose. Nevertheless, the much improved characteristics of the chopper-stabilized op-amp made it the dominant way to use op-amps. Techniques that used the non-inverting input regularly would not be very popular until the 1960s when op-amp ICs started to show up in the field.

1953: A commercially available op-amp. In 1953, vacuum tube op-amps became commercially available with the release of the model K2-W from George A. Philbrick Researches, Incorporated. The designation on the devices shown, GAP/R, is an acronym for the complete company name. Two nine-pin 12AX7 vacuum tubes were mounted in an octal package and had a model K2-P chopper add-on available that would effectively "use up" the non-inverting input. This op-amp was based on a descendant of Loebe Julie's 1947 design and, along with its successors, would start the widespread use of op-amps in industry.

Figure: *GAP/R's model P45: a solid-state, discrete op-amp (1961).*

1961: A discrete IC op-amp. With the birth of the transistor in 1947, and the silicon transistor in 1954, the concept of ICs became a reality. The introduction of the planar process in 1959 made transistors and ICs stable enough to be commercially useful. By 1961, solid-state, discrete op-amps were being produced. These op-amps were effectively small circuit boards with packages such as edge connectors. They usually had hand-selected resistors in order to improve things such as voltage offset and drift. The P45 (1961) had a gain of 94 dB and ran on ±15 V rails. It was intended to deal with signals in the range of ±10 V.

1961: A varactor bridge op-amp. There have been many different directions taken in op-amp design. Varactor bridge op-amps started to be produced in the early 1960s. They were designed to have extremely small input current and are still amongst the best op-amps available in terms of common-mode rejection with the ability to correctly deal with hundreds of volts at their inputs.

Figure: *GAP/R's model PP65: a solid-state op-amp in a potted module (1962)*

1962: An op-amp in a potted module. By 1962, several companies were producing modular potted packages that could be plugged into printed circuit boards. These packages were crucially important as they made the operational amplifier into a single black box which could be easily treated as a component in a larger circuit.

1963: A monolithic IC op-amp. In 1963, the first monolithic IC op-amp, the μA702 designed by Bob Widlar at Fairchild Semiconductor, was released. Monolithic ICs consist of a single chip as opposed to a chip and discrete parts (a discrete IC) or multiple chips bonded and connected on a circuit board (a hybrid IC). Almost all modern op-amps are monolithic ICs; however, this first IC did not meet with much success. Issues such as an uneven supply voltage, low gain and a small dynamic range held off the dominance of monolithic op-amps until 1965 when the μA709 (also designed by Bob Widlar) was released.

1968: Release of the μA741. The popularity of monolithic op-amps was further improved upon the release of the LM101 in 1967, which solved a variety of issues, and the subsequent release of the μA741 in 1968. The μA741 was extremely similar to the LM101 except that Fairchild's facilities allowed them to include a 30 pF compensation capacitor inside the chip instead of requiring external compensation. This simple difference has made the 741 *the* canonical op-amp and many modern amps base their pinout on the 741s. The μA741 is still in production, and has become ubiquitous in electronics—many manufacturers produce a version of this classic chip, recognizable by part numbers containing *741*. The same part is manufactured by several companies.

1970: First high-speed, low-input current FET design. In the 1970s high speed, low-input current designs started to be made by using FETs. These would be largely replaced by op-amps made with MOSFETs in the 1980s. During the 1970s single sided supply op-amps also became available.

Figure: *ADI's HOS-050: a high speed hybrid IC op-amp (1979)*

1972: Single sided supply op-amps being produced. A single sided supply op-amp is one where the input and output voltages can be as low as the negative power supply voltage instead of needing to be at least two volts above it. The result is that it can operate in many applications with the negative supply pin on the op-amp being connected to the signal ground, thus eliminating the need for a separate negative power supply.

Figure: *An op-amp in a modern mini DIP*

The LM324 (released in 1972) was one such op-amp that came in a quad package (four separate op-amps in one package) and became an industry standard. In addition to packaging multiple op-amps in a single package, the 1970s also saw the birth of op-amps in hybrid packages. These op-amps were generally improved versions of existing monolithic op-amps. As the properties of monolithic op-amps improved, the more complex hybrid ICs were quickly relegated to systems that are required to have extremely long service lives or other speciality systems.

Recent trends. Recently supply voltages in analog circuits have decreased (as they have in digital logic) and low-voltage op-amps have been introduced reflecting this. Supplies of ±5 V and increasingly 3.3 V (sometimes as low as 1.8 V) are common. To maximize the signal range modern op-amps commonly have rail-to-rail output (the output signal can range from the lowest supply voltage to the highest) and sometimes rail-to-rail inputs.

Differential Amplifier Configurations

A differential amplifier is a type of electronic amplifier that amplifies the difference between two voltages but does not amplify the particular voltages.

Theory

Many electronic devices use differential amplifiers internally. The output of an ideal differential amplifier is given by:

$$V_{\text{out}} = A_{\text{d}}(V_{\text{in}}^{+} - V_{\text{in}}^{-})$$

Where V_{in}^{+} and V_{in}^{-} are the input voltages and A_{d} is the differential gain.

In practice, however, the gain is not quite equal for the two inputs. This means, for instance, that if V_{in}^{+} and V_{in}^{-} are equal, the output will not be zero, as it would be in the ideal case. A more realistic expression for the output of a differential amplifier thus includes a second term.

$$V_{\text{out}} = A_{\text{d}}(V_{\text{in}}^{+} - V_{\text{in}}^{-}) + A_{\text{c}}\left(\frac{V_{\text{in}}^{+} + V_{\text{in}}^{-}}{2}\right)$$

A_{c} is called the common-mode gain of the amplifier. As differential amplifiers are often used to null out noise or bias-voltages that appear at both inputs, a low common-mode gain is usually desired.

The common-mode rejection ratio (CMRR), usually defined as the ratio between differential-mode gain and common-mode gain, indicates the ability of the amplifier to accurately cancel voltages that are common to both inputs. The common-mode rejection ratio is defined as:

$$\text{CMRR} = 10\log_{10}\left(\frac{A_{\text{d}}}{A_{\text{cm}}}\right)^{2} = 20\log_{10}\left(\frac{A_{\text{d}}}{|A_{\text{cm}}|}\right)$$

In a perfectly symmetrical differential amplifier, A_c is zero and the CMRR is infinite. Note that a differential amplifier is a more general form of amplifier than one with a single input; by grounding one input of a differential amplifier, a single-ended amplifier results.

Long-tailed Pair

Historical Background: The long-tailed pair was originally implemented using a pair of vacuum tubes. The circuit works the same way for all three-terminal devices with current gain. The long-tail resistor circuit bias points are largely determined by Ohm's Law and less so by active component characteristics.

The long-tailed pair was developed from earlier knowledge of push-pull circuit techniques and measurement bridges. The earliest circuit that is truly recognizable as a long-tailed pair in its conventional form is given by Matthews (1934) and the same circuit form appears in a patent submitted by Alan Blumlein in 1936. By the end of the 1930s the topology was well established and had been described by various authors including Offner (1937),Schmitt (1937) and Toennies (1938) and It was particularly used for detection and measurement of physiological impulses.

The long-tailed pair was very successfully used in early British computing, most notably the Pilot ACE Model and descendants, Wilkes' EDSAC, and probably others designed by people who worked with Blumlein or his peers. The long-tailed pair has many attributes as a switch: largely immune to tube (transistor) variations (of great importance when machines contained 1,000 or more tubes), high gain, gain stability, high input impedance, medium/low output impedance, good clipper (with not-too-long tail), non-inverting (EDSAC contained no inverters!) and large output voltage swings. One disadvantage is that the output voltage swing (typically ±10–20 V) was imposed upon a high DC voltage (200 V or so), requiring care in signal coupling, usually some form of wide-band DC coupling. Many computers of this time tried to avoid this problem by using only AC-coupled pulse logic, which made them very large and overly complex (ENIAC: 18,000 tubes for a 20 digit calculator) or unreliable. DC-coupled circuitry became the norm after the first generation of vacuum tube computers.

Configurations

A differential (long-tailed, emitter-coupled) pair amplifier consists of two amplifying stages with common (emitter, source or cathode) degeneration.

Differential Output

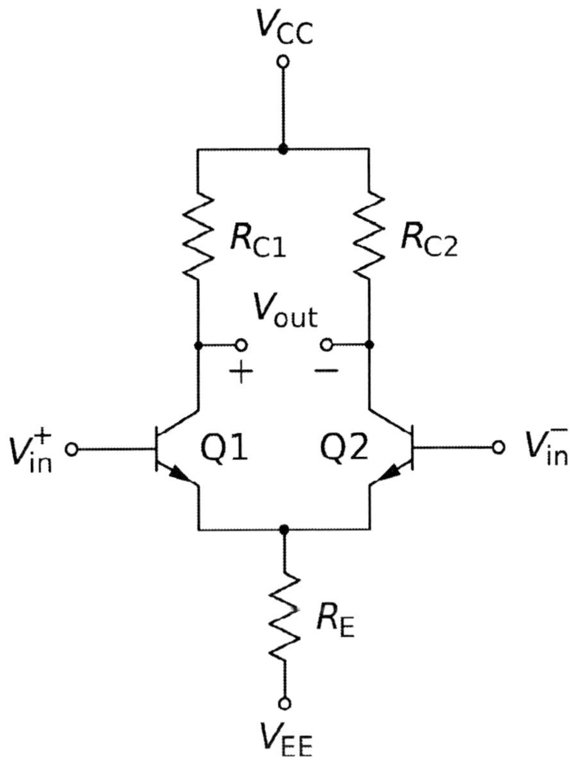

Figure: *A classic long-tailed pair*

With two inputs and two outputs, this forms a differential amplifier stage. The two bases (or grids or gates) are inputs which are differentially amplified (subtracted and multiplied) by the pair; they can be fed with a differential (balanced) input signal, or one input could be grounded to form a phase splitter circuit. An amplifier with differential output can drive floating load or another stage with differential input.

Single-ended Output

If the differential output is not desired, then only one output can be used (taken from just one of the collectors (or anodes or drains), disregarding the other output without a collector inductor; this configuration is referred to as *single-ended output*. The gain is half that of the stage with differential output. To avoid sacrificing gain, a differential to single-ended converter can be utilized. This is often implemented as a current mirror.

Operation

To explain the circuit operation, four particular modes are isolated below although, in practice, some of them act simultaneously and their effects are superimposed.

Biasing

In contrast with classic amplifying stages that are biased from the side of the base (and so they are highly β-dependent), the differential pair is directly biased from the side of the emitters by sinking/injecting the total quiescent current. The series negative feedback (the emitter degeneration) makes the transistors act as voltage stabilizers; it forces them to adjust their V_{BE} voltages (base currents) to pass the quiescent current through their collector-emitter junctions. So, due to the negative feedback, the quiescent current depends slightly on the transistor's β.

The biasing base currents needed to evoke the quiescent collector currents usually come from the ground, pass through the input sources and enter the bases. So, the sources have to be galvanic (DC) to ensure paths for the biasing currents and low resistive enough to not create significant voltage drops across them. Otherwise, additional DC elements should be connected between the bases and the ground (or the positive power supply).

Common Mode

At common mode (the two input voltages change in the same directions), the two voltage (emitter) followers cooperate with each other working together on the common high-resistive emitter load (the "long tail"). They all together increase or decrease the voltage of the common emitter point (figuratively speaking, they together "pull up" or "pull down" it so that it moves). In addition, the dynamic load "helps" them by changing its instant ohmic resistance in the same direction as the input voltages (it increases when the voltage increases and vice versa.) thus keeping up constant total resistance between the two supply rails. There is a full (100%) negative feedback; the two input base voltages and the emitter voltage change simultaneously while the collector currents and the total current do not change. As a result, the output collector voltages do not change as well.

Differential Mode

Normal. At differential mode (the two input voltages change in opposite directions), the two voltage (emitter) followers oppose each other - while one of them tries to increase the voltage of the common emitter point, the other tries to decrease it (figuratively speaking, one of them "pulls up" the common point while the other "pulls down" it so that it stays immovable) and v.v. So, the common point does not change its voltage; it behaves like a virtual ground with a magnitude determined by the common-mode input voltages. The high-resistive emitter element does not play any role since it is shunted by the other low-resistive emitter follower.

There is no negative feedback since the emitter voltage does not change at all when the input base voltages change. Òhe common quiescent current vigorously steers between the two transistors and the output collector voltages vigorously change. The two transistors mutually ground their emitters; so, although they are common-collector stages, they actually act as common-emitter stages with maximum gain. Bias stability and independence from variations in device parameters can be improved by negative feedback introduced via cathode/emitter resistors with relatively small resistances.

Overdriven.If the input differential voltage changes significantly (more than about a hundred millivolts), the base-emitter junction of the transistor driven by the lower input voltage becomes backward biased and its collector voltage reaches the positive supply rail. The other transistor (driven by the higher input voltage) saturates and its collector voltage begins following the input one. This mode is used in differential switches and ECL gates.

Breakdown. If the input voltage continues increasing and exceeds the base-emitter breakdown voltage, the base-emitter junction of the transistor driven by the lower input voltage breaks down. If the input sources are low resistive, an unlimited current will flow directly through the "diode bridge" between the two input sources and will damage them.

At common mode, the emitter voltage follows the input voltage variations; there is a full negative feedback and the gain is minimum. At differential mode, the emitter voltage is fixed (equal to the instant common input voltage); there is no negative feedback and the gain is maximum.

Single-ended Input

The differential pair can be used as an amplifier with a single-ended input if one of the inputs is grounded or fixed to a reference voltage (usually, the other collector is used as a single-ended output) This arrangement can be thought as of cascaded common-collector and common-base stages or as a buffered common-base stage.

The emitter-coupled amplifier is compensated for temperature drifts, V_{BE} is cancelled, and the Miller effect and transistor saturation are avoided. That is why it is used to form emitter-coupled amplifiers (avoiding Miller effect), phase splitter circuits (obtaining two inverse voltages), ECL gates and switches (avoiding transistor saturation), etc.

Improvements

Emitter constant current source:

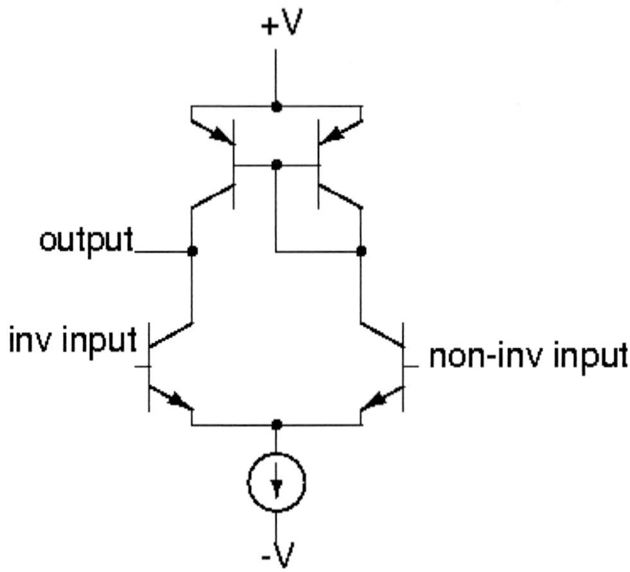

Figure: *An improved long-tailed pair with current-mirror load and constant-current biasing*

The quiescent current has to be constant to ensure constant collector voltages at common mode. This requirement is not so important in the case of a differential output since the two collector voltages will vary simultaneously but their difference (the output voltage) will not vary; only the output range will decrease. But in the case of a single-ended output, it is extremely important to keep a constant current since the output collector voltage will vary.

Thus the higher the resistance of the current source R_c, the lower A_c is, and the better the CMRR. The constant current needed can be produced by connecting an element (resistor) with very high resistance between the shared emitter node and the supply rail (negative for NPN and positive for PNP transistors) but this will require high supply voltage. That is why, in more sophisticated designs, an element with high differential (dynamic) resistance approximating a constant current source/sink is substituted for the long tail. It is usually implemented by a current mirror because of its high compliance voltage (small voltage drop across the output transistor).

The same arrangement is widely used in cascode circuits as well. It can be generalized by an equivalent circuit consisting of a constant current source loaded by two connected in parallel voltage sources with equal voltages. The current source determines the common current flowing through the voltage sources while the voltage sources fix the voltage across the current source. The emitter current source is usually implemented as a common-emitter transistor stage with constant base voltage driving with current the two common-base transistor stages. So, this arrangement can be considered as a cascode consisting of cascaded common-emitter and common-base stages.

Collector Current Mirror

The collector resistors can be replaced by a current mirror, whose output part acts as an active load. Thus the differential collector current signal is converted to a single ended voltage signal without the intrinsic 50% losses and the gain is extremely increased. This is achieved by copying the input collector current from the right to the left side where the magnitudes of the two input signals add. For this purpose, the input of the current mirror is connected to the right output and the output of the current mirror is connected to the left output of the differential amplifier.

The current mirror inverts the right collector current and tries to pass it through the left transistor that produces the left collector current. In the middle point between the two left transistors, the two signal currents (current changes) are subtracted. In this case (differential input signal), they are equal and opposite. Thus, the difference is twice the individual signal currents (ΔI - $(-\Delta I) = 2\Delta I$) and the differential to single ended conversion is completed without gain losses.

Interfacing Considerations

Floating Input Source: It is possible to connect a floating source between the two bases, but it is necessary to ensure paths for the biasing base currents. In the case of galvanic source, only one resistor has to be connected between one of the bases and the ground. The biasing current will enter directly this base and indirectly (through the input source) the other one. If the source is capacitive, two resistors have to be connected between the two bases and the ground to ensure different paths for the base currents.

Input/output Impedance

The input impedance of the differential pair highly depends on the input mode. At common mode, the two parts behave as common-collector stages with high emitter loads; so, the input impedances are extremely high. At differential mode, they behave as common-emitter stages with grounded emitters; so, the input impedances are low.

The output impedance of the differential pair is high (especially for the improved differential pair).

Input/output Range

The common-mode input voltage can vary between the two supply rails but cannot closely reach them since some voltage drops (minimum 1 volt) have to remain across the output transistors of the two current mirrors.

Other Differential Amplifiers

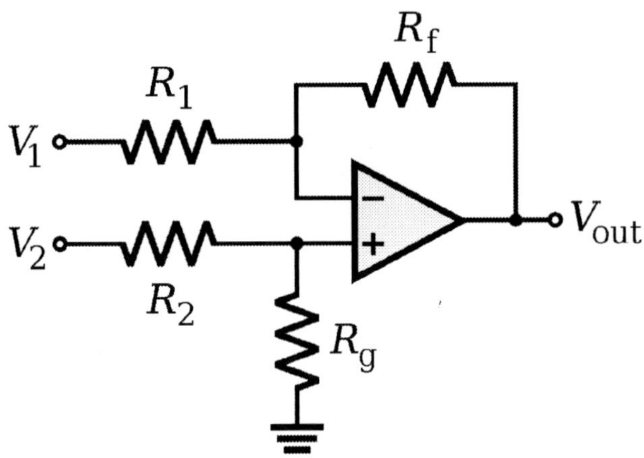

Figure: *Op-amp differential amplifier*

An operational amplifier, or op-amp, is a differential amplifier with very high differential-mode gain, very high input impedances, and a low output impedance. By applying negative feedback, an op-amp differential amplifier with predictable and stable gain can be built. Some kinds of differential amplifier usually include several simpler differential amplifiers. For example, an instrumentation amplifier, a fully differential amplifier, an instrument amplifier, or an isolation amplifier are often built from several op-amps.

Applications

Differential amplifiers are found in many circuits that utilize series negative feedback (op-amp follower, non-inverting amplifier, etc.), where one input is used for the input signal, the other for the feedback signal (usually implemented by operational amplifiers). For comparison, the old-fashioned inverting single-ended op-amps from the early 1940s could realize only parallel negative feedback by connecting additional resistor networks (an op-amp inverting amplifier is the most popular example).

A common application is for the control of motors or servos, as well as for signal amplification applications. In discrete electronics, a common arrangement for implementing a differential amplifier is the long-tailed pair, which is also usually found as the differential element in most op-amp integrated circuits. A long-tailed pair can be used as an analog multiplier with the differential voltage as one input and the biasing current as another.

A differential amplifier is used as the input stage emitter coupled logic gates and as switch. When used as a switch, the "left" base/grid is used as signal input and the "right" base/grid is grounded; output is taken from the right collector/plate. When the input is zero or negative, the output is close to zero (but can be not saturated); when the input is positive, the output is most-positive, dynamic operation being the same as the amplifier use described above.

Slew Rate

In electronics, slew rate is defined as the maximum rate of change of output voltage per unit of time and is expressed as volt per micro second. Limitations in slew rate capability can give rise to non linear effects in electronic amplifiers. For a sinusoidal waveform not to be subject to slew rate limitation, the slew rate capability (in volts per second) at all points in an amplifier must satisfy the following condition:

$$\mathrm{SR} \geq 2\pi f \times V_{\mathrm{pk}},$$

where f is the operating frequency, and V_{pk} is the peak amplitude of the waveform.

In mechanics the slew rate is given in dimensions $1/T$ and is associated with the change in position over time of an object which orbits around the observer. Slew rate can also be measured in degrees per second.

Definition

The slew rate of an electronic circuit is defined as the maximum rate of change of the output voltage. Slew rate is usually expressed in units of V/μs.

$$\text{SR} = \max\left(\left|\frac{dv_{\text{out}}(t)}{dt}\right|\right)$$

where $v_{\text{out}}(t)$ is the output produced by the amplifier as a function of time t.

Measurement

The slew rate can be measured using a function generator (usually square wave) and oscilloscope. The unit of slew rate is typically V/μs. The slew rate is same for both when feedback is considered or not considered.

Slew Rate Limiting in Amplifiers

There are slight differences between different amplifier designs in how the slewing phenomenon occurs. However, the general principles are the same as in this illustration. The input stage of modern amplifiers is usually a differential amplifier with a transconductance characteristic. This means the input stage takes a differential input voltage and produces an output current into the second stage.

The transconductance is typically very high — this is where the large open loop gain of the amplifier is generated. This also means that a fairly small input voltage can cause the input stage to saturate. In saturation, the stage produces a nearly constant output current. The second stage of modern power amplifiers is, amongst other things, where frequency compensation is accomplished. The low pass characteristic of this stage approximates an integrator. A constant current input will therefore produce a linearly increasing output. If the second stage has an effective input capacitance C and voltage gain A_2, then slew rate in this example can be expressed as:

$$\text{SR} = \frac{I_{\text{sat}}}{C} A_2$$

where I_{sat} is the output current of the first stage in saturation.

Slew rate helps us to identify what is the maximum input frequency and amplitude applicable to the amplifier such that the output is not distorted. Thus it becomes imperative to check the datasheet for the device's slew rate before using it for high-frequency applications.

Pin Configuration of 741 Op-Amp

Generally, operational amplifiers are extremely high voltage gain op-amps and they are standard building blocks of analogue circuits.

The most commonly used op-amp is IC741. The 741 op-amp is a voltage amplifier, it inverts the input voltage at the output, can be found almost everywhere in electronic circuits. Let's see the pin configuration and testing of 741 op-amps. Usually, this is a numbered counter clockwise around the chip.

It is an 8 pin IC. They provide superior performance in integrator, summing amplifier and general feedback applications. These are high gain op-amp; the voltage on the inverting input can be maintained almost equal to Vin.

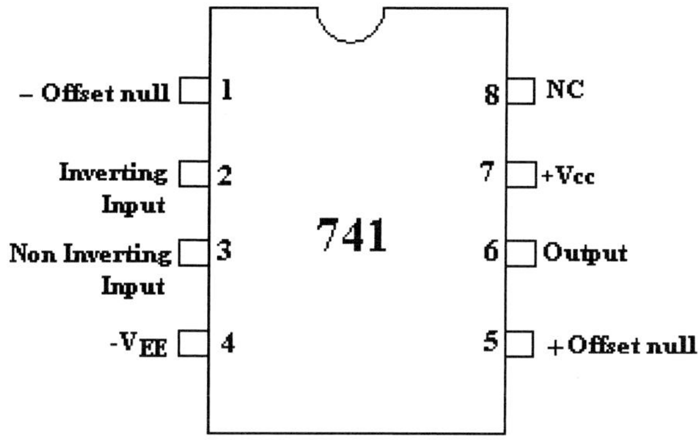

It is a 8-pin dual-in-line package with a pinout shown above.

Pin 1: Offset null.

Pin 2: Inverting input terminal.

Pin 3: Non-inverting input terminal.

Pin 4: –VCC (negative voltage supply).

Pin 5: Offset null.

Pin 6: Output voltage.

Pin 7: +VCC (positive voltage supply).

Pin 8: No Connection.

The important pins are 2, 3 and 6. In inverting amplifier, a positive voltage is applied to pin2 of the op-amp; we get output as negative voltage through pin 6. The polarity has been inverted. In a non-inverting amplifier, a positive voltage is applied to pin3 of the op-amp; we get output as positive voltage through pin 6. Polarity remains the same in non-inverting amplifier. Vcc is usually in the range from 12 to 15 volts.

When two supplies (+Vcc/-Vcc) are used, they are the same voltage and of opposite sign in almost all cases. Remember that the operational amplifier is a high gain, differential voltage amplifier. For a 741 operational amplifier, the gain is at least 100,000 and can be more than a million (1,000,000). That's an important fact you'll need to remember as you put the 741 into a circuit. There are many common application circuits using IC741 op-amp, they are adder, comparator, subtractor, integrator, differentiator and voltage follower.

Below is some example of 741 IC based circuits. However, the 741 is used as a comparator and not an amplifier. The difference between the two is small but significant. Even if used as a comparator the 741 still detects weak signals so that they can be recognized more easily. It is important to understand these circuits.

A comparator is a circuit that compares two input voltages. One voltage is called the reference voltage and the other is called the input voltage. It is a circuit which compares a signal voltage applied at one input of an op-amp with a known reference voltage at the other input. It is basically an open loop op-amp with output ±Vsat as in the ideal transfer characteristics. It is clear that the change in the output state takes place with an increment in input Vi of only 2mv.

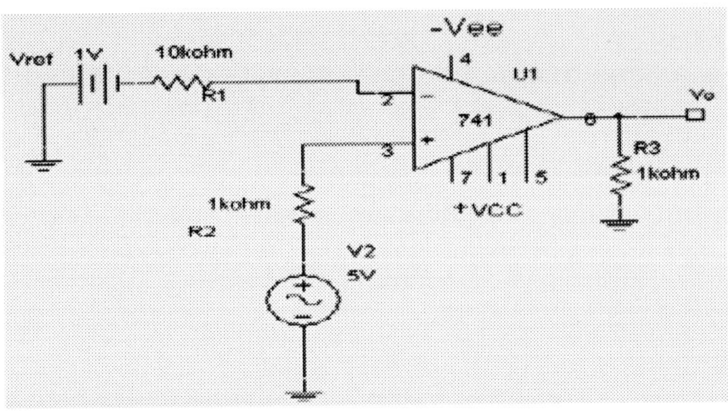

Characteristics of Ideal Op Amp

What is an Ideal Op Amp?

An ideal op amp is an op amp that has perfect conditions to allow it to function as an op amp with 100% efficiency.

An ideal op amp will display the following characeristics, of which are all explained in detail below. Ideal op amps will have infinite voltage gain, infinitely high impedance, zero output impedance, its gain is independent of input frequency, it has zero voltage offset, its output can swing positive or negative to the same voltages as the supply rails, and its output swings instantly to the correct value.

Below is a table that charts each of the major characteristics of an op amp and how they differ in ideal and real op amps.

***Table:** Ideal Op Amp Characteristics*

Characteristics	Ideal Op Amp	Real Op Amp
Infinite Voltage Gain	An ideal op amp will have infinite voltage gain. Op amps are devices that many times are used to function as amplifiers. A voltage is input into the op amp and as output, it produces the voltage amplified. An ideal op amp will produce mega-gain, practically, it will be able to produce infinite gain. It will amplify the signal infinite times over so that we can have as much gain as we'd ever need.	A real op amp can only produce a finite gain.
Infinitely high input impedance	An ideal op amp will have infintely high input impedance. This will ensure that the op amp causes no loading in the circuit. The lower the input impedance, the more current that an op amp will draw. The higher the impedance, the lower the current that an op will draw. We want high input impedance so that the op amp doesn't disturb the original circuit by pulling current from it. To do this, we need infinitely high input impedance.	A real op amp has finite input impedance. Even though many types of op amps, such as MOSFETs, have extremely high input impedance, in the order of teraohms, it is still finite.
Zero Output Impedance	An ideal op amp will have zero output impedance. When an op amp produces its output signal, we want the op amp to have zero voltage so that the maximum voltage will be transferred to the output load. Voltage is divided in a circuit according to the amount of impedance present in a circuit. Voltage drops across a component of higher impedance. In order for the voltage to drop across the output load, that load must be of greater impedance than the output of the op amp. This is why, ideally, we want the output impedance of the op amp to be zero	A real op amp will always have some output impedance, though it is low. A typical value can be 75? .
Gain Independent of Frequency	In an ideal op amp, the gain that the op amp produces will be independent of frequency. This means that regardless of the frequency of the input signal going into the op amp, the gain that is produced will be constant and good across all frequencies.	In real op amps, the gain that is produced is only for a certain bandwidth of frequencies. Outside of this bandwidth, the gain that the op amp produces will decline.
Zero Input Voltage Offset	In an ideal op amp, if no voltage is applied to the inverting and noninverting input pins, the op amp will output a voltage of 0, since there is no difference at all of the voltage applied to the 2 input pins.	A real op amp will have slight offset even if the voltage applied to the pins are the same. To correct this offset, voltage must be applied to the offset pin.

In real life, as with all ideal components, an ideal op amp does not exist. However, if we can get an op amp to display as close as possible the characteristics of an ideal op amp as closely, we can make a more efficient op amp that has better output in real-world conditions.

Now we will go over all of these characteristics of an ideal op amp mentioned above, so that you can know what each is and so the difference between an ideal case and a real case.

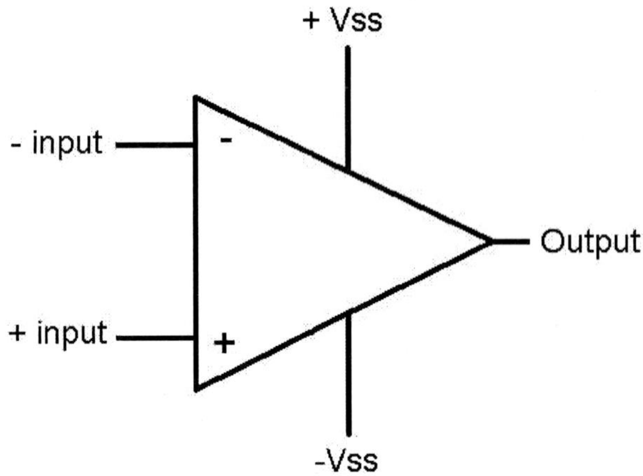

Again, ideal op amps can't exist because op amps, as all electronic components, will have some internal resistance, which won't allow maximum efficiency. However, if we can get real op amps as closely as possible to ideal conditions, we will have very efficient, useful op amps.

Concept of Virtual Ground

In electronics, a virtual ground (or virtual earth) is a node of a circuit that is maintained at a steady reference potential, without being connected directly to the reference potential. In some cases the reference potential is considered to be that of the surface of the earth, and the reference node is called "ground" or "earth" as a consequence.

The virtual ground concept aids circuit analysis in operational amplifier and other circuits and provides useful practical circuit effects that would be difficult to achieve in other ways. In circuit theory, a node may have any value of current or voltage but physical implementations of a virtual ground will have limitations of current handling ability and a non-zero impedance which may have practical side effects.

Construction

A voltage divider, using two resistors, can be used to create a virtual ground node. If two voltage sources are connected in series with two resistors, it can be shown that the midpoint becomes a virtual ground if

$$\frac{V_2}{V_1} = -\frac{R_2}{R_1}$$

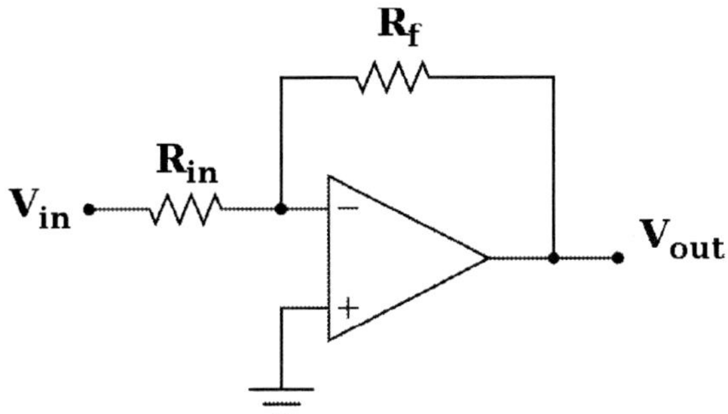

Figure: *Op-amp inverting amplifier*

An active virtual ground circuit is sometimes called a rail splitter. Such a circuit uses an op-amp or some other circuit element that has gain. Since an operational amplifier has very high open-loop gain, the potential difference between its inputs tend to zero when a feedback network is implemented. To achieve a reasonable voltage at the output (and thus equilibrium in the system), the output supplies the inverting input (via the feedback network) with enough voltage to reduce the potential difference between the inputs to microvolts. The non-inverting (+) input of the operational amplifier is grounded; therefore, its inverting (-) input, although not connected to ground, will assume a similar potential, becoming a *virtual ground* if the opamp is working in its linear region (i.e., outputs have not saturated).

Applications

Voltage is a differential quantity, which appears between two points. In order to deal only with a voltage (an electrical potential) of a single point, the second point has to be connected to a reference point (ground). Usually, the power supply terminals serve as steady grounds; when the internal points of compound power sources are accessible, they can also serve as real grounds.

If there are not accessible source internal points, external circuit points having steady voltage with respect to the source terminals can serve as artificial *virtual grounds*. Such a point has to have steady potential, which does not vary when a load is attached.

Op-Amp Applications

Practical Considerations

Input offset Problems: The equations shown below, pertaining to each type of circuit, assume that an ideal op amp is used. Those interested in construction of any of these circuits for practical use should consult a more detailed reference. Resistors used in practical solid-state op-amp circuits are typically in the kΩ range. Resistors much greater than 1 MΩ cause excessive thermal noise and make the circuit operation susceptible to significant errors due to bias or leakage currents.

Practical operational amplifiers draw a small current from each of their inputs due to bias requirements and leakage. These currents flow through the resistances connected to the inputs and produce small voltage drops across those resistances. In AC signal applications this seldom matters. If high-precision DC operation is required, however, these voltage drops need to be considered. The design technique is to try to ensure that these voltage drops are equal for both inputs, and therefore cancel. If these voltage drops are equal and the common-mode rejection ratio of the operational amplifier is good, there will be considerable cancellation and improvement in DC accuracy.

If the input currents into the operational amplifier are equal, to reduce offset voltage the designer must ensure that the DC resistance looking out of each input is also matched. In general, input currents differ, the difference being called the *input offset current*, I_{os}. Matched external input resistances R_{in} will still produce an input voltage error of $R_{in} \cdot I_{os}$. Most manufacturers provide a method for tuning the operational amplifier to balance the input currents (e.g., "offset null" or "balance" pins that can interact with an external voltage source attached to a potentiometer). Otherwise, a tunable external voltage can be added to one of the inputs in order to balance out the offset effect. In cases where a design calls for one input to be short-circuited to ground, that short circuit can be replaced with a variable resistance that can be tuned to mitigate the offset problem.

Many operational amplifiers that have MOSFET-based input stages have input leakage currents that will truly be negligible to most designs.

Power Supply Effects

Although the power supplies are not shown in the operational amplifier designs below, they can be critical in operational amplifier design. Power supply imperfections (e.g., power signal ripple, non-zero source impedance) may lead to noticeable deviations from ideal operational amplifier behaviour. For example, operational amplifiers have a specified power supply rejection ratio that indicates how well the output can reject signals that appear on the power supply inputs.

Power supply inputs are often noisy in large designs because the power supply is used by nearly every component in the design, and inductance effects prevent current from being instantaneously delivered to every component at once. As a consequence, when a component requires large injections of current (e.g., a digital component that is frequently switching from one state to another), nearby components can experience sagging at their connection to the power supply. This problem can be mitigated with copious use of bypass capacitors connected across each power supply pin and ground. When bursts of current are required by a component, the component can *bypass* the power supply by receiving the current directly from the nearby capacitor (which is then slowly recharged by the power supply).

Additionally, current drawn into the operational amplifier from the power supply can be used as inputs to external circuitry that augment the capabilities of the operational amplifier. For example, an operational amplifier may not be fit for a particular high-gain application because its output would be required to generate signals outside of the safe range generated by the amplifier. In this case, an external push–pull amplifier can be controlled by the current into and out of the operational amplifier. Thus, the operational amplifier may itself operate within its factory specified bounds while still allowing the negative feedback path to include a large output signal well outside of those bounds.

Amplifiers

Inverting amplifier:

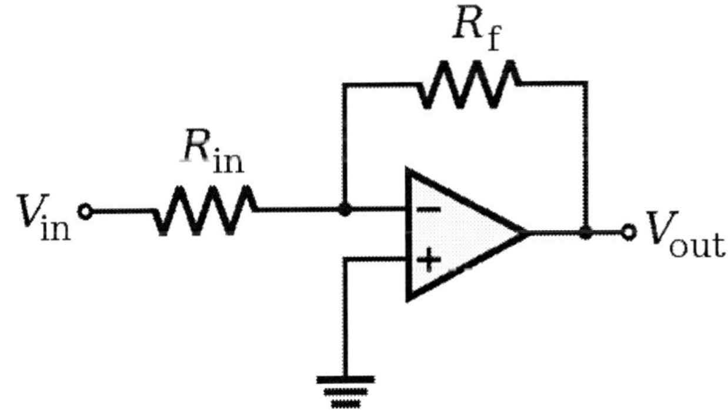

An inverting amplifier inverts and scales the input signal. As long as the op-amp gain is very large, the amplifier gain is determined by two stable external resistors (the feedback resistor R_f and the input resistor R_{in}) and not by op-amp parameters which are highly temperature dependent. In particular, the R_{in}–R_f resistor network acts as an electronic seesaw (i.e., a class-1 lever) where the inverting (i.e., “) input of the operational amplifier is like a fulcrum about which the seesaw pivots. That is, because the operational amplifier is in a negative-feedback configuration, its internal high gain effectively fixes the inverting (i.e., “) input at the same 0 V (*ground*) voltage of the non-inverting (i.e., +) input, which is similar to the stiff mechanical support provided by the fulcrum of the seesaw. Continuing the analogy,

- Just as the movement of one end of the seesaw is opposite the movement of the other end of the seesaw, positive movement away from 0 V at the input of the R_{in}–R_f network is matched by negative movement away from 0 V at the output of the network; thus, the amplifier is said to be *inverting*.
- In the seesaw analogy, the mechanical moment or torque from the force on one side of the fulcrum is balanced exactly by the force on the other side of the fulcrum; consequently, asymmetric lengths in the seesaw allow for small forces on one side of the seesaw to generate large forces on the other side of the seesaw. In the inverting amplifier, electrical current, like torque, is conserved across the R_{in}–R_f network and relative differences between the R_{in} and R_f resistors allow small voltages on one side of the network to generate large voltages (with opposite sign) on the other side of the network. Thus, the device *amplifies* (and inverts) the input voltage. However, in this analogy, it is the reciprocals of the resistances (i.e., the conductances or admittances) that play the role of lengths in the seesaw.

Hence, the amplifier output is related to the input as in

$$V_{out} = -\frac{R_f}{R_{in}} V_{in} \ .$$

So the voltage gain of the amplifier is $A = -R_f / R_{in}$ where the negative sign is a convention indicating that the output is negated. For example, if R_f is 10 kΩ and R_{in} is 1 kΩ, then the gain is –10 kΩ/1 kΩ, or –10 (or –10 V/V). Moreover, the input impedance of the device is R_{in} because the operational amplifier's inverting (i.e., –) input is a virtual ground.

In a real operational amplifier, the current into its two inputs is small but non-zero (e.g., due to input bias currents). The current into the inverting (i.e., –) input of the operational amplifier is drawn across the R_{in} and R_f resistors in parallel, which appears like a small parasitic voltage difference between the inverting (i.e., –) and non-inverting (i.e., +) inputs of the operational amplifier. To mitigate this practical problem, a third resistor of value $R_f \parallel R_{in} \triangleq R_f R_{in} / (R_f + R_{in})$ can be added between the non-inverting (i.e., +) input and the true ground. This resistor does not affect the idealized operation of the device because no current enters the ideal non-inverting input.

However, in the practical case, if input currents are roughly equivalent, the voltage added at the inverting input will match the voltage at the non-inverting input, and so this common-mode signal will be ignored by the operational amplifier (which operates on differences between its inputs).

Non-inverting Amplifier

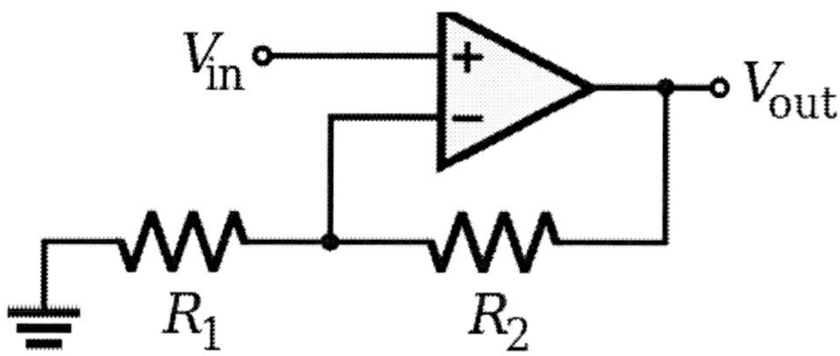

Amplifies a voltage (multiplies by a constant greater than 1)

$$V_{out} = V_{in}\left(1 + \frac{R_2}{R_1}\right)$$

- Input impedance $Z_{in} \approx \infty$
  - The input impedance is *at least* the impedance between non-inverting (+) and inverting (–) inputs, which is typically 1 MΩ to 10 TΩ, plus the impedance of the path from the inverting (–) input to ground (i.e., R_1 in parallel with R_2).
  - Because negative feedback ensures that the non-inverting and inverting inputs match, the input impedance is actually much higher.
- Although this circuit has a large input impedance, it suffers from error of input bias current.
  - The non-inverting (+) and inverting (–) inputs draw small leakage currents into the operational amplifier.
  - These input currents generate voltages that act like unmodelled input offsets. These unmodelled effects can lead to noise on the output (e.g., offsets or drift).

- Assuming that the two leaking currents are matched, their effect can be mitigated by ensuring the DC impedance looking out of each input is the same.
  - The voltage produced by each bias current is equal to the product of the bias current with the equivalent DC impedance looking out of each input. Making those impedances equal makes the offset voltage at each input equal, and so the non-zero bias currents will have no impact on the difference between the two inputs.
  - A resistor of value

$$R_1 \parallel R_2 \triangleq \left(\frac{1}{R_1}+\frac{1}{R_2}\right)^{-1}=\frac{R_1R_2}{R_1+R_2},$$

- which is the equivalent resistance of R_1 in parallel with R_2, between the V_{in} source and the non-inverting (+) input will ensure the impedances looking out of each input will be matched.
  - The matched bias currents will then generate matched offset voltages, and their effect will be hidden to the operational amplifier (which acts on the difference between its inputs) so long as the CMRR is good.
- Very often, the input currents are *not* matched.
  - Most operational amplifiers provide some method of balancing the two input currents (e.g., by way of an external potentiometer).
  - Alternatively, an external offset can be added to the operational amplifier input to nullify the effect.
  - Another solution is to insert a variable resistor between the V_{in} source and the non-inverting (+) input. The resistance can be tuned until the offset voltages at each input are matched.
  - Operational amplifiers with MOSFET-based input stages have input currents that are so small that they often can be neglected.

Differential Amplifier (Difference Amplifier)

The name "differential amplifier" should not be confused with the "differentiator," which is also shown on this page.

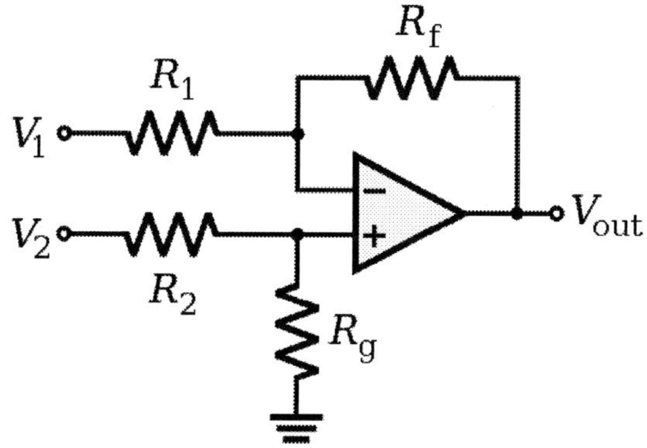

The "instrumentation amplifier," which is also shown on this page, is a modification of the differential amplifier that also provides high input impedance.

The circuit shown computes the difference of two voltages multiplied by some constant. In particular, the output voltage is:

$$V_{out} = \frac{(R_f + R_1)R_g}{(R_g + R_2)R_1}V_2 - \frac{R_f}{R_1}V_1 = \left(\frac{R_1 + R_f}{R_1}\right)\cdot\left(\frac{R_g}{R_g + R_2}\right)V_2 - \frac{R_f}{R_1}V_1.$$

The differential input impedance Z_{in} (i.e., the impedance between the two input pins) is approximately $R_1 + R_2$. The input currents vary with the operating point of the circuit. Consequently, if the two sources feeding this circuit have appreciable output impedance, then non-linearities can appear in the output. An instrumentation amplifier mitigates these problems.

- Under the condition that the $R_f/R_1 = R_g/R_2$, the output expression becomes:

  $$V_{out} = A(V_2 - V_1) \quad \text{where} \quad A \triangleq \frac{R_f}{R_1}$$

  is the *differential gain* of the circuit.

Moreover, the amplifier synthesized with this choice of parameters has good common-mode rejection in theory because components of the signals that have $V_1 = V_2$ are not expressed on the output. Although this property is described here with resistances, it is a more general property of the impedances in the circuit. So, for example, if a compensation capacitor is added across any resistor (e.g., to improve phase margin and ensure closed-loopstability of the operational amplifier), similar changes need to be made in the rest of the circuit to maintain the ratio balance. Otherwise, high-frequency components common to both V_1 and V_2 can express themselves on the output. Additionally, because of leakage or bias currents in a real operational amplifier, it is usually desirable for the impedance looking out each input to the operational amplifier to be equal to the impedance looking out of the other input of the operational amplifier.

Otherwise, the same current into each operational amplifier input will generate a parasitic differential signal and thus a parasitic output component. Consequently, choosing $R_1 = R_2$ and $R_f = R_g$ is common in practice.

- In the special case when $R_f/R_1 = R_g/R_2$, as before, and $R_f = R_1$, the differential gain $A = 1$, and the circuit is a differential follower with:

$$V_{out} = V_2 - V_1.$$

Voltage follower (Unity Buffer Amplifier)

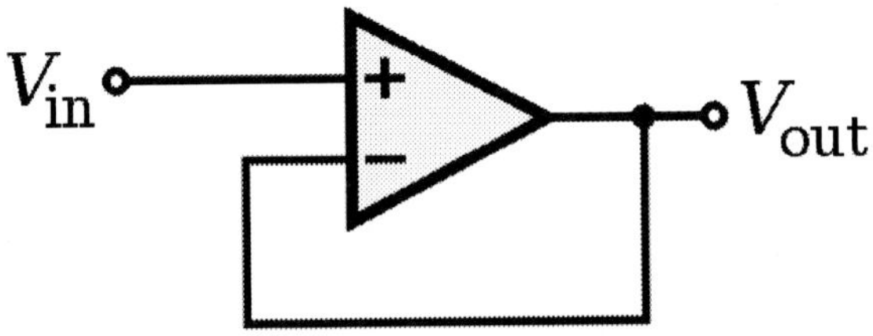

Used as a buffer amplifier to eliminate loading effects (e.g., connecting a device with a high source impedance to a device with a low input impedance).

$$V_{out} = V_{in}$$

$Z_{in} = \infty$ (realistically, the differential input impedance of the op-amp itself, 1 MΩ to 1 TΩ)

Due to the strong (i.e., unity gain) feedback and certain non-ideal characteristics of real operational amplifiers, this feedback system is prone to have poor stability margins. Consequently, the system may be unstable when connected to sufficiently capacitive loads.

In these cases, a lag compensation network (e.g., connecting the load to the voltage follower through a resistor) can be used to restore stability. The manufacturer data sheet for the operational amplifier may provide guidance for the selection of components in external compensation networks.

Alternatively, another operational amplifier can be chosen that has more appropriate internal compensation.

Summing Amplifier

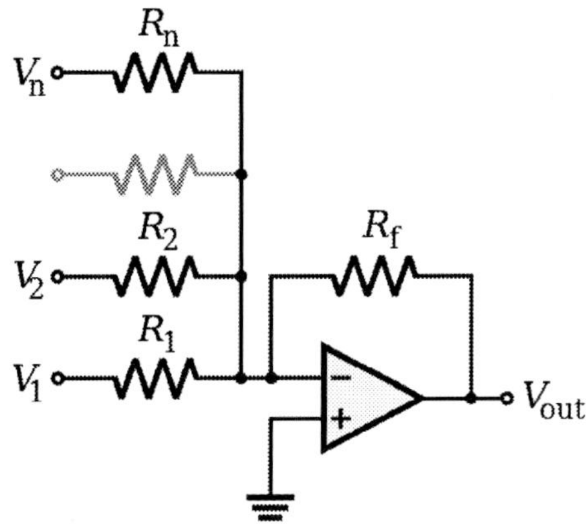

Figure: *A summing amplifier sums several (weighted) voltages:*

$$V_{out} = -R_f\left(\frac{V_1}{R_1} + \frac{V_2}{R_2} + \cdots + \frac{V_n}{R_n}\right)$$

- When $R_1 = R_2 = \cdots = R_n$, and R_f independent

$$V_{out} = -\frac{R_f}{R_1}(V_1 + V_2 + \cdots + V_n)$$

- When $R_1 = R_2 = \cdots = R_n = R_f$

$$V_{out} = -(V_1 + V_2 + \cdots + V_n)$$

- Output is inverted
- Input impedance of the *n*th input is $Z_n = R_n$ (V_- is a virtual ground)

Instrumentation Amplifier

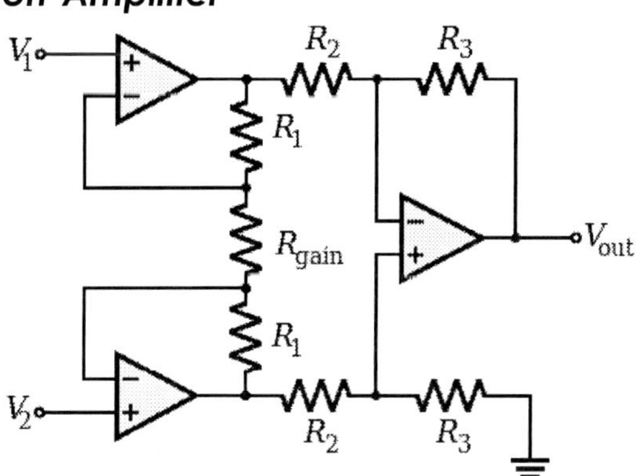

Combines very high input impedance, high common-mode rejection, low DC offset, and other properties used in making very accurate, low-noise measurements

- Is made by adding a non-invertingbuffer to each input of the differential amplifier to increase the input impedance.

Oscillators

Wien bridge oscillator:

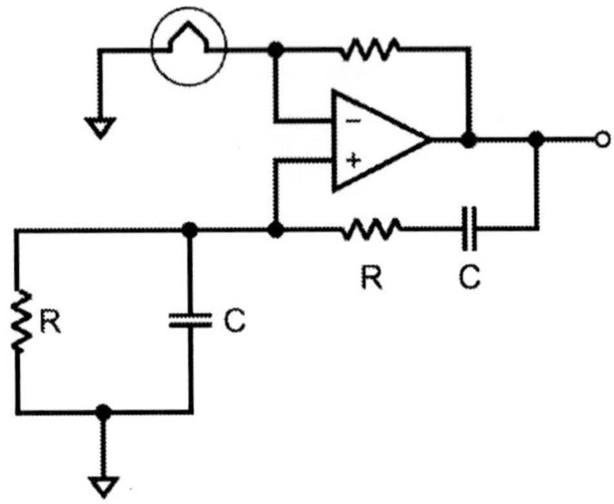

Produces a very low distortion sine wave. Uses negative temperature compensation in the form of a light bulb or diode.

Relaxation Oscillator

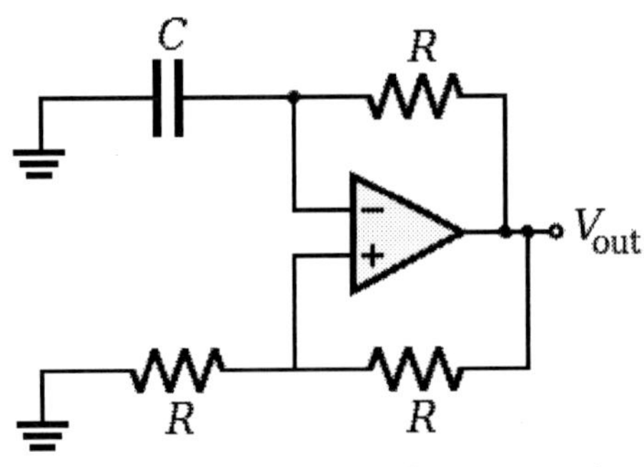

By using an RC network to add slow negative feedback to the inverting Schmitt trigger, a relaxation oscillator is formed. The feedback through the RC network causes the Schmitt trigger output to oscillate in an endless symmetric square wave (i.e., the Schmitt trigger in this configuration is an astablemultivibrator).

Filters

Operational amplifiers can be used in construction of active filters, providing high-pass, low-pass, band-pass, reject and delay functions.

The high input impedance and gain of an op-amp allow straightforward calculation of element values, allowing accurate implementation of any desired filter topology with little concern for the loading effects of stages in the filter or of subsequent stages. However, the frequencies at which active filters can be implemented is limited; when the behaviour of the amplifiers departs significantly from the ideal behaviour assumed in elementary design of the filters, filter performance is degraded.

Comparators and Detectors

Comparator:

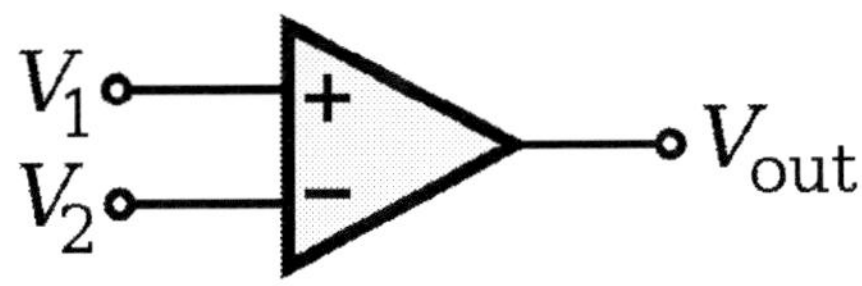

Bistable output that indicates which of the two inputs has a higher voltage. That is,

$$V_{out} = \begin{cases} V_{S+} & \text{if } V_1 > V_2, \\ V_{S-} & \text{if } V_1 < V_2, \\ 0 & \text{if } V_1 = V_2, \end{cases}$$

where V_{S+} and V_{S-} are nominally the positive and negative supply voltages (which are not shown in the diagram).

Threshold Detector

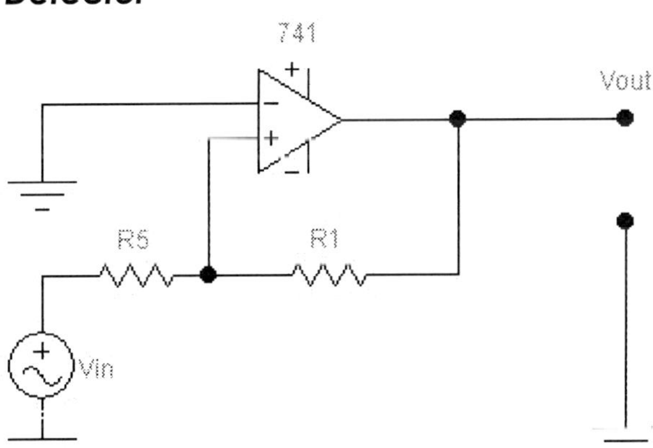

The threshold detector with hysteresis consists of an operational amplifier and a series of resistors that provide hysteresis. Like other detectors, this device functions as a voltage switch, but with an important difference. The state of the detector output is not directly affected by input voltage, but instead by the voltage drop across its input terminals (here, referred to as Va). From Kirchhoff's Current Law, this value depends both on Vin and the output voltage of the threshold detector itself, both multiplied by a resistor ratio.

$$V_a = \frac{R_1}{R_1 + R_5} V_{in} + \frac{R_5}{R_1 + R_5} V_{THRout}$$

Unlike the zero crossing detector, the detector with hysteresis does not switch when Vin is zero, rather the output becomes Vsat+ when Va becomes positive and Vsat- when Va becomes negative. Further examination of the Va equation reveals that Vin can exceed zero (positive or negative) by a certain magnitude before the output of the detector is caused to switch.

By adjusting the value of R1, the magnitude of Vin that will cause the detector to switch can be increased or decreased. Hysteresis is useful in various applications. It has better noise immunity than the level detector, so it is used in interface circuits. Its positive feedback has a faster transition, so it is used in timing applications such as frequency counters. It is also used in astablemultivibrators found in instruments such as function generators.

Zero Level Detector

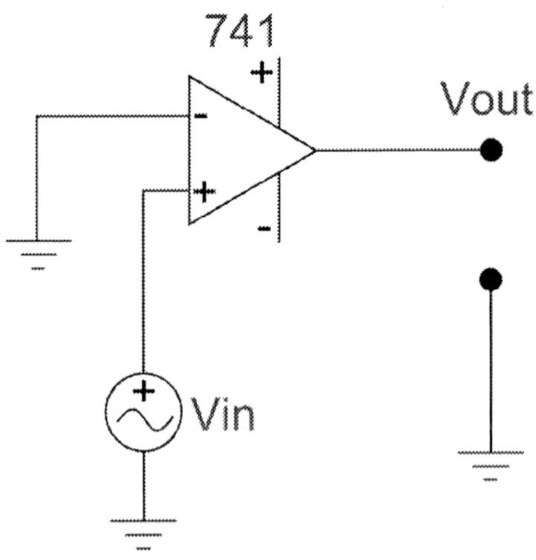

Circuit Diagram: A *zero crossing detector* is a comparator with the reference level set at zero. It is used for detecting the zero crossings of AC signals. It can be made from an operational amplifier with an input voltage at its positive input.

When the input voltage is positive, the output voltage is a positive value, when the input voltage is negative, the output voltage is a negative value. The magnitude of the output voltage is a property of the operational amplifier and its power supply.

Applications include converting an analog signal into a form suitable to use for frequency measurements, in phase locked loops, or for controlling power electronics circuits that must switch with a defined relationship to an alternating current waveform.

This detector exploits the property that the instantaneous frequency of an FM wave is approximately given by $f_i = \frac{\Delta t}{2}$ where Δt is the time difference between adjacent zero crossings of FM wave

Schmitt Trigger

A bistablemultivibrator implemented as a comparator with hysteresis.

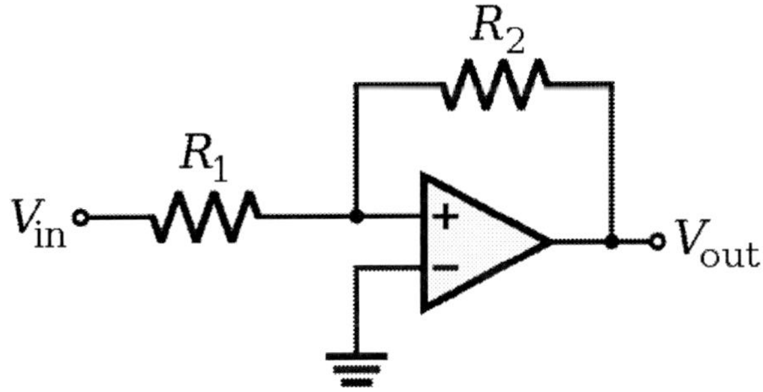

In this configuration, the input voltage is applied through the resistor R_1 (which may be the source internal resistance) to the non-inverting input and the inverting input is grounded or referenced. The hysteresis curve is non-inverting and the switching thresholds are $\pm\frac{R_1}{R_2}V_{sat}$ where V_{sat} is the greatest output magnitude of the operational amplifier.

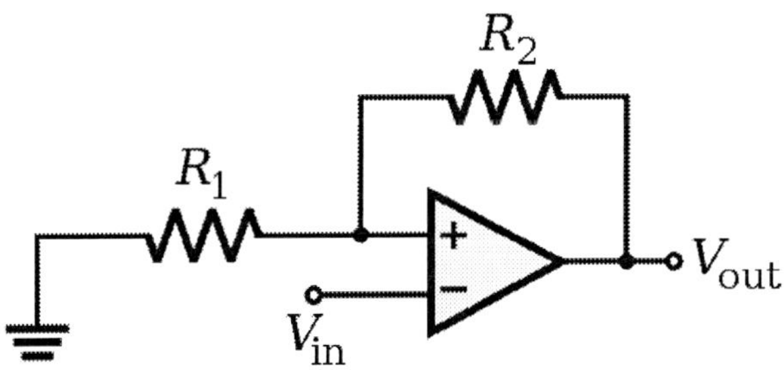

Alternatively, the input source and the ground may be swapped. Now the input voltage is applied directly to the inverting input and the non-inverting input is grounded or referenced. The hysteresis curve is inverting and the switching thresholds are $\pm\frac{R_1}{R_1+R_2}V_{sat}$. Such a configuration is used in the relaxation oscillator shown above.

Integration and Differentiation

Inverting Integrator

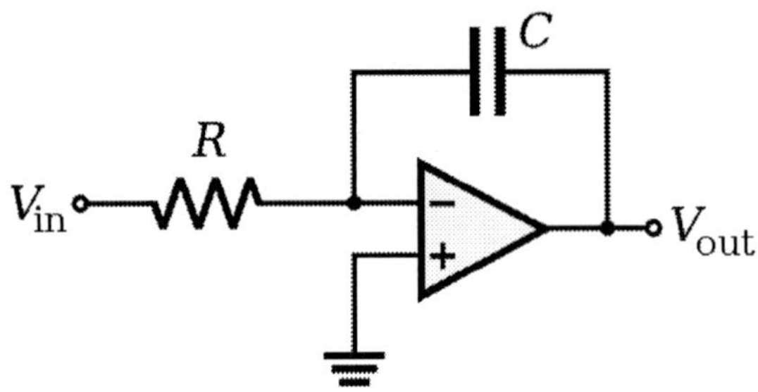

Integrates the (inverted) signal over time

$$V_{out} = -\int_0^t \frac{V_{in}}{RC}\mathrm{d}t + V_{initial}$$

(where V_{in} and V_{out} are functions of time, $V_{initial}$ is the output voltage of the integrator at time t = 0.)

- This can also be viewed as a low-passelectronic filter. It is a filter with a single pole at DC (i.e., where $\omega = 0$) and gain.
- There are several potential problems with this circuit.
  - o It is usually assumed that the input V_{in} has zero DC component (i.e., has a zero average value). Otherwise, unless the capacitor is periodically discharged, the output will drift outside of the operational amplifier's operating range.
  - o Even when V_{in} has no offset, the leakage or bias currents into the operational amplifier inputs can add an unexpected offset voltage to V_{in} that causes the output to drift. Balancing input currents and replacing the non-inverting (+) short-circuit to ground with a resistor with resistance R can reduce the severity of this problem.
  - o Because this circuit provides no DC feedback (i.e., the capacitor appears like an open circuit to signals with $\omega = 0$),

the offset of the output may not agree with expectations (i.e., V_{initial} may be out of the designer's control with the present circuit).

Many of these problems can be made less severe by adding a *large* resistor R_F in parallel with the feedback capacitor. At significantly high frequencies, this resistor will have negligible effect. However, at low frequencies where there are drift and offset problems, the resistor provides the necessary feedback to hold the output steady at the correct value. In effect, this resistor reduces the DC gain of the "integrator" – it goes from infinite to some finite value R_F / R.

Inverting Differentiator

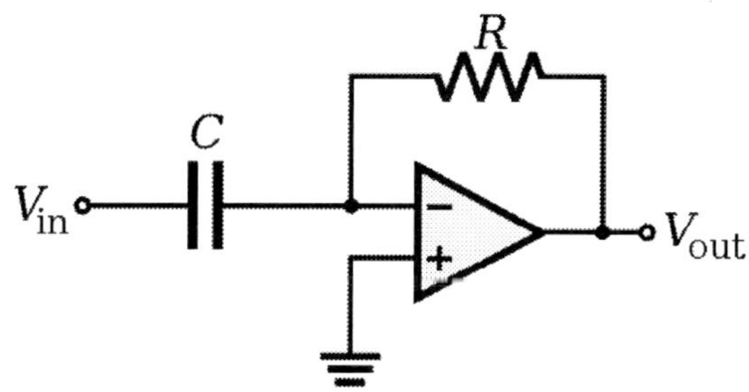

Differentiates the (inverted) signal over time.

$$V_{\text{out}} = -RC\frac{\mathrm{d}V_{\text{in}}}{\mathrm{d}t} \qquad \text{where } V_{\text{in}} \text{ and } V_{\text{out}} \text{ are functions of time.}$$

- This can also be viewed as a high-passelectronic filter. It is a filter with a single zero at DC (i.e., where angular frequency $\omega = 0$ radians) and gain. The high-pass characteristics of a differentiating amplifier (i.e., the low-frequency zero) can lead to stability challenges when the circuit is used in an analog servo loop (e.g., in a PID controller with a significant derivative gain). In particular, as a root locus analysis would show, increasing feedback gain will drive a closed-loop pole toward marginal stability at the DC zero introduced by the differentiator.

Synthetic Elements

Simulates an inductor (i.e., provides inductance without the use of a possibly costly inductor). The circuit exploits the fact that the current flowing through a capacitor behaves through time as the voltage across an inductor. The capacitor used in this circuit is smaller than the inductor it simulates and its capacitance is less subject to changes in value due to environmental changes.

This circuit is unsuitable for applications relying on the back EMF property of an inductor as this will be limited in a gyrator circuit to the voltage supplies of the op-amp.

Inductance gyrator:

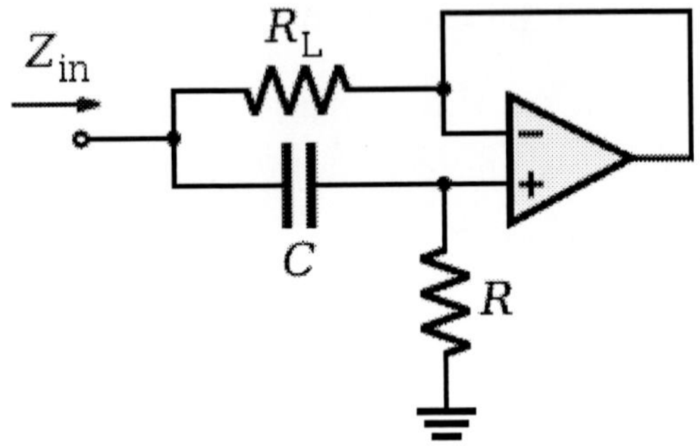

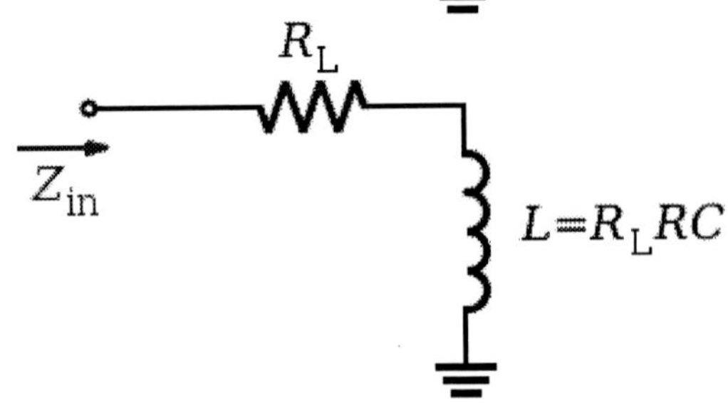

Negative Impedance Converter (NIC)

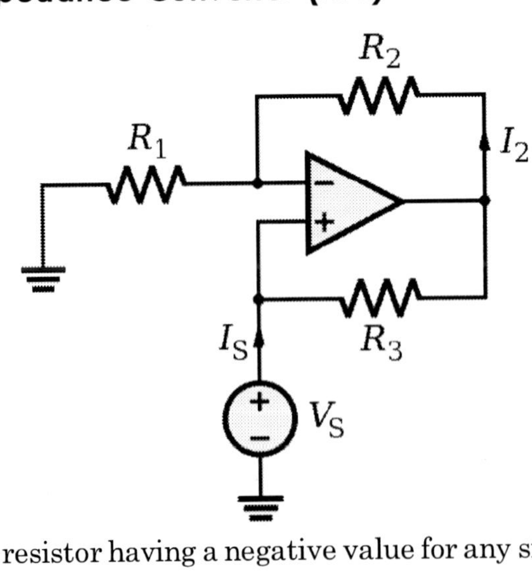

Creates a resistor having a negative value for any signal generator

- In this case, the ratio between the input voltage and the input current (thus the input resistance) is given by:

$$R_{in} = -R_3 \frac{R_1}{R_2}$$

In general, the components R_1, R_2, and R_3 need not be resistors; they can be any component that can be described with an impedance.

Non-linear

Precision rectifier:

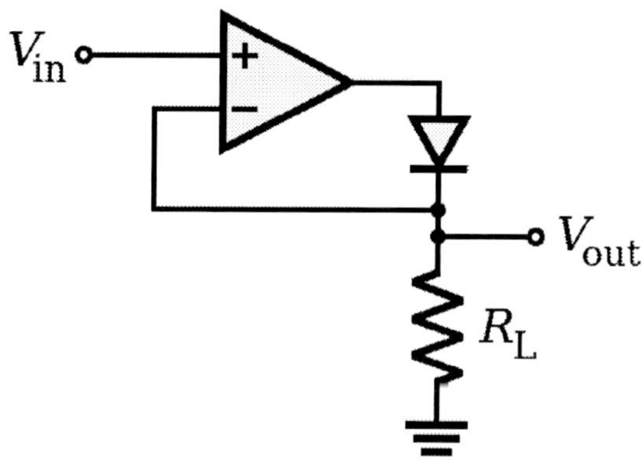

The voltage drop V_F across the forward biased diode in the circuit of a passive rectifier is undesired. In this active version, the problem is solved by connecting the diode in the negative feedback loop. The op-amp compares the output voltage across the load with the input voltage and increases its own output voltage with the value of V_F. As a result, the voltage drop V_F is compensated and the circuit behaves very nearly as an ideal (*super*) diode with $V_F = 0$ V.

The circuit has speed limitations at high frequency because of the slow negative feedback and due to the low slew rate of many non-ideal op-amps.

Logarithmic Output

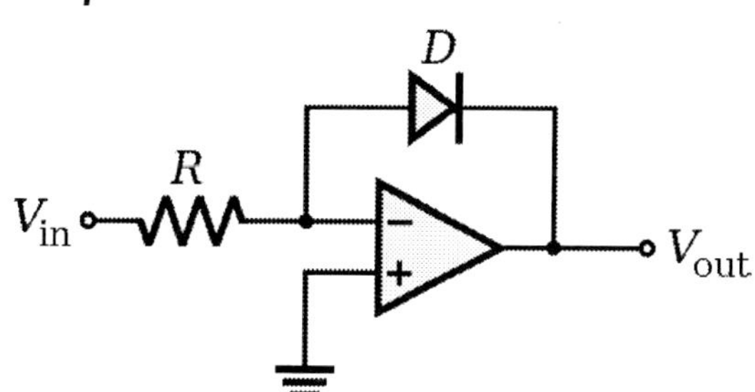

- The relationship between the input voltage v_{in} and the output voltage v_{out} is given by:

$$v_{out} = -V_T \ln\left(\frac{v_{in}}{I_S R}\right)$$

where I_S is the *saturation current* and V_T is the *thermal voltage.*

- If the operational amplifier is considered ideal, the negative pin is virtually grounded, so the current flowing into the resistor from the source (and thus through the diode to the output, since the op-amp inputs draw no current) is:

$$\frac{v_{in}}{R} = I_R = I_D$$

where I_D is the current through the diode. As known, the relationship between the current and the voltage for a diode is:

$$I_D = I_S\left(e^{\frac{V_D}{V_T}} - 1\right).$$

This, when the voltage is greater than zero, can be approximated by:

$$I_D \simeq I_S e^{\frac{V_D}{V_T}}.$$

Putting these two formulae together and considering that the output voltage is the negative of the voltage across the diode ($V_{out} = -V_D$), the relationship is proven.

This implementation does not consider temperature stability and other non-ideal effects.

Exponential Output

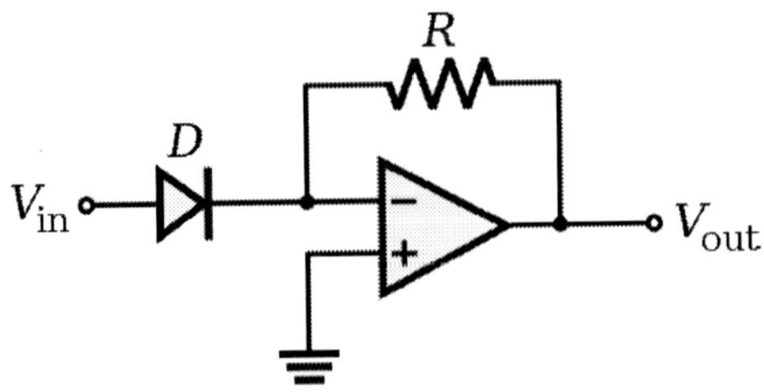

- The relationship between the input voltage and the output voltage is given by:

$$v_{out} = -RI_S e^{\frac{v_{in}}{V_T}}$$

whereis I_S the saturation current and V_T is the thermal voltage.

- Considering the operational amplifier ideal, then the negative pin is virtually grounded, so the current through the diode is given by:

$$I_D = I_S\left(e^{\frac{V_D}{V_T}} - 1\right)$$

when the voltage is greater than zero, it can be approximated by:

$$I_D \simeq I_S e^{\frac{V_D}{V_T}}.$$

The output voltage is given by:

$$v_{out} = -RI_D.$$

Chapter

5

Timers and Data Converters

IC 555 Timer

The 555 timer IC is an integrated circuit (chip) used in a variety of timer, pulse generation, and oscillator applications. The 555 can be used to provide time delays, as an oscillator, and as a flip-flop element. Derivatives provide up to four timing circuits in one package.

Introduced in 1972 by Signetics, the 555 is still in widespread use due to its ease of use, low price, and good stability. It is now made by many companies in the original bipolar and also in low-power CMOS types. As of 2003, it was estimated that 1 billion units are manufactured every year.

Design

The IC was designed in 1971 by Hans Camenzind under contract to Signetics, which was later acquired by Philips (now NXP). Depending on the manufacturer, the standard 555 package includes 25 transistors, 2 diodes and 15 resistors on a silicon chip installed in an 8-pin mini dual-in-line package (DIP-8). Variants available include the 556 (a 14-pin DIP combining two 555s on one chip), and the two 558 & 559s (both a 16-pin DIP combining four slightly modified 555s with DIS & THR connected internally, and TR is falling edge sensitive instead of level sensitive).

The NE555 parts were commercial temperature range, 0 °C to +70 °C, and the SE555 part number designated the military temperature range, –55 °C to +125 °C.

These were available in both high-reliability metal can (T package) and inexpensive epoxy plastic (V package) packages.

Thus the full part numbers were NE555V, NE555T, SE555V, and SE555T. It has been hypothesized that the 555 got its name from the three 5 kΩ resistors used within, but Hans Camenzind has stated that the number was arbitrary.

Low-power versions of the 555 are also available, such as the 7555 and CMOS TLC555.

The 7555 is designed to cause less supply noise than the classic 555 and the manufacturer claims that it usually does not require a "control" capacitor and in many cases does not require a decoupling capacitor on the power supply.

Such a practice should nevertheless be avoided, because noise produced by the timer or variation in power supply voltage might interfere with other parts of a circuit or influence its threshold voltages.

Pins

The connection of the pins for a DIP package is as follows:

Pin	*Name*	*Purpose*
1	GND	Ground reference voltage, low level (0 V)
2	TRIG	The OUT pin goes high and a timing interval starts when this input falls below 1/2 of CTRL voltage (which is typically 1/3 of V_{CC}, when CTRL is open).
3	OUT	This output is driven to approximately 1.7V below $+V_{CC}$ or GND.
4	RESET	A timing interval may be reset by driving this input to GND, but the timing does not begin again until RESET rises above approximately 0.7 volts. Overrides TRIG which overrides THR.
5	CTRL	Provides "control" access to the internal voltage divider (by default, 2/3 V_{CC}).
6	THR	The timing (OUT high) interval ends when the voltage at THR is greater than that at CTRL (2/3 V_{CC} if CTRL is open).
7	DIS	Open collector output which may discharge a capacitor between intervals. In phase with output.
8	V_{CC}	Positive supply voltage, which is usually between 3 and 15 V depending on the variation.

Pin 5 is also sometimes called the CONTROL VOLTAGE pin. By applying a voltage to the CONTROL VOLTAGE input one can alter the timing characteristics of the device. In most applications, the CONTROL VOLTAGE input is not used. It is usual to connect a 10 nF capacitor between pin 5 and 0 V to prevent interference. The CONTROL VOLTAGE input can be used to build an astable with a frequency modulated output.

Modes

The 555 has three operating modes:

- Monostable mode: In this mode, the 555 functions as a "one-shot" pulse generator. Applications include timers, missing pulse detection, bouncefree switches, touch switches, frequency divider, capacitance measurement, pulse-width modulation (PWM) and so on.
- Astable (free-running) mode: The 555 can operate as an oscillator. Uses include LED and lamp flashers, pulse generation, logic clocks, tone generation, security alarms, pulse position modulation and so on. The 555 can be used as a simple ADC, converting an analog value to a pulse length. E.g. selecting a thermistor as timing resistor allows the use of the 555 in a temperature sensor: the period of the output pulse is determined by the temperature. The use of a microprocessor based circuit can then convert the pulse period to temperature, linearize it and even provide calibration means.
- Bistable mode or Schmitt trigger: The 555 can operate as a flip-flop, if the DIS pin is not connected and no capacitor is used. Uses include bounce-free latched switches.

Monostable

In the monostable mode, the 555 timer acts as a "one-shot" pulse generator. The pulse begins when the 555 timer receives a signal at the trigger input that falls below a third of the voltage supply. The width of the output pulse is determined by the time constant of an RC network, which consists of a capacitor (C) and a resistor (R). The output pulse ends when the voltage on the capacitor equals 2/3 of the supply voltage. The output pulse width can be lengthened or shortened to the need of the specific application by adjusting the values of R and C.

The output pulse width of time t, which is the time it takes to charge C to 2/3 of the supply voltage, is given by

$$t = RC\ln(3) \approx 1.1RC$$

where t is in seconds, R is in ohms and C is in farads. While using the timer IC in monostable mode, the main disadvantage is that the time span between any two triggering pulses must be greater than the RC time constant.

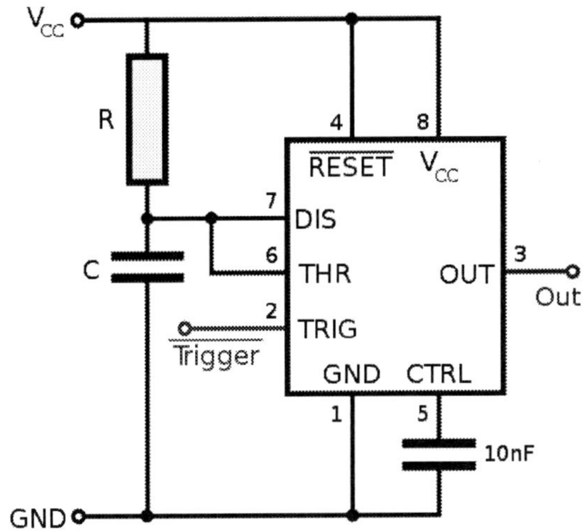

Figure: *Schematic of a 555 in monostable mode*

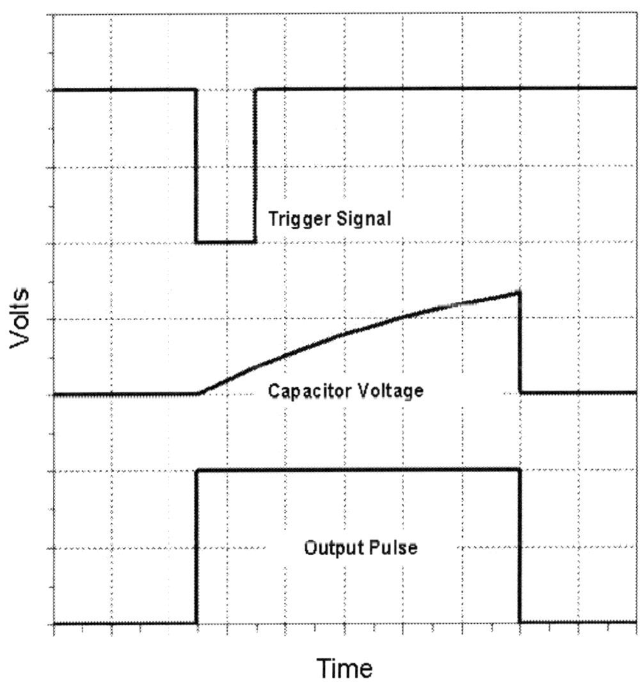

Figure: *The relationships of the trigger signal, the voltage on C and the pulse width in monostable mode*

Bistable

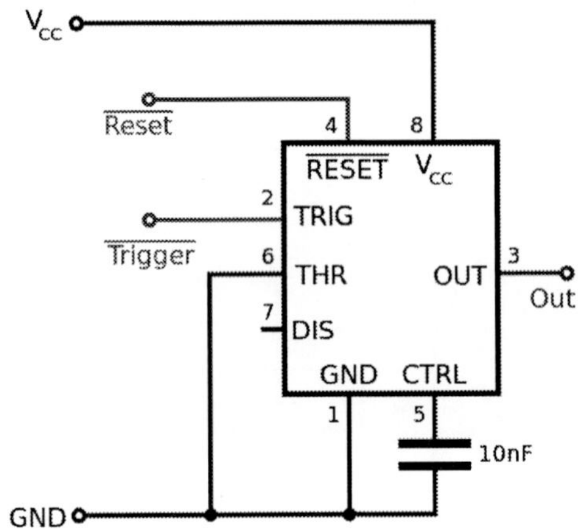

Figure: *Schematic of a 555 in bistable mode*

In bistable mode, the 555 timer acts as a basic flip-flop. The trigger and reset inputs (pins 2 and 4 respectively on a 555) are held high via Pull-up resistors while the threshold input (pin 6) is simply grounded. Thus configured, pulling the trigger momentarily to ground acts as a 'set' and transitions the output pin (pin 3) to Vcc (high state). Pulling the reset input to ground acts as a 'reset' and transitions the output pin to ground (low state). No capacitors are required in a bistable configuration. Pin 5 (control) is connected to ground via a small-value capacitor (usually 0.01 to 0.1 uF); pin 7 (discharge) is left floating.

Astable

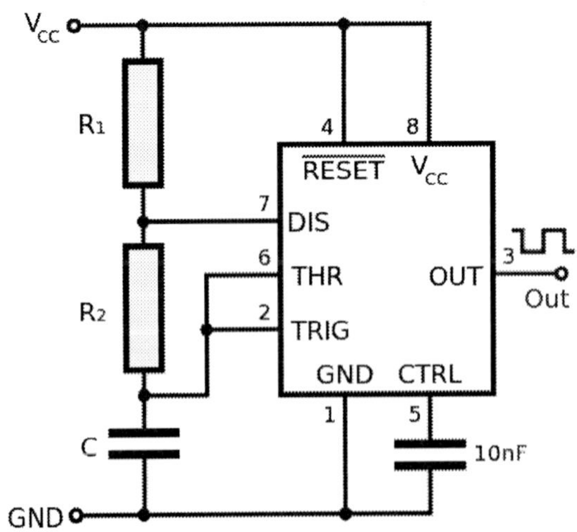

Figure: *Standard 555 astable circuit*

In astable mode, the 555 timer puts out a continuous stream of rectangular pulses having a specified frequency. Resistor R_1 is connected between V_{CC} and the discharge pin (pin 7) and another resistor (R_2) is connected between the discharge pin (pin 7), and the trigger (pin 2) and threshold (pin 6) pins that share a common node.

Hence the capacitor is charged through R_1 and R_2, and discharged only through R_2, since pin 7 has low impedance to ground during output low intervals of the cycle, therefore discharging the capacitor.

In the astable mode, the frequency of the pulse stream depends on the values of R_1, R_2 and C:

$$f = \frac{1}{\ln(2) \cdot C \cdot (R_1 + 2R_2)}$$

The high time from each pulse is given by:

$$\text{high} = \ln(2) \cdot (R_1 + R_2) \cdot C$$

and the low time from each pulse is given by:

$$\text{low} = \ln(2) \cdot R_2 \cdot C$$

where R_1 and R_2 are the values of the resistors in ohms and C is the value of the capacitor in farads.

The power capability of R_1 must be greater than $\frac{V_{cc}^2}{R_1}$.

Particularly with bipolar 555s, low values of R1 must be avoided so that the output stays saturated near zero volts during discharge, as assumed by the above equation.

Otherwise the output low time will be greater than calculated above. It should be noted that the first cycle will take appreciably longer than the calculated time, as the capacitor must charge from 0V to 2/3 of V_{CC} from power-up, but only from 1/3 of V_{CC} to 2/3 of V_{CC} on subsequent cycles.

The circuit configuration above does not permit a duty cycle of less than 50%, because the time-constant for charging C1 is always greater than for discharging. To achieve any arbitrary duty cycle, R_2 can be moved to be in series with pin 7, the discharge pin. The duration of the high-output interval (during the charging of C1) is then 0.693(R_1C_1), and the low-output interval (while discharging C1) is 0.693(R_2C_1). The total time period, T, is 0.693(R1+R2)C1.

A more complicated method to achieve a duty cycle of less than 50% is to use a small diode (that is fast enough for the application) in parallel with R_2 (instead of placing it on pin 7), with the anode on the capacitor side. This bypasses R_2 during the high part of the cycle so that the high interval depends approximately only on R_1 and C.

The presence of the diode is a voltage drop that slows charging on the capacitor so that the high time is longer than the expected and often-cited ln(2)*R_1C = 0.693 R_1C.

The low time will be the same as without the diode as shown above. With a diode, the high time is

$$\text{high} = R_1 C \cdot \ln\left(\frac{2V_{cc} - 3V_{diode}}{V_{cc} - 3V_{diode}}\right)$$

where V_{diode} is when the diode has a current of 1/2 of V_{cc}/R_1 which can be determined from its datasheet or by testing. As an extreme example, when V_{cc} = 5 and V_{diode} = 0.7, high time = 1.00 R_1C which is 45% longer than the "expected" 0.693 R_1C. At the other extreme, when V_{cc} = 15 and V_{diode} = 0.3, the high time = 0.725 R_1C which is closer to the expected 0.693 R_1C. The equation reduces to the expected 0.693 R_1C if V_{diode} = 0.

The operation of RESET in this mode is not well defined, some manufacturers' parts will hold the output state to what it was when RESET is taken low, others will send the output either high or low.

Specifications

These specifications apply to the NE555. Other 555 timers can have different specifications depending on the grade (military, medical, etc.).

Supply voltage (V_{CC})	4.5 to 15 V
Supply current (V_{CC} = +5 V)	3 to 6 mA
Supply current (V_{CC} = +15 V)	10 to 15 mA
Output current (maximum)	200 mA
Maximum Power dissipation	600 mW
Power consumption (minimum operating)	30 mW@5V, 225 mW@15V
Operating temperature	0 to 70 °C

Derivatives

Many pin-compatible variants, including CMOS versions, have been built by various companies. Bigger packages also exist with two or four timers on the same chip. The 555 is also known under the following type numbers:

Manufacturer	*Model*	*Remark*
Custom Silicon Solutions	CSS555/CSS555C	CMOS from 1.2 V, IDD < 5 µA
CEMI	ULY7855	
ECG Philips	ECG955M	
Exar	XR-555	
Fairchild Semiconductor	NE555/KA555	
GoldStar	GSC555	CMOS
Harris	HA555	
IK Semicon	ILC555	CMOS from 2 V
Intersil	SE555/NE555	
Intersil	ICM7555	CMOS
Lithic Systems	LC555	
Maxim	ICM7555	CMOS from 2 V
Motorola	MC1455/MC1555	
National Semiconductor	LM1455/LM555/LM555C	
National Semiconductor	LMC555	CMOS from 1.5 V
NTE Sylvania	NTE955M	
Raytheon	RM555/RC555	
RCA	CA555/CA555C	
STMicroelectronics	NE555N/ K3T647	
Texas Instruments	SN52555/SN72555	
Texas Instruments	TLC555	CMOS from 2 V
USSR	K1006??1	
X-REL Semiconductor	XTR655	Operation from -60°C to 250+°C
Zetex	ZSCT1555	down to 0.9 V
NXP Semiconductors	ICM7555	CMOS
HFO / East Germany	B555	

556 Dual Timer

The dual version is called 556. It features two complete 555s in a 14 pin DIL package.

558 Quad Timer

The quad version is called 558 and has 16 pins. To fit four 555s into a 16 pin package the control, voltage, and reset lines are shared by all four modules. Each module's discharge and threshold circuits are wired together internally.

XTR651 Extended Functionality HiRelHiTemp (-60°C to 250+°C)

This version includes non-overlapped complementary outputs, coarse temperature sensor and on-chip 200pF timing capacitance.

Example Applications

Joystick Interface Circuit Using the 558 Quad Timer: The Apple IImicrocomputer used a quad timer 558 in monostable (or "one-shot") mode to interface up to four "game paddles" or two joysticks to the host computer. It also used a single 555 for flashing the display cursor. A similar circuit was used in the IBM PC. In the joystick interface circuit of the IBM PC, the capacitor (C) of the RC network was generally a 10 nF capacitor. The resistor (R) of the RC network consisted of the potentiometer inside the joystick along with an external resistor of 2.2 kilohms. The joystick potentiometer acted as a variable resistor. By moving the joystick, the resistance of the joystick increased from a small value up to about 100 kilohms. The joystick operated at 5 V.

Software running in the host computer started the process of determining the joystick position by writing to a special address (ISA bus I/O address 201h). This would result in a trigger signal to the quad timer, which would cause the capacitor (C) of the RC network to begin charging and cause the quad timer to output a pulse. The width of the pulse was determined by how long it took the C to charge up to 2/3 of 5 V (or about 3.33 V), which was in turn determined by the joystick position. Software running in the host computer measured the pulse width to determine the joystick position. A wide pulse represented the full-right joystick position, for example, while a narrow pulse represented the full-left joystick position.

Astableand Monostable Multivibrator Configurations

A multivibrator is an electronic circuit used to implement a variety of simple two-state systems such as oscillators, timers and flip-flops. It is characterized by two amplifying devices (transistors, electron tubes or other devices) cross-coupled by resistors or capacitors. The name "multivibrator" was initially applied to the free-running oscillator version of the circuit because its output waveform was rich in harmonics. There are three types of multivibrator circuits depending on the circuit operation:

- astable, in which the circuit is not stable in either state —it continually switches from one state to the other. It functions as a relaxation oscillator.
- monostable, in which one of the states is stable, but the other state is unstable (transient). A trigger pulse causes the circuit to enter the unstable state. After entering the unstable state,

the circuit will return to the stable state after a set time. Such a circuit is useful for creating a timing period of fixed duration in response to some external event. This circuit is also known as a one shot.

- bistable, in which the circuit is stable in either state. It can be flipped from one state to the other by an external trigger pulse. This circuit is also known as a flip flop. It can be used to store one bit of information.

Multivibrators find applications in a variety of systems where square waves or timed intervals are required. For example, before the advent of low-cost integrated circuits, chains of multivibrators found use as frequency dividers. A free-running multivibrator with a frequency of one-half to one-tenth of the reference frequency would accurately lock to the reference frequency. This technique was used in early electronic organs, to keep notes of different octaves accurately in tune. Other applications included early television systems, where the various line and frame frequencies were kept synchronized by pulses included in the video signal.

History

The classic multivibrator circuit (also called a *plate-coupled multivibrator*) is first described by H. Abraham and E. Bloch in *Publication 27* of the French *Ministère de la Guerre*, and in *Annales de Physique 12, 252 (1919)*. It is a predecessor of Eccles-Jordan trigger derived from this circuit a year later.

The terminology of multivibrators has been somewhat variable, historically. For example:

- 1942 – multivibrator implies astable: "The multivibrator circuit is somewhat similar to the flip-flop circuit, but the coupling from the anode of one valve to the grid of the other is by a condenser only, so that the coupling is not maintained in the steady state."
- 1942 – multivibrator as a particular flip-flop circuit: "Such circuits were known as 'trigger' or 'flip-flop' circuits and were of very great importance. The earliest and best known of these circuits was the multivibrator."
- 1943 – flip-flop as one-shot pulse generator: "It should be noted that an essential difference between the two-valve flip-flop and the multivibrator is that the flip-flop has one of the valves biased to cutoff."

- 1949 – monostable as flip-flop: "Monostablemultivibrators have also been called 'flip-flops'."
- 1949 – monostable as flip-flop: "... a flip-flop is a monostablemultivibrator and the ordinary multivibrator is an astablemultivibrator."

Astablemultivibrator

An astablemultivibrator is a regenerative circuit consisting of two amplifying stages connected in a positive feedback loop by two capacitive-resistive coupling networks. The amplifying elements may be junction or field-effect transistors, vacuum tubes, operational amplifiers, or other types of amplifier. The example diagram shows bipolar junction transistors.

The circuit is usually drawn in a symmetric form as a cross-coupled pair. Two output terminals can be defined at the active devices, which will have complementary states; one will have high voltage while the other has low voltage, (except during the brief transitions from one state to the other).

Operation

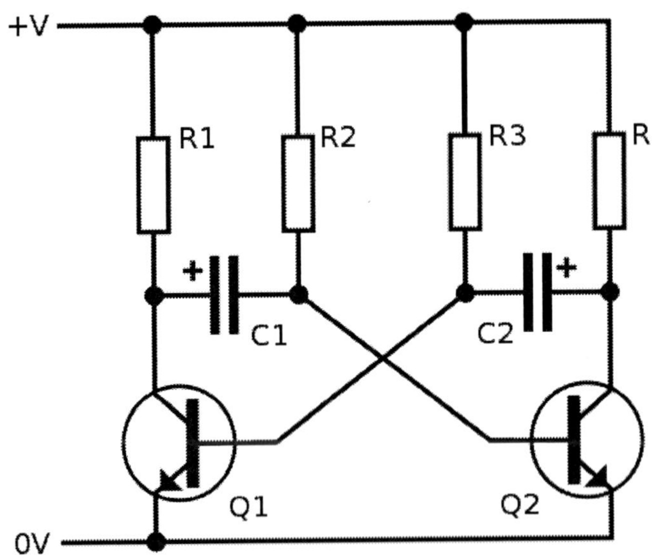

Figure: *Basic BJTastablemultivibrator*

The circuit has two stable states that change alternatively with maximum transition rate because of the "accelerating" positive feedback. It is implemented by the coupling capacitors that instantly transfer voltage changes because the voltage across a capacitor cannot suddenly change. In each state, one transistor is switched on and the other is switched off. Accordingly, one fully charged capacitor discharges (reverse charges) slowly thus converting the time into an exponentially changing voltage. At the same time, the other empty capacitor quickly charges thus restoring its charge (the first capacitor acts as a time-setting capacitor and the second prepares to play this role in the next state). The circuit operation is based on the fact that the forward-biased base-emitter junction of the switched-on bipolar transistor can provide a path for the capacitor restoration.

State 1 (Q1 is Switched on, Q2 is Switched off):

In the beginning, the capacitor C1 is fully charged (in the previous State 2) to the power supply voltage *V* with the polarity. Q1 is *on* and connects the left-hand positive plate of C1 to ground. As its right-hand negative plate is connected to Q2 base, a maximum negative voltage (-*V*) is applied to Q2 base that keeps Q2 firmly *off*. C1 begins discharging (reverse charging) via the high-value base resistor R2, so that the voltage of its right-hand plate (and at the base of Q2) is rising from below ground (-*V*) toward +*V*. As Q2 base-emitter junction is backward-biased, it does not conduct, so all the current from R2 goes into C1. Simultaneously, C2 that is fully discharged and even slightly charged to 0.6 V (in the previous State 2) quickly charges via the low-value collector resistor R4 and Q1 forward-biased base-emitter junction (because R4 is less than R2, C2 charges faster than C1). Thus C2 restores its charge and prepares for the next State 2 when it will act as a time-setting capacitor. Q1 is firmly saturated in the beginning by the "forcing" C2 charging current added to R3 current; in the end, only R3 provides the needed input base current. The resistance R3 is chosen small enough to keep Q1 (not deeply) saturated after C2 is fully charged.

When the voltage of C1 right-hand plate (Q2 base voltage) becomes positive and reaches 0.6 V, Q2 base-emitter junction begins diverting a part of R2 charging current. Q2 begins conducting and this starts the avalanche-like positive feedback process as follows. Q2 collector voltage begins falling; this change transfers through the fully charged C2 to Q1 base and Q1 begins cutting off. Its collector voltage begins rising; this change transfers back through the almost empty C1 to Q2 base and makes Q2 conduct more thus sustaining the initial input impact on Q2 base. Thus the initial input change circulates along the feedback loop and grows in an avalanche-like manner until finally Q1 switches off and Q2 switches on. The forward-biased Q2 base-emitter junction fixes the voltage of C1 right-hand plate at 0.6 V and does not allow it to continue rising toward +*V*.

State 2 (Q1 is switched off, Q2 is switched on):

Now, the capacitor C2 is fully charged (in the previous State 1) to the power supply voltage *V* with the polarity. Q2 is *on* and connects the right-hand positive plate of C2 to ground. As its left-hand negative plate is connected to Q1 base, a maximum negative voltage (-*V*) is applied to Q1 base that keeps Q1 firmly *off*. C2 begins discharging (reverse charging) via the high-value base resistor R3, so that the voltage of its left-hand plate (and at the base of Q1) is rising from below ground (-*V*) toward +*V*. Simultaneously, C1 that is fully discharged and even slightly charged to 0.6 V (in the previous State 1) quickly charges via the low-value collector resistor R1 and Q2 forward-biased base-emitter junction (because R1 is less than R3, C1 charges faster than C2). Thus C1 restores its charge and prepares for the next State 1 when it will act again as a time-setting capacitor...and so on... (the next explanations are a mirror copy of the second part of Step 1).

Multivibrator Drequency

Derivation: The duration of state 1 (low output) will be related to the time constant R_2C_1 as it depends on the charging of C1, and the duration of state 2 (high output) will be related to the time constant R_3C_2 as it depends on the charging of C2. Because they do not need to be the same, an asymmetric duty cycle is easily achieved.

The voltage on a capacitor with non-zero initial charge is:

$$V_{\text{cap}}(t) = \left[\left(V_{\text{capinit}} - V_{\text{charging}} \right) \times e^{-\frac{t}{RC}} \right] + V_{\text{charging}}$$

Looking at C2, just before Q2 turns on the left terminal of C2 is at the base-emitter voltage of Q1 (V_{BE_Q1}) and the right terminal is at V_{CC} ("V_{CC}" is used here instead of "+*V*" to ease notation). The voltage across C2 is V_{CC} minus V_{BE_Q1} . The moment after Q2 turns on, the right terminal of C2 is now at 0 V which drives the left terminal of C2 to 0 V minus (V_{CC} - V_{BE_Q1}) or V_{BE_Q1} - V_{CC}. From this instant in time, the left terminal of C2 must be charged back up to V_{BE_Q1}. How long this takes is half our multivibrator switching time (the other half comes from C1). In the charging capacitor equation above, substituting:

V_{BE_Q1} for $V_{\text{cap}}(t)$

$(V_{BE\_Q1}$ - $V_{CC})$ for $V_{capinit}$

V_{CC} for $V_{charging}$

results in:

$$V_{BE\_Q1} = \left(\left[\left(V_{BE\_Q1} - V_{CC} \right) - V_{CC} \right] \times e^{-\frac{t}{RC}} \right) + V_{CC}$$

Solving for t results in:

$$t = -RC \times \ln \left(\frac{V_{BE\_Q1} - V_{CC}}{V_{BE\_Q1} - 2V_{CC}} \right)$$

For this circuit to work, $V_{CC} >> V_{BE\_Q1}$ (for example: V_{CC}=5 V, V_{BE_Q1}=0.6 V), therefore the equation can be simplified to:

$$t = -RC \times \ln \left(\frac{-V_{CC}}{-2V_{CC}} \right)$$

or

$$t = -RC \times \ln \left(\frac{1}{2} \right)$$

or

$$t = RC \times \ln(2)$$

The period of each *half* of the multivibrator is therefore given by $t = \ln(2)RC$.

The total period of oscillation is given by:

$T = t_1 + t_2 = \ln(2)R_2C_1 + \ln(2)R_3C_2$

$$f = \frac{1}{T} = \frac{1}{\ln(2) \cdot (R_2C_1 + R_3C_2)} \approx \frac{1}{0.693 \cdot (R_2C_1 + R_3C_2)}$$

where...

- f is frequency in hertz.
- R_2 and R_3 are resistor values in ohms.
- C_1 and C_2 are capacitor values in farads.
- T is the period (In this case, the sum of two period durations).

For the special case where

- $t_1 = t_2$ (50% duty cycle)
- $R_2 = R_3$

- $C_1 = C_2$

$$f = \frac{1}{T} = \frac{1}{\ln(2)\cdot 2RC} \approx \frac{0.72}{RC}$$ Output pulse shape

The output voltage has a shape that approximates a square waveform. It is considered below for the transistor Q1.

During State 1, Q2 base-emitter junction is backward-biased and the capacitor C1 is "unhooked" from ground. The output voltage of the switched-on transistor Q1 changes rapidly from high to low since this low-resistive output is loaded by a high impedance load (the series connected capacitor C1 and the high-resistive base resistor R2).

During State 2, Q2 base-emitter junction is forward-biased and the capacitor C1 is "hooked" to ground. The output voltage of the switched-off transistor Q1 changes exponentially from low to high since this relatively high resistive output is loaded by a low impedance load (the capacitance C1). This is the output voltage of R_1C_1 integrating circuit.

To approach the needed square waveform, the collector resistors have to be low resistance. The base resistors have to be low enough to make the transistors saturate in the end of the restoration ($R_B < \beta.R_C$).

Initial Power-up

When the circuit is first powered up, neither transistor will be switched on. However, this means that at this stage they will both have high base voltages and therefore a tendency to switch on, and inevitable slight asymmetries will mean that one of the transistors is first to switch on. This will quickly put the circuit into one of the above states, and oscillation will ensue. In practice, oscillation always occurs for practical values of *R* and *C*.

However, if the circuit is temporarily held with both bases high, for longer than it takes for both capacitors to charge fully, then the circuit will remain in this stable state, with both bases at 0.6 V, both collectors at 0 V, and both capacitors charged backwards to "0.6 V. This can occur at startup without external intervention, if *R* and *C* are both very small.

Frequency Divider

An astablemultivibrator can be synchronized to an external chain of pulses. A single pair of active devices can be used to divide a reference by a large ratio, however, the stability of the technique is poor owing to the variability of the power supply and the circuit elements; a division ratio of 10, for example, is easy to obtain but not dependable. Chains of bistable flip-flops provide more predictable division, at the cost of more active elements.

Protective Components

While not fundamental to circuit operation, diodes connected in series with the base or emitter of the transistors are required to prevent the base-emitter junction being driven into reverse breakdown when the supply voltage is in excess of the V_{eb} breakdown voltage, typically around 5-10 volts for general purpose silicon transistors. In the monostable configuration, only one of the transistors requires protection.

Monostablemultivibrator Circuit

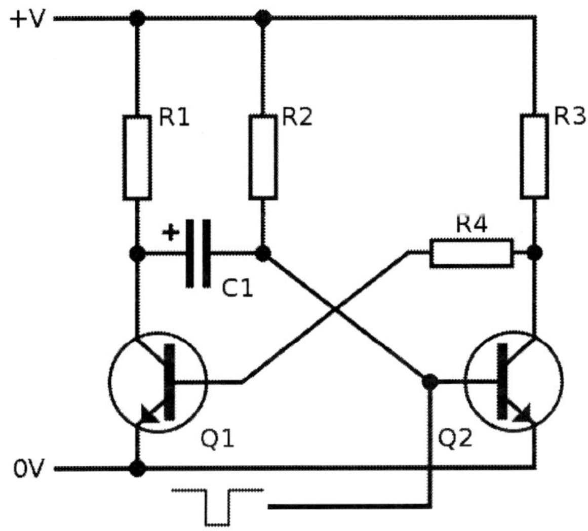

Figure: Basic BJTmonostablemultivibrator.

In the monostable-multivibrator, the one resistive-capacitive network (C_2-R_3) is replaced by a resistive network (just a resistor). The circuit can be thought as a 1/2 astablemultivibrator. Q2 collector voltage is the output of the circuit (in contrast to the astable circuit, it has a perfect square waveform since the output is not loaded by the capacitor).

When triggered by an input pulse, a monostable-multivibrator will switch to its unstable position for a period of time, and then return to its stable state. The time period monostablemultivibrator remains in unstable state is given by $t = \ln(2)R_2C_1$. If repeated application of the input pulse maintains the circuit in the unstable state, it is called a *retriggerable*monostable. If further trigger pulses do not affect the period, the circuit is a *non-retriggerable*multivibrator.

For the circuit in Figure, in the stable state Q1 is turned off and Q2 is turned on. It is triggered by zero or negative input signal applied to Q2 base (with the same success it can be triggered by applying a positive input signal through a resistor to Q1 base). As a result, the circuit goes in State 1 described above. After elapsing the time, it returns to its stable initial state.

Bistablemultivibrator Circuit

In the bistablemultivibrator, both the resistive-capacitive network are replaced by resistive networks (just resistors or direct coupling).

This latch circuit is similar to an astablemultivibrator, except that there is no charge or discharge time, due to the absence of capacitors. Hence, when the circuit is switched on, if Q1 is on, its collector is at 0 V. As a result, Q2 gets switched off. This results in more than half $+V$ volts being applied to R4 causing current into the base of Q1, thus keeping it on. Thus, the circuit remains stable in a single state continuously. Similarly, Q2 remains on continuously, if it happens to get switched on first.

Switching of state can be done via Set and Reset terminals connected to the bases. For example, if Q2 is on and Set is grounded momentarily, this switches Q2 off, and makes Q1 on. Thus, Set is used to "set" Q1 on, and Reset is used to "reset" it to off state.

Data Converters

Data conversion is the conversion of computer data from one format to another. Throughout a computer environment, data is encoded in a variety of ways. For example, computer hardware is built on the basis of certain standards, which requires that data contains, for example, parity bit checks. Similarly, the operating system is predicated on certain standards for data and file handling. Furthermore, each computer programme handles data in a different manner. Whenever any one of these variable is changed, data must be converted in some way before it can be used by a different computer, operating system or programme. Even different versions of these elements usually involve different data structures. For example, the changing of bits from one format to another, usually for the purpose of application interoperability or of capability of using new features, is merely a data conversion. Data conversions may as simple as the conversion of a text file from one character encoding system to another;

or more complex, such as the conversion of office file formats, or the conversion of image and audio file formats.

There are many ways in which data is converted within the computer environment. This may be seamless, as in the case of upgrading to a newer version of a computer programme. Alternatively, the conversion may require processing by the use of a special conversion programme, or it may involve a complex process of going through intermediary stages, or involving complex "exporting" and "importing" procedures, which may converting to and from a tab-delimited or comma-separated text file. In some cases, a programme may recognise several data file formats at the data input stage and then is also capable of storing the output data in a number of different formats. Such a programme may be used to convert a file format. If the source format or target format is not recognised, then at times third programme may be available which permits the conversion to an intermediate format, which can then be reformatted using the first programme. There are many possible scenarios.

Information Basics

Before any data conversion is carried out, the user or application programmer should keep a few basics of computing and information theory in mind. These include:

- Information can easily be discarded by the computer, but adding information takes effort.
- The computer can add information only in a rule-based fashion.
- Upsampling the data or converting to a more feature-rich format does not add information; it merely makes room for that addition, which usually a human must do.
- By storing data electronically, the chances of data loss are greatly lowered and it is easier to perform data mining and other conversions.
- Data stored in an electronic format can be quickly modified and analyzed.

For example, a true colour image can easily be converted to grayscale, while the opposite conversion is a painstaking process. Converting a Unix text file to a Microsoft (DOS/Windows) text file involves adding characters, but this does not increase the entropy since it is rule-based; whereas the addition of colour information to a grayscale image cannot be done programmetically, since only a human knows which colours are needed for each section of the picture–

there are no rules that can be used to automate that process. Converting a 24-bit PNG to a 48-bit one does not add information to it, it only pads existing RGB pixel values with zeroes, so that a pixel with a value of FF C3 56, for example, becomes FF00 C300 5600. The conversion makes it possible to change a pixel to have a value of, for instance, FF80 C340 56A0, but the conversion itself does not do that, only further manipulation of the image can. Converting an image or audio file in a lossy format (like JPEG or Vorbis) to a lossless (like PNG or FLAC) or uncompressed (like BMP or WAV) format only wastes space, since the same image with its loss of original information (the artifacts of lossy compression) becomes the target. A JPEG image can never be restored to the quality of the original lossless image from which it was made, no matter how much the user tries the "JPEG Artifact Removal" feature of his or her image manipulation programme.

Automatic restorage of information that was lost through a lossy compression process would probably require important advances in artificial intelligence.

Because of these realities of computing and information theory, data conversion is more often than not a complex and error-prone process that requires the help of experts.

Pivotal Conversion

Data conversion can occur directly from one format to another, but many applications that convert between multiple formats use a pivotal encoding by way of which any source format is converted to its target. For example, it is possible to convert Cyrillic text from KOI8-R to Windows-1251 using a lookup table between the two encodings, but the modern approach is to convert the KOI8-R file to Unicode first and from that to Windows-1251. This is a more manageable approach: an application specializing in character encoding conversion would have to keep hundreds of lookup tables, for all the permutations of character encoding conversions available, while keeping lookup tables just for each character set to Unicode scales down the number to a few tens.

Pivotal conversion is similarly used in other areas. Office applications, when employed to convert between office file formats, use their internal, default file format as a pivot. For example, a word processor may convert an RTF file to a WordPerfect file by converting the RTF to OpenDocument and then that to WordPerfect format. An image conversion programme does not convert a PCX image to PNG directly; instead, when loading the PCX image, it decodes it to a simple bitmap format for internal use in memory, and when commanded to convert to PNG, that memory image is converted to the target format. An audio converter that converts from FLAC to AAC decodes the source file to raw PCM data in memory first, and then performs the lossy AAC compression on that memory image to produce the target file.

Lost and Inexact Data Conversion

The objective of data conversion is to maintain all of the data, and as much of the embedded information as possible. This can only be done if the target format supports the same features and data structures present in the source file. Conversion of a word processing document to a plain text file necessarily involves loss of formatting information, because plain text format does not support word processing constructs such as marking a word as boldface. For this reason, conversion from one format to one which does not support a feature which is important to the user is rarely carried out, though it may be necessary for interoperability, e.g. converting a file from one version of Microsoft Word to an earlier version to enable transfer and use by other users who do not have the same later version of Word installed on their computer.

Loss of information can be mitigated by approximation in the target format. There is no way of converting a character like *δ* to ASCII, since the ASCII standard lacks it, but the information may be retained by approximating the character as *ae*. Of course, this is not an optimal solution, and can impact operations like searching and copying; and if a language makes a distinction between *δ* and *ae*, then that approximation does involve loss of information.

Data conversion can also suffer from inexactitude, the result of converting between formats that are conceptually different. The WYSIWYG paradigm, extant in word processors and desktop publishing applications, versus the structural-descriptive paradigm, found in SGML, XML and many applications derived therefrom, like HTML and MathML, is one example. Using a WYSIWYG HTML editor conflates the two paradigms, and the result is HTML files with suboptimal, if not nonstandard, code. In the WYSIWYG paradigm a double linebreak signifies a new paragraph, as that is the visual cue for such a construct, but a WYSIWYG HTML editor will usually convert such a sequence to

, which is structurally no new paragraph at all. As another example, converting from PDF to an editable word processor format is a tough chore, because PDF records the textual information like engraving on stone, with each character given a fixed position and linebreaks hard-coded, whereas word processor formats accommodate text reflow. PDF does not know of a word space character—the space between two letters and the space between two words differ only in quantity. Therefore, a title with ample letter-spacing for effect will usually end up with spaces in the word processor file, for example INTRODUCTION with spacing of 1 em as I N T R O D U C T I O N on the word processor.

Open vs. Secret Specifications

Successful data conversion requires thorough knowledge of the workings of both source and target formats. In the case where the specification of a format is unknown, reverse engineering will be needed to carry out conversion. Reverse engineering can achieve close approximation of the original specifications, but errors and missing features can still result.

Electronics

Data format conversion can also occur at the physical layer of an electronic communication system. Conversion between line codes such as NRZ and RZ can be accomplished when necessary.

Analog-to-digital Converter

An analog-to-digital converter (abbreviated ADC, A/D or A to D) is a device that converts a continuous physical quantity (usually voltage) to a digital number that represents the quantity's amplitude.

The conversion involves quantization of the input, so it necessarily introduces a small amount of error. Instead of doing a single conversion, an ADC often performs the conversions ("samples" the input) periodically. The result is a sequence of digital values that have converted a continuous-time and continuous-amplitude analog signal to a discrete-time and discrete-amplitude digital signal.

An ADC is defined by its bandwidth (the range of frequencies it can measure) and its signal to noise ratio (how accurately it can measure a signal relative to the noise it introduces). The actual bandwidth of an ADC is characterized primarily by its sampling rate, and to a lesser extent by how it handles errors such as aliasing. The dynamic range of an ADC is influenced by many factors, including the resolution (the number of output levels it can quantize a signal to), linearity and accuracy (how well the quantization levels match the true analog signal) and jitter (small timing errors that introduce additional noise). The dynamic range of an ADC is often summarized in terms of its effective number of bits (ENOB), the number of bits of each measure it returns that are on average not noise. An ideal ADC has an ENOB equal to its resolution. ADCs are chosen to match the bandwidth and required signal to noise ratio of the signal to be quantized. If an ADC operates at a sampling rate greater than twice the bandwidth of the signal, then perfect reconstruction is possible given an ideal ADC and neglecting quantization error. The presence of quantization error limits the dynamic range of even an ideal ADC, however, if the dynamic range of the ADC exceeds that of the input signal, its effects may be neglected resulting in an essentially perfect digital representation of the input signal.

An ADC may also provide an isolated measurement such as an electronic device that converts an input analogvoltage or current to a digital number proportional to the magnitude of the voltage or current. However, some non-electronic or only partially electronic devices, such as rotary encoders, can also be considered ADCs. The digital output may use different coding schemes. Typically the digital output will be a two's complement binary number that is proportional to the input, but there are other possibilities. An encoder, for example, might output a Gray code. The inverse operation is performed by a digital-to-analog converter (DAC).

Concepts

Resolution:

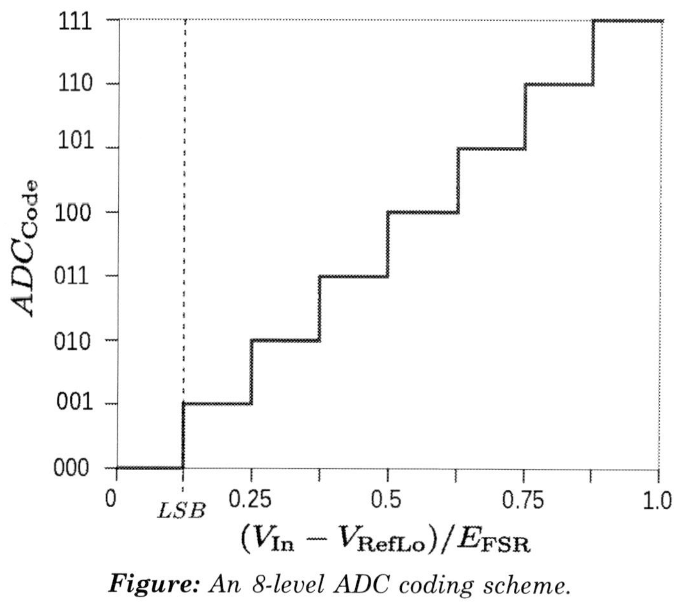

Figure: *An 8-level ADC coding scheme.*

The resolution of the converter indicates the number of discrete values it can produce over the range of analog values. The resolution determines the magnitude of the quantization error and therefore determines the maximum possible average signal to noise ratio for an ideal ADC without the use of oversampling.

The values are usually stored electronically in binary form, so the resolution is usually expressed in bits. In consequence, the number of discrete values available, or "levels", is assumed to be a power of two. For example, an ADC with a resolution of 8 bits can encode an analog input to one in 256 different levels, since $2^8 = 256$. The values can represent the ranges from 0 to 255 (i.e. unsigned integer) or from "128 to 127 (i.e. signed integer), depending on the application.

Resolution can also be defined electrically, and expressed in volts. The minimum change in voltage required to guarantee a change in the output code level is called the least significant bit (LSB) voltage. The resolution Q of the ADC is equal to the LSB voltage. The voltage resolution of an ADC is equal to its overall voltage measurement range divided by the number of discrete values:

$$Q = \frac{E_{\text{FSR}}}{2^M - 1},$$

where M is the ADC's resolution in bits and E_{FSR} is the full scale voltage range (also called 'span'). E_{FSR} is given by

$$E_{\text{FSR}} = V_{\text{RefHi}} - V_{\text{RefLow}},$$

where V_{RefHi} and V_{RefLow} are the upper and lower extremes, respectively, of the voltages that can be coded.

Normally, the number of voltage intervals is given by

$$N = 2^M - 1,$$

where M is the ADC's resolution in bits.

That is, one voltage interval is assigned in between two consecutive code levels.

Example:

- Coding scheme as in figure (assume input signal x(t) = Acos(t), A = 5V)
- Full scale measurement range = -5 to 5 volts
- ADC resolution is 8 bits: $2^8 - 1 = 256 - 1 = 255$ quantization levels (codes)

- ADC voltage resolution, $Q = (10\text{ V} - 0\text{ V}) / 255 = 10\text{ V} / 255 \approx 0.039\text{ V} \approx 39\text{ mV}$.

In practice, the useful resolution of a converter is limited by the best signal-to-noise ratio (SNR) that can be achieved for a digitized signal. An ADC can resolve a signal to only a certain number of bits of resolution, called the effective number of bits (ENOB). One effective bit of resolution changes the signal-to-noise ratio of the digitized signal by 6 dB, if the resolution is limited by the ADC. If a preamplifier has been used prior to A/D conversion, the noise introduced by the amplifier can be an important contributing factor towards the overall SNR.

Quantization Error

Quantization error is the noise introduced by quantization in an ideal ADC. It is a rounding error between the analog input voltage to the ADC and the output digitized value. The noise is non-linear and signal-dependent.

In an ideal analog-to-digital converter, where the quantization error is uniformly distributed between "1/2 LSB and +1/2 LSB, and the signal has a uniform distribution covering all quantization levels, the Signal-to-quantization-noise ratio (SQNR) can be calculated from

$$\text{SQNR} = 20\log_{10}(2^{Q}) \approx 6.02 \cdot Q \text{ dB}$$

Where Q is the number of quantization bits. For example, a 16-bit ADC has a maximum signal-to-noise ratio of 6.02 × 16 = 96.3 dB, and therefore the quantization error is 96.3 dB below the maximum level. Quantization error is distributed from DC to the Nyquist frequency, consequently if part of the ADC's bandwidth is not used (as in oversampling), someS of the quantization error will fall out of band, effectively improving the SQNR. In an oversampled system, noise shaping can be used to further increase SQNR by forcing more quantization error out of the band.

Dither

In ADCs, performance can usually be improved using dither. This is a very small amount of random noise (white noise), which is added to the input before conversion.

Its effect is to cause the state of the LSB to randomly oscillate between 0 and 1 in the presence of very low levels of input, rather than sticking at a fixed value. Rather than the signal simply getting cut off altogether at this low level (which is only being quantized to a resolution of 1 bit), it extends the effective range of signals that the ADC can convert, at the expense of a slight increase in noise – effectively the quantization error is diffused across a series of noise values which is far less objectionable than a hard cutoff. The result is an accurate representation of the signal over time. A suitable filter at the output of the system can thus recover this small signal variation.

An audio signal of very low level (with respect to the bit depth of the ADC) sampled without dither sounds extremely distorted and unpleasant. Without dither the low level may cause the least significant bit to "stick" at 0 or 1. With dithering, the true level of the audio may be calculated by averaging the actual quantized sample with a series of other samples [the dither] that are recorded over time.

A virtually identical process, also called dither or dithering, is often used when quantizing photographic images to a fewer number of bits per pixel—the image becomes noisier but to the eye looks far more realistic than the quantized image, which otherwise becomes banded. This analogous process may help to visualize the effect of dither on an analogue audio signal that is converted to digital.

Dithering is also used in integrating systems such as electricity metres. Since the values are added together, the dithering produces results that are more exact than the LSB of the analog-to-digital converter.

Note that dither can only increase the resolution of a sampler, it cannot improve the linearity, and thus accuracy does not necessarily improve.

Accuracy

An ADC has several sources of errors. Quantization error and (assuming the ADC is intended to be linear) non-linearity are intrinsic to any analog-to-digital conversion.

These errors are measured in a unit called the least significant bit (LSB). In the above example of an eight-bit ADC, an error of one LSB is 1/256 of the full signal range, or about 0.4%.

Non-linearity

All ADCs suffer from non-linearity errors caused by their physical imperfections, causing their output to deviate from a linear function (or some other function, in the case of a deliberately non-linear ADC) of their input. These errors can sometimes be mitigated by calibration, or prevented by testing.

Important parameters for linearity are integral non-linearity (INL) and differential non-linearity (DNL). These non-linearities reduce the dynamic range of the signals that can be digitized by the ADC, also reducing the effective resolution of the ADC.

Jitter

When digitizing a sine wave $x(t) = A\sin(2\pi f_0 t)$, the use of a non-ideal sampling clock will result in some uncertainty in when samples are recorded. Provided that the actual sampling time *uncertainty* due to the *clock jitter*is Δt, the error caused by this phenomenon can be estimated as $E_{ap} \leq |x'(t)\Delta t| \leq 2A\pi f_0 \Delta t$.

This will result in additional recorded noise that will reduce the ENOB below that predicted by quantization error alone.

The error is zero for DC, small at low frequencies, but significant when high frequencies have high amplitudes. This effect can be ignored if it is drowned out by the *quantizing error*. Jitter requirements can be calculated using the following formula: $\Delta t < \dfrac{1}{2^q \pi f_0}$, where q is the number of ADC bits.

Output size (bits)	Signal Frequency						
	1 Hz	1 kHz	10 kHz	1 MHz	10 MHz	100 MHz	1 GHz
8	1,243 μs	1.24 μs	124 ns	1.24 ns	124 ps	12.4 ps	1.24 ps
10	311 μs	311 ns	31.1 ns	311 ps	31.1 ps	3.11 ps	0.31 ps
12	77.7 μs	77.7 ns	7.77 ns	77.7 ps	7.77 ps	0.78 ps	0.08 ps
14	19.4 μs	19.4 ns	1.94 ns	19.4 ps	1.94 ps	0.19 ps	0.02 ps
16	4.86 μs	4.86 ns	486 ps	4.86 ps	0.49 ps	0.05 ps	–
18	1.21 ns	121 ps	6.32 ps	1.21 ps	0.12 ps	–	–
20	304 ps	30.4 ps	1.58 ps	0.16 ps	–	–	–

Clock jitter is caused by phase noise. The resolution of ADCs with a digitization bandwidth between 1 MHz and 1 GHz is limited by jitter.

When sampling audio signals at 44.1 kHz, the anti-aliasing filter should have eliminated all frequencies above 22 kHz.

The input frequency (in this case, < 22kHz kHz), not the ADC clock frequency, is the determining factor with respect to jitter performance.

Sampling Rate

The analog signal is continuous in time and it is necessary to convert this to a flow of digital values. It is therefore required to define the rate at which new digital values are sampled from the analog signal. The rate of new values is called the *sampling rate* or *sampling frequency* of the converter.

A continuously varying bandlimited signal can be sampled (that is, the signal values at intervals of time T, the sampling time, are measured and stored) and then the original signal can be *exactly* reproduced from the discrete-time values by an interpolation formula. The accuracy is limited by quantization error. However, this faithful reproduction is only possible if the sampling rate is higher than twice the highest frequency of the signal. This is essentially what is embodied in the Shannon-Nyquist sampling theorem.

Since a practical ADC cannot make an instantaneous conversion, the input value must necessarily be held constant during the time that the converter performs a conversion (called the *conversion time*). An input circuit called a sample and hold performs this task—in most cases by using a capacitor to store the analog voltage at the input, and using an electronic switch or gate to disconnect the capacitor from the input. Many ADC integrated circuits include the sample and hold subsystem internally.

Aliasing

An ADC works by sampling the value of the input at discrete intervals in time. Provided that the input is sampled above the Nyquist rate, defined as twice the highest frequency of interest, then all frequencies in the signal can be reconstructed. If frequencies above half the Nyquist rate are sampled, they are incorrectly detected as lower frequencies, a process referred to as aliasing. Aliasing occurs because instantaneously sampling a function at two or fewer times per cycle results in missed cycles, and therefore the appearance of an incorrectly lower frequency. For example, a 2 kHz sine wave being sampled at 1.5 kHz would be reconstructed as a 500 Hz sine wave.

To avoid aliasing, the input to an ADC must be low-pass filtered to remove frequencies above half the sampling rate. This filter is called an anti-aliasing filter, and is essential for a practical ADC system that is applied to analog signals with higher frequency content. In applications where protection against aliasing is essential, oversampling may be used to greatly reduce or even eliminate it.

Although aliasing in most systems is unwanted, it should also be noted that it can be exploited to provide simultaneous down-mixing of a band-limited high frequency signal. The alias is effectively the lower heterodyne of the signal frequency and sampling frequency.

Oversampling

Signals are often sampled at the minimum rate required, for economy, with the result that the quantization noise introduced is white noise spread over the whole pass band of the converter. If a signal is sampled at a rate much higher than the Nyquist frequency and then digitally filtered to limit it to the signal bandwidth there are the following advantages:

- digital filters can have better properties (sharper rolloff, phase) than analogue filters, so a sharper anti-aliasing filter can be realised and then the signal can be downsampled giving a better result
- a 20-bit ADC can be made to act as a 24-bit ADC with 256× oversampling
- thesignal-to-noise ratio due to quantization noise will be higher than if the whole available band had been used. With this technique, it is possible to obtain an effective resolution larger than that provided by the converter alone
- The improvement in SNR is 3 dB (equivalent to 0.5 bits) per octave of oversampling which is not sufficient for many applications. Therefore, oversampling is usually coupled with noise shaping. With noise shaping, the improvement is 6L+3 dB per octave where L is the order of loop filter used for noise shaping. e.g. – a 2nd order loop filter will provide an improvement of 15 dB/octave.

Oversampling is typically used in audio frequency ADCs where the required sampling rate (typically 44.1 or 48kHz) is very low compared to the clock speed of typical transistor circuits (>1 MHz). In this case, by using the extra bandwidth to distribute quantization error onto out of band frequencies, the accuracy of the ADC can be greatly increased at no cost. Furthermore, as any aliased signals are also typically out of band, aliasing can often be completely eliminated using very low cost filters.

Relative Speed and Precision

The speed of an ADC varies by type. The Wilkinson ADC is limited by the clock rate which is processable by current digital circuits. Currently, frequencies up to 300 MHz are possible. For a successive-approximation ADC, the conversion time scales with the logarithm of the resolution, e.g. the number of bits. Thus for high resolution, it is possible that the successive-approximation ADC is faster than the Wilkinson.

However, the time consuming steps in the Wilkinson are digital, while those in the successive-approximation are analog. Since analog is inherently slower than digital, as the resolution increases, the time required also increases. Thus there are competing processes at work. Flash ADCs are certainly the fastest type of the three. The conversion is basically performed in a single parallel step. For an 8-bit unit, conversion takes place in a few tens of nanoseconds.

There is, as expected, somewhat of a tradeoff between speed and precision. Flash ADCs have drifts and uncertainties associated with the comparator levels. This results in poor linearity. For successive-approximation ADCs, poor linearity is also present, but less so than for flash ADCs. Here, non-linearity arises from accumulating errors from the subtraction processes. Wilkinson ADCs have the highest linearity of the three. These have the best differential non-linearity. The other types require channel smoothing to achieve the level of the Wilkinson.

The Sliding Scale Principle

The sliding scale or randomizing method can be employed to greatly improve the linearity of any type of ADC, but especially flash and successive approximation types. For any ADC the mapping from input voltage to digital output value is not exactly a floor or ceiling function as it should be. Under normal conditions, a pulse of a particular amplitude is always converted to a digital value. The problem lies in that the ranges of analog values for the digitized values are not all of the same width, and the differential linearity decreases proportionally with the divergence from the average width.

The sliding scale principle uses an averaging effect to overcome this phenomenon. A random, but known analog voltage is added to the sampled input voltage. It is then converted to digital form, and the equivalent digital amount is subtracted, thus restoring it to its original value. The advantage is that the conversion has taken place at a random point. The statistical distribution of the final levels is decided by a weighted average over a region of the range of the ADC. This in turn desensitizes it to the width of any specific level.

ADC Types

These are the most common ways of implementing an electronic ADC:

- A direct-conversion ADC or flash ADC has a bank of comparators sampling the input signal in parallel, each firing for their decoded voltage range. The comparator bank feeds a logic circuit that generates a code for each voltage range. Direct conversion is very fast, capable of gigahertz sampling rates, but usually has only 8 bits of resolution or fewer, since the number of comparators needed, $2^N - 1$, doubles with each additional bit, requiring a large, expensive circuit. ADCs of this type have a large die size, a high input capacitance, high power dissipation, and are prone to produce glitches at the output (by outputting an out-of-sequence code). Scaling to newer submicrometre technologies does not help as the device mismatch is the dominant design limitation. They are often used for video, wideband communications or other fast signals in optical storage.
- A successive-approximation ADC uses a comparator to successively narrow a range that contains the input voltage. At each successive step, the converter compares the input voltage to the output of an internal digital to analog converter which might represent the midpoint of a selected voltage range. At each step in this process, the approximation is stored in a successive approximation register (SAR). For example, consider an input voltage of 6.3 V and the initial range is 0 to 16 V. For the first step, the input 6.3 V is compared to 8 V (the midpoint of the 0–16 V range). The comparator reports that the input voltage is less than 8 V, so the SAR is updated to narrow the range to 0–8 V. For the second step, the input voltage is compared to 4 V (midpoint of 0–8). The comparator reports the input voltage is above 4 V, so the SAR is updated to reflect the input voltage is in the range 4–8 V. For the third step, the input voltage is compared with 6 V (halfway between 4 V and 8 V); the comparator reports the input voltage is greater than 6 volts, and search range becomes 6–8 V. The steps are continued until the desired resolution is reached.
- A ramp-compare ADC produces a saw-tooth signal that ramps up or down then quickly returns to zero. When the ramp starts, a timer starts counting. When the ramp voltage matches

the input, a comparator fires, and the timer's value is recorded. Timed ramp converters require the least number of transistors. The ramp time is sensitive to temperature because the circuit generating the ramp is often just some simple oscillator. There are two solutions: use a clocked counter driving a DAC and then use the comparator to preserve the counter's value, or calibrate the timed ramp. A special advantage of the ramp-compare system is that comparing a second signal just requires another comparator, and another register to store the voltage value. A very simple (non-linear) ramp-converter can be implemented with a microcontroller and one resistor and capacitor. Vice versa, a filled capacitor can be taken from an integrator, time-to-amplitude converter, phase detector, sample and hold circuit, or peak and hold circuit and discharged. This has the advantage that a slow comparator cannot be disturbed by fast input changes.

- The Wilkinson ADC was designed by D. H. Wilkinson in 1950. The Wilkinson ADC is based on the comparison of an input voltage with that produced by a charging capacitor. The capacitor is allowed to charge until its voltage is equal to the amplitude of the input pulse (a comparator determines when this condition has been reached). Then, the capacitor is allowed to discharge linearly, which produces a ramp voltage. At the point when the capacitor begins to discharge, a gate pulse is initiated. The gate pulse remains on until the capacitor is completely discharged. Thus the duration of the gate pulse is directly proportional to the amplitude of the input pulse. This gate pulse operates a linear gate which receives pulses from a high-frequency oscillator clock. While the gate is open, a discrete number of clock pulses pass through the linear gate and are counted by the address register. The time the linear gate is open is proportional to the amplitude of the input pulse, thus the number of clock pulses recorded in the address register is proportional also. Alternatively, the charging of the capacitor could be monitored, rather than the discharge.
- An integrating ADC (also dual-slope or multi-slope ADC) applies the unknown input voltage to the input of an integrator and allows the voltage to ramp for a fixed time period (the run-up period). Then a known reference voltage of opposite polarity is applied to the integrator and is allowed to ramp until the integrator output returns to zero (the run-down period). The

input voltage is computed as a function of the reference voltage, the constant run-up time period, and the measured run-down time period. The run-down time measurement is usually made in units of the converter's clock, so longer integration times allow for higher resolutions. Likewise, the speed of the converter can be improved by sacrificing resolution. Converters of this type (or variations on the concept) are used in most digital voltmeters for their linearity and flexibility.

- A delta-encoded ADC or counter-ramp has an up-down counter that feeds a digital to analog converter (DAC). The input signal and the DAC both go to a comparator. The comparator controls the counter. The circuit uses negative feedback from the comparator to adjust the counter until the DAC's output is close enough to the input signal. The number is read from the counter. Delta converters have very wide ranges and high resolution, but the conversion time is dependent on the input signal level, though it will always have a guaranteed worst-case. Delta converters are often very good choices to read real-world signals. Most signals from physical systems do not change abruptly. Some converters combine the delta and successive approximation approaches; this works especially well when high frequencies are known to be small in magnitude.
- A pipeline ADC (also called subrangingquantizer) uses two or more steps of subranging. First, a coarse conversion is done. In a second step, the difference to the input signal is determined with a digital to analog converter (DAC). This difference is then converted finer, and the results are combined in a last step. This can be considered a refinement of the successive-approximation ADC wherein the feedback reference signal consists of the interim conversion of a whole range of bits (for example, four bits) rather than just the next-most-significant bit. By combining the merits of the successive approximation and flash ADCs this type is fast, has a high resolution, and only requires a small die size.
- A sigma-delta ADC (also known as a delta-sigma ADC) oversamples the desired signal by a large factor and filters the desired signal band. Generally, a smaller number of bits than required are converted using a Flash ADC after the filter. The resulting signal, along with the error generated by the discrete levels of the Flash, is fed back and subtracted from the input to the filter. This negative feedback has the effect of noise

shaping the error due to the Flash so that it does not appear in the desired signal frequencies. A digital filter (decimation filter) follows the ADC which reduces the sampling rate, filters off unwanted noise signal and increases the resolution of the output (sigma-delta modulation, also called delta-sigma modulation).

- A time-interleaved ADC uses M parallel ADCs where each ADC samples data every M:th cycle of the effective sample clock. The result is that the sample rate is increased M times compared to what each individual ADC can manage. In practice, the individual differences between the M ADCs degrade the overall performance reducing the SFDR. However, technologies exist to correct for these time-interleaving mismatch errors.
- An ADC with intermediate FM stage first uses a voltage-to-frequency converter to convert the desired signal into an oscillating signal with a frequency proportional to the voltage of the desired signal, and then uses a frequency counter to convert that frequency into a digital count proportional to the desired signal voltage. Longer integration times allow for higher resolutions. Likewise, the speed of the converter can be improved by sacrificing resolution. The two parts of the ADC may be widely separated, with the frequency signal passed through an opto-isolator or transmitted wirelessly. Some such ADCs use sine wave or square wave frequency modulation; others use pulse-frequency modulation. Such ADCs were once the most popular way to show a digital display of the status of a remote analog sensor.

There can be other ADCs that use a combination of electronics and other technologies:

- A time-stretch analog-to-digital converter (TS-ADC) digitizes a very wide bandwidth analogsignal, that cannot be digitized by a conventional electronic ADC, by time-stretching the signal prior to digitization. It commonly uses a photonic preprocess or frontend to time-stretch the signal, which effectively slows the signal down in time and compresses its bandwidth. As a result, an electronic backend ADC, that would have been too slow to capture the original signal, can now capture this slowed down signal. For continuous capture of the signal, the frontend also divides the signal into multiple segments in addition to time-stretching. Each segment is individually digitized by a separate electronic ADC. Finally, a digital signal processor rearranges the samples and removes any distortions added by the frontend to yield the binary data that is the digital representation of the original analog signal.

Commercial Analog-to-digital Converters

Commercial ADCs are usually implemented as integrated circuits. Most converters sample with 6 to 24 bits of resolution, and produce fewer than 1 megasample per second. Thermal noise generated by passive components such as resistors masks the measurement when higher resolution is desired. For audio applications and in room temperatures, such noise is usually a little less than 1 μV (microvolt) of white noise. If the MSB corresponds to a standard 2 V of output signal, this translates to a noise-limited performance that is less than 20~21 bits, and obviates the need for any dithering. As of February 2002, Mega- and giga-sample per second converters are available. Mega-sample converters are required in digital video cameras, video capture cards, and TV tuner cards to convert full-speed analog video to digital video files.

Commercial converters usually have ±0.5 to ±1.5 LSB error in their output. In many cases, the most expensive part of an integrated circuit is the pins, because they make the package larger, and each pin has to be connected to the integrated circuit's silicon. To save pins, it is common for slow ADCs to send their data one bit at a time over a serial interface to the computer, with the next bit coming out when a clock signal changes state, say from 0 to 5 V. This saves quite a few pins on the ADC package, and in many cases, does not make the overall design any more complex (even microprocessors which use memory-mapped I/O only need a few bits of a port to implement a serial bus to an ADC).

Commercial ADCs often have several inputs that feed the same converter, usually through an analogmultiplexer. Different models of ADC may include sample and hold circuits, instrumentation amplifiers or differential inputs, where the quantity measured is the difference between two voltages.

Applications

Music recording: Analog-to-digital converters are integral to current music reproduction technology. People produce much music on computers using an analog recording and therefore need analog-to-digital converters to create the pulse-code modulation (PCM) data streams that go onto compact discs and digital music files.

The current crop of analog-to-digital converters utilized in music can sample at rates up to 192 kilohertz. Considerable literature exists on these matters, but commercial considerations often play a significant role. Most high-profile recording studios record in 24-bit/192-176.4 kHz pulse-code modulation (PCM) or in Direct Stream Digital (DSD) formats, and then downsample or decimate the signal for Red-Book CD production (44.1 kHz) or to 48 kHz for commonly used for radio and television broadcast applications.

Digital Signal Processing

People must use ADCs to process, store, or transport virtually any analog signal in digital form. TV tuner cards, for example, use fast video analog-to-digital converters. Slow on-chip 8, 10, 12, or 16 bit analog-to-digital converters are common in microcontrollers. Digital storage oscilloscopes need very fast analog-to-digital converters, also crucial for software defined radio and their new applications.

Scientific Instruments

Digital imaging systems commonly use analog-to-digital converters in digitizingpixels. Some radar systems commonly use analog-to-digital converters to convert signal strength to digital values for subsequent signal processing. Many other in situ and remote sensing systems commonly use analogous technology.

The number of binary bits in the resulting digitized numeric values reflects the resolution, the number of unique discrete levels of quantization (signal processing).

The correspondence between the analog signal and the digital signal depends on the quantization error. The quantization process must occur at an adequate speed, a constraint that may limit the resolution of the digital signal.

Many sensors produce an analog signal; temperature, pressure, pH, light intensity etc. All these signals can be amplified and fed to an ADC to produce a digital number proportional to the input signal.

Electrical Symbol

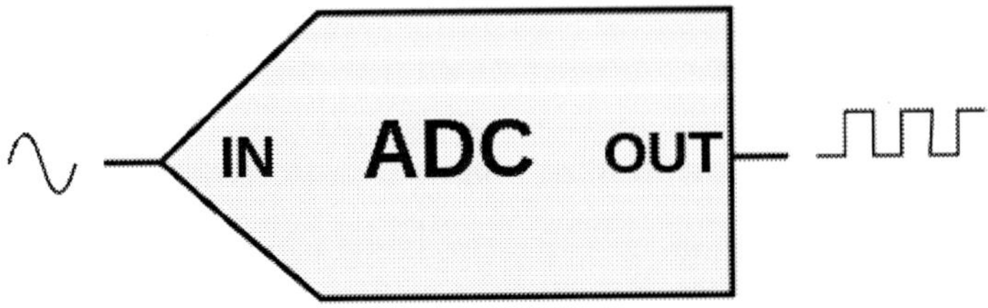

Testing

Testing an Analog to Digital Converter requires an analog input source, hardware to send control signals and capture digital data output. Some ADCs also require an accurate source of reference signal.

The key parameters to test a SAR ADC are following:

1. DC Offset Error
2. DC Gain Error
3. Signal to Noise Ratio (SNR)
4. Total Harmonic Distortion (THD)
5. Integral Non Linearity (INL)
6. Differential Non Linearity (DNL)
7. Spurious Free Dynamic Range
8. Power Dissipation

Digital-to-analog Converter

In electronics, a digital-to-analog converter (DAC or D-to-A) is a device that converts a digital (usually binary) code to an analog signal (current, voltage, or electric charge). An analog-to-digital converter (ADC) performs the reverse operation. Signals are easily stored and transmitted in digital form, but a DAC is needed for the signal to be recognized by human senses or other non-digital systems.

A common use of digital-to-analog converters is generation of audio signals from digital information in music players. Digital video signals are converted to analog in televisions and mobile phones to display colours and shades. Digital-to-analog conversion can degrade a signal, so conversion details are normally chosen so that the errors are negligible. Due to cost and the need for matched components, DACs are almost exclusively manufactured on integrated circuits (ICs). There are many DAC architectures which have different advantages and disadvantages. The suitability of a particular DAC for an application is determined by a variety of measurements including speed and resolution.

Overview

A DAC converts an abstract finite-precision number (usually a fixed-pointbinary number) into a physical quantity (e.g., a voltage or a pressure). In particular, DACs are often used to convert finite-precision time series data to a continually varying physical signal.

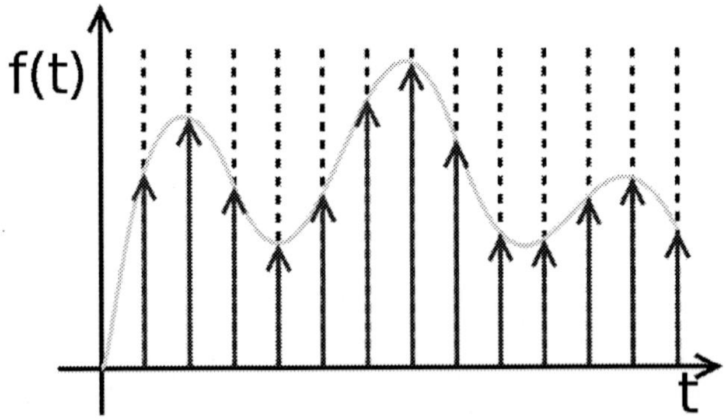

***Figure:** Ideally sampled signal.*

A typical DAC converts the abstract numbers into a concrete sequence of impulses that are then processed by a reconstruction filter using some form of interpolation to fill in data between the impulses. Other DAC methods (e.g., methods based on delta-sigma modulation) produce a pulse-density modulated signal that can then be filtered in a similar way to produce a smoothly varying signal.

As per the Nyquist–Shannon sampling theorem, a DAC can reconstruct the original signal from the sampled data provided that its bandwidth meets certain requirements (e.g., a baseband signal with bandwidth less than the Nyquist frequency). Digital sampling introduces quantization error that manifests as low-level noise added to the reconstructed signal.

Practical Operation

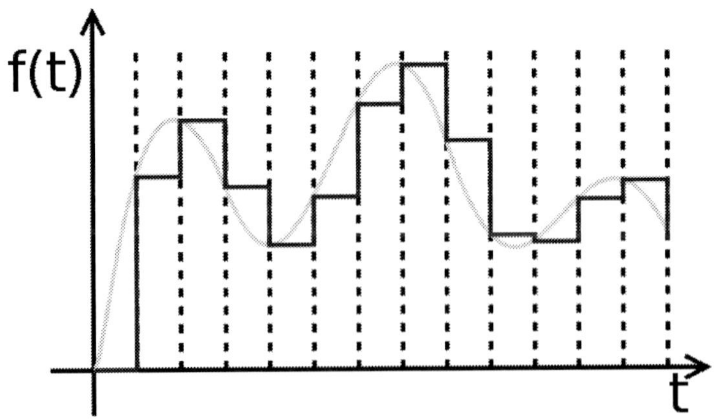

***Figure:** Piecewise constant output of an idealized DAC lacking a reconstruction filter. In a practical DAC, a filter or the finite bandwidth of the device smooths out the step response into a continuous curve.*

Instead of impulses, usually the sequence of numbers update the analog voltage at uniform sampling intervals which are often then interpolated via a reconstruction filter to continuously varied levels.

These numbers are written to the DAC, typically with a clock signal that causes each number to be latched in sequence, at which time the DAC output voltage changes rapidly from the previous value to the value represented by the currently latched number. The effect of this is that the output voltage is *held* in time at the current value until the next input number is latched resulting in a piecewise constant or 'staircase' shaped output. This is equivalent to a zero-order hold operation and has an effect on the frequency response of the reconstructed signal.

The fact that DACs output a sequence of piecewise constant values (known as zero-order hold in sample data textbooks) or rectangular pulses causes multiple harmonics above the Nyquist frequency. Usually, these are removed with a low pass filter acting as a reconstruction filter in applications that require it.

Applications

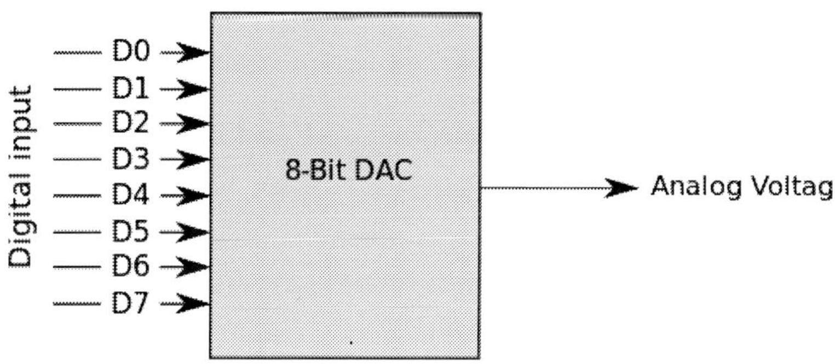

Figure: *A simplified functional diagram of an 8-bit DAC*

Audio

Most modern audio signals are stored in digital form (for example MP3s and CDs) and in order to be heard through speakers they must be converted into an analog signal. DACs are therefore found in CD players, digital music players, and PC sound cards.

Specialist standalone DACs can also be found in high-end hi-fi systems. These normally take the digital output of a compatible CD player or dedicated transport (which is basically a CD player with no internal DAC) and convert the signal into an analogline-level output that can then be fed into an amplifier to drive speakers.

Similar digital-to-analog converters can be found in digital speakers such as USB speakers, and in sound cards.

In VoIP (Voice over IP) applications, the source must first be digitized for transmission, so it undergoes conversion via an Analog-to-Digital Converter, and is then reconstructed into analog using a DAC on the receiving party's end.

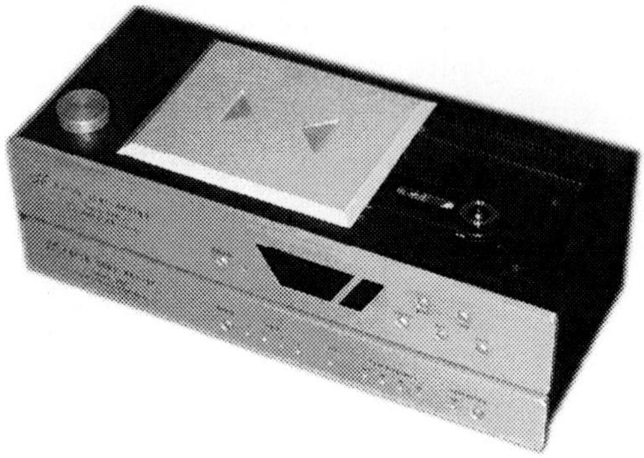

Figure: *Top-loading CD player and external digital-to-analog converter.*

Video

Video sampling tends to work on a completely different scale altogether thanks to the highly nonlinear response both of cathode ray tubes (for which the vast majority of digital video foundation work was targeted) and the human eye, using a "gamma curve" to provide an appearance of evenly distributed brightness steps across the display's full dynamic range - hence the need to use RAMDACs in computer video applications with deep enough colour resolution to make engineering a hardcoded value into the DAC for each output level of each channel impractical (e.g. an Atari ST or Sega Genesis would require 24 such values; a 24-bit video card would need 768...). Given this inherent distortion, it is not unusual for a television or video projector to truthfully claim a linear contrast ratio (difference between darkest and brightest output levels) of 1000:1 or greater, equivalent to 10 bits of audio precision even though it may only accept signals with 8-bit precision and use an LCD panel that only represents 6 or 7 bits per channel.

Video signals from a digital source, such as a computer, must be converted to analog form if they are to be displayed on an analog monitor. As of 2007, analog inputs were more commonly used than digital, but this changed as flat panel displays with DVI and/or HDMI connections became more widespread. A video DAC is, however, incorporated in any digital video player with analog outputs. The DAC is usually integrated with some memory (RAM), which contains conversion tables for gamma correction, contrast and brightness, to make a device called a RAMDAC.

A device that is distantly related to the DAC is the digitally controlled potentiometer, used to control an analog signal digitally.

Mechanical

An unusual application of digital-to-analog conversion was the whiffletree electromechanical digital-to-analog converter linkage in the IBM Selectric typewriter.

DAC Types

The most common types of electronic DACs are:

- The pulse-width modulator, the simplest DAC type. A stable current or voltage is switched into a low-pass analog filter with a duration determined by the digital input code. This technique is often used for electric motor speed control, but has many other applications as well.
- Oversampling DACs or interpolating DACs such as the delta-sigma DAC, use a pulse density conversion technique. The oversampling technique allows for the use of a lower resolution DAC internally. A simple 1-bit DAC is often chosen because the oversampled result is inherently linear. The DAC is driven with a pulse-density modulated signal, created with the use of a low-pass filter, step nonlinearity (the actual 1-bit DAC), and negative feedback loop, in a technique called delta-sigma modulation. This results in an effective high-pass filter acting on the quantization (signal processing) noise, thus steering this noise out of the low frequencies of interest into the megahertz frequencies of little interest, which is called noise shaping. The quantization noise at these high frequencies is removed or greatly attenuated by use of an analog low-pass filter at the output (sometimes a simple RC low-pass circuit is sufficient). Most very high resolution DACs (greater than 16 bits) are of this type due to its high linearity and low cost. Higher oversampling rates can relax the specifications of the output low-pass filter and enable further suppression of quantization noise. Speeds of greater than 100 thousand samples per second (for example, 192 kHz) and resolutions of 24 bits are attainable with delta-sigma DACs. A short comparison with pulse-width modulation shows that a 1-bit

DAC with a simple first-order integrator would have to run at 3 THz (which is physically unrealizable) to achieve 24 meaningful bits of resolution, requiring a higher-order low-pass filter in the noise-shaping loop. A single integrator is a low-pass filter with a frequency response inversely proportional to frequency and using one such integrator in the noise-shaping loop is a first order delta-sigma modulator. Multiple higher order topologies (such as MASH) are used to achieve higher degrees of noise-shaping with a stable topology.

- The binary-weighted DAC, which contains individual electrical components for each bit of the DAC connected to a summing point. These precise voltages or currents sum to the correct output value. This is one of the fastest conversion methods but suffers from poor accuracy because of the high precision required for each individual voltage or current. Such high-precision components are expensive, so this type of converter is usually limited to 8-bit resolution or less.
  - o Switched resistor DAC contains of a parallel resistor network. Individual resistors are enabled or bypassed in the network based on the digital input.
  - o Switched current source DAC, from which different current sources are selected based on the digital input.
  - o Switched capacitor DAC contains a parallel capacitor network. Individual capacitors are connected or disconnected with switches based on the input.
- The R-2R ladder DAC which is a binary-weighted DAC that uses a repeating cascaded structure of resistor values R and 2R. This improves the precision due to the relative ease of producing equal valued-matched resistors (or current sources). However, wide converters perform slowly due to increasingly large RC-constants for each added R-2R link.
- The Successive-Approximation or Cyclic DAC, which successively constructs the output during each cycle. Individual bits of the digital input are processed each cycle until the entire input is accounted for.
- The thermometer-coded DAC, which contains an equal resistor or current-source segment for each possible value of DAC output. An 8-bit thermometer DAC would have 255 segments, and a 16-bit thermometer DAC would have 65,535 segments. This is perhaps the fastest and highest precision DAC

architecture but at the expense of high cost. Conversion speeds of >1 billion samples per second have been reached with this type of DAC.

- Hybrid DACs, which use a combination of the above techniques in a single converter. Most DAC integrated circuits are of this type due to the difficulty of getting low cost, high speed and high precision in one device.
    - o The segmented DAC, which combines the thermometer-coded principle for the most significant bits and the binary-weighted principle for the least significant bits. In this way, a compromise is obtained between precision (by the use of the thermometer-coded principle) and number of resistors or current sources (by the use of the binary-weighted principle). The full binary-weighted design means 0% segmentation, the full thermometer-coded design means 100% segmentation.
- Most DACs, shown earlier in this list, rely on a constant reference voltage to create their output value. Alternatively, a *multiplying DAC* takes a variable input voltage for their conversion. This puts additional design constraints on the bandwidth of the conversion circuit.

DAC Performance

DACs are very important to system performance. The most important characteristics of these devices are:

Resolution: The number of possible output levels the DAC is designed to reproduce. This is usually stated as the number of bits it uses, which is the base two logarithm of the number of levels. For instance a 1 bit DAC is designed to reproduce 2 (2^1) levels while an 8 bit DAC is designed for 256 (2^8) levels. Resolution is related to the effective number of bits which is a measurement of the actual resolution attained by the DAC. Resolution determines colour depth in video applications and audio bit depth in audio applications.

Maximum Sampling Rate: A measurement of the maximum speed at which the DACs circuitry can operate and still produce the correct output. As stated in the Nyquist–Shannon sampling theorem defines a relationship between the sampling frequency and bandwidth of the sampled signal.

Monotonicity: The ability of a DAC's analog output to move only in the direction that the digital input moves (i.e., if the input increases, the output doesn't dip before asserting the correct output.) This characteristic is very important for DACs used as a low frequency signal source or as a digitally programmable trim element.

Total Harmonic Distortion and Noise (THD+N): A measurement of the distortion and noise introduced to the signal by the DAC. It is expressed as a percentage of the total power of unwanted harmonicdistortion and noise that accompany the desired signal. This is a very important DAC characteristic for dynamic and small signal DAC applications.

Dynamic Range: A measurement of the difference between the largest and smallest signals the DAC can reproduce expressed in decibels. This is usually related to resolution and noise floor. Other measurements, such as phase distortion and jitter, can also be very important for some applications, some of which (e.g. wireless data transmission, composite video) may even *rely* on accurate production of phase-adjusted signals.

Linear PCM audio sampling usually works on the basis of each bit of resolution being equivalent to 6 decibels of amplitude (a 2x increase in volume or precision). Non-linear PCM encodings (A-law / μ-law, ADPCM, NICAM) attempt to improve their effective dynamic ranges by a variety of methods - logarithmic step sizes between the output signal strengths represented by each data bit (trading greater quantisation distortion of loud signals for better performance of quiet signals)

DAC Figures of Merit

- Static performance:
  - o Differential nonlinearity (DNL) shows how much two adjacent code analog values deviate from the ideal 1 LSB step.
  - o Integral nonlinearity (INL) shows how much the DAC transfer characteristic deviates from an ideal one. That is, the ideal characteristic is usually a straight line; INL shows how much the actual voltage at a given code value differs from that line, in LSBs (1 LSB steps).
  - o Gain
  - o Offset
  - o Noise is ultimately limited by the thermal noise generated by passive components such as resistors. For audio applications and in room temperatures, such noise is usually

a little less than 1 μV (microvolt) of white noise. This limits performance to less than 20~21 bits even in 24-bit DACs.

- Frequency domain performance
  - Spurious-free dynamic range (SFDR) indicates in dB the ratio between the powers of the converted main signal and the greatest undesired spur.
  - Signal-to-noise and distortion ratio (SNDR) indicates in dB the ratio between the powers of the converted main signal and the sum of the noise and the generated harmonic spurs
  - i-th harmonic distortion (HDi) indicates the power of the i-th harmonic of the converted main signal
  - Total harmonic distortion (THD) is the sum of the powers of all HDi
  - If the maximum DNL error is less than 1 LSB, then the D/A converter is guaranteed to be monotonic. However, many monotonic converters may have a maximum DNL greater than 1 LSB.
- Time domain performance:
  - Glitch impulse area (glitch energy)
  - Response uncertainty
  - Time nonlinearity (TNL)

Chapter

6

Basic Digital Electronics

Binary Number Systems and Codes

In mathematics and computer science, the binary numeral system, or base-2 numeral system, represents numeric values using two symbols: typically 0 and 1. More specifically, the usual base-2 system is a positional notation with a radix of 2. Numbers represented in this system are commonly called binary numbers. Because of its straightforward implementation in digital electronic circuitry using logic gates, the binary system is used internally by almost all modern computers and computer-based devices such as mobile phones.

History

The Indian scholar Pingala (around 5th–2nd centuries BC) developed mathematical concepts for describing prosody, and in doing so presented the first known description of a binary numeral system. He used binary numbers in the form of short and long syllables (the latter equal in length to two short syllables), making it similar to Morse code. Pingala's Hindu classic titled Chandahsâstra (8.23) describes the formation of a matrix in order to give a unique value to each metre. An example of such a matrix is as follows (note that these binary representations are "backwards" compared to modern, Western positional notation):

0 0 0 0 numerical value 1_{10}
1 0 0 0 numerical value 2_{10}
0 1 0 0 numerical value 3_{10}
1 1 0 0 numerical value 4_{10}

A set of eight trigrams (Bagua) and a set of 64 hexagrams ("sixty-four" gua), analogous to the three-bit and six-bit binary numerals, were in usage at least as early as the Zhou Dynasty of ancient China through the classic text *Yijing*.

In the 11th century, scholar and philosopher Shao Yong developed a method for arranging the hexagrams which corresponds, albeit unintentionally, to the sequence 0 to 63, as represented in binary, with yin as 0, yang as 1 and the least significant bit on top. The ordering is also the lexicographical order on sextuples of elements chosen from a two-element set. Similar sets of binary combinations have also been used in traditional African divination systems such as Ifα as well as in medieval Western geomancy. The base-2 system utilized in geomancy had long been widely applied in sub-Saharan Africa.

In 1605 Francis Bacon discussed a system whereby letters of the alphabet could be reduced to sequences of binary digits, which could then be encoded as scarcely visible variations in the font in any random text. Importantly for the general theory of binary encoding, he added that this method could be used with any objects at all: "provided those objects be capable of a twofold difference only; as by Bells, by Trumpets, by Lights and Torches, by the report of Muskets, and any instruments of like nature". The modern binary number system was discovered by Gottfried Leibniz in 1679. Leibniz's system uses 0 and 1, like the modern binary numeral system. As a Sinophile, Leibniz was aware of the Yijing (or I-Ching) and noted with fascination how its hexagrams correspond to the binary numbers from 0 to 111111, and concluded that this mapping was evidence of major Chinese accomplishments in the sort of philosophical mathematics he admired.

In 1854, British mathematician George Boole published a landmark paper detailing an algebraic system of logic that would become known as Boolean algebra. His logical calculus was to become instrumental in the design of digital electronic circuitry.

In 1937, Claude Shannon produced his master's thesis at MIT that implemented Boolean algebra and binary arithmetic using electronic relays and switches for the first time in history. Entitled *A Symbolic Analysis of Relay and Switching Circuits*, Shannon's thesis essentially founded practical digital circuit design.

In November 1937, George Stibitz, then working at Bell Labs, completed a relay-based computer he dubbed the "Model K" (for "Kitchen", where he had assembled it), which calculated using binary addition. Bell Labs thus authorized a full research programme in late 1938 with Stibitz at the helm. Their Complex Number Computer, completed 8 January 1940, was able to calculate complex numbers. In a demonstration to the American Mathematical Society conference at Dartmouth College on 11 September 1940, Stibitz was able to send the Complex Number Calculator remote commands over telephone lines by a teletype. It was the first computing machine ever used remotely over a phone line. Some participants of the conference who witnessed the demonstration were John Von Neumann, John Mauchly and Norbert Wiener, who wrote about it in his memoirs.

Representation

Any number can be represented by any sequence of bits (binary digits), which in turn may be represented by any mechanism capable of being in two mutually exclusive states. The following sequence of symbols could all be interpreted as the binary numeric value of 667:

1 0 1 0 0 1 1 0 1 1

| – | – – | | – | |

x o x o o x x o x x

y n y n n y y n y y

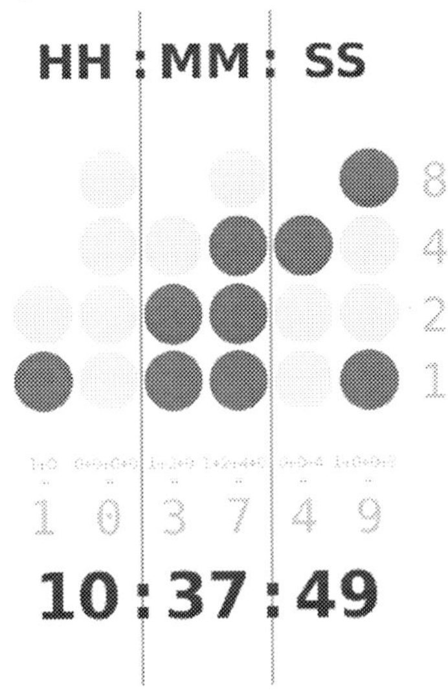

Figure: *A binary clock might use LEDs to express binary values. In this clock, each column of LEDs shows a binary-coded decimal numeral of the traditional sexagesimal time.*

The numeric value represented in each case is dependent upon the value assigned to each symbol. In a computer, the numeric values may be represented by two different voltages; on a magnetic disk, magnetic polarities may be used. A "positive", "yes", or "on" state is not necessarily equivalent to the numerical value of one; it depends on the architecture in use.

In keeping with customary representation of numerals using Arabic numerals, binary numbers are commonly written using the symbols 0 and 1. When written, binary numerals are often subscripted, prefixed or suffixed in order to indicate their base, or radix. The following notations are equivalent:

100101 binary (explicit statement of format)

100101b (a suffix indicating binary format)

100101B (a suffix indicating binary format)

bin 100101 (a prefix indicating binary format)

100101_2 (a subscript indicating base-2 (binary) notation)

%100101 (a prefix indicating binary format)

0b100101 (a prefix indicating binary format, common in programming languages)

6b100101 (a prefix indicating number of bits in binary format, common in programming languages)

When spoken, binary numerals are usually read digit-by-digit, in order to distinguish them from decimal numerals. For example, the binary numeral 100 is pronounced *one zero zero*, rather than *one hundred*, to make its binary nature explicit, and for purposes of correctness. Since the binary numeral 100 represents the value four, it would be confusing to refer to the numeral as *one hundred* (a word that represents a completely different value, or amount). Alternatively, the binary numeral 100 can be read out as "four" (the correct *value*), but this does not make its binary nature explicit.

Counting in Binary

Counting in binary is similar to counting in any other number system. Beginning with a single digit, counting proceeds through each symbol, in increasing order. Before examining binary counting, it is useful to briefly discuss the more familiar decimal counting system as a frame of reference.

Decimal Counting: Decimal counting uses the ten symbols 0 through 9. Counting primarily involves incremental manipulation of the "low-order" digit, or the rightmost digit, often called the "first digit". When the available symbols for the low-order digit are exhausted, the next-higher-order digit (located one position to the left) is incremented, and counting in the low-order digit starts over at 0. In decimal, counting proceeds like so:

000, 001, 002, ... 007, 008, 009, (rightmost digit starts over, and next digit is incremented)

010, 011, 012, ...

...

090, 091, 092, ... 097, 098, 099, (rightmost two digits start over, and next digit is incremented)

100, 101, 102, ...

After a digit reaches 9, an increment resets it to 0 but also causes an increment of the next digit to the left.

Binary Counting

In binary, counting follows similar procedure, except that only the two symbols 0 and 1 are used. Thus, after a digit reaches 1 in binary, an increment resets it to 0 but also causes an increment of the next digit to the left:

0000,

0001, (rightmost digit starts over, and next digit is incremented)

0010, 0011, (rightmost two digits start over, and next digit is incremented)

0100, 0101, 0110, 0111, (rightmost three digits start over, and the next digit is incremented)

1000, 1001, 1010, 1011, 1100, 1101, 1110, 1111 ...

Since binary is a base-2 system, each digit represents an increasing power of 2, with the rightmost digit representing 2^0, the next representing 2^1, then 2^2, and so on. To determine the decimal representation of a binary number simply take the sum of the products of the binary digits and the powers of 2 which they represent. For example, the binary number 100101 is converted to decimal form as follows:

$$100101_2 = [\,(1) \times 2^5\,] + [\,(0) \times 2^4\,] + [\,(0) \times 2^3\,] + [\,(1) \times 2^2\,] + [\,(0) \times 2^1\,] + [\,(1) \times 2^0\,]$$

$$100101_2 = [\,1 \times 32\,] + [\,0 \times 16\,] + [\,0 \times 8\,] + [\,1 \times 4\,] + [\,0 \times 2\,] + [\,1 \times 1\,]$$

$$100101_2 = 37_{10}$$

To create higher numbers, additional digits are simply added to the left side of the binary representation.

Fractions

Fractions in binary only terminate if the denominator has 2 as the only prime factor. As a result, 1/10 does not have a finite binary representation, and this causes 10×0.1 not to be precisely equal to 1 in floating point arithmetic. As an example, to interpret the binary expression for 1/3 = .010101..., this means: $1/3 = 0 \times 2^{-1} + 1 \times 2^{-2} + 0 \times 2^{-3} + 1 \times 2^{-4} + \ldots = 0.3125 + \ldots$ An exact value cannot be found with a sum of a finite number of inverse powers of two, the zeros and ones in the binary representation of 1/3 alternate forever.

Fraction	*Decimal*	*Binary*	*Fractional approximation*
1/1	1 or 0.999...	1 or 0.111...	1/2 + 1/4 + 1/8...
1/2	0.5 or 0.4999...	0.1 or 0.0111...	1/4 + 1/8 + 1/16 . . .
1/3	0.333...	0.010101...	1/4 + 1/16 + 1/64 . . .
1/4	0.25 or 0.24999...	0.01 or 0.00111...	1/8 + 1/16 + 1/32 . . .
1/5	0.2 or 0.1999...	0.00110011...	1/8 + 1/16 + 1/128 . . .
1/6	0.1666...	0.0010101...	1/8 + 1/32 + 1/128 . . .
1/7	0.142857142857...	0.001001...	1/8 + 1/64 + 1/512 . . .
1/8	0.125 or 0.124999...	0.001 or 0.000111...	1/16 + 1/32 + 1/64 . . .
1/9	0.111...	0.000111000111...	1/16 + 1/32 + 1/64 . . .
1/10	0.1 or 0.0999...	0.000110011...	1/16 + 1/32 + 1/256 . . .
1/11	0.090909...	0.00010111010001011101...	1/16 + 1/64 + 1/128 . . .
1/12	0.08333...	0.00010101...	1/16 + 1/64 + 1/256 . . .
1/13	0.076923076923...	0.000100111011000100111011...	1/16 + 1/128 + 1/256 . . .
1/14	0.0714285714285...	0.0001001001...	1/16 + 1/128 + 1/1024 . . .
1/15	0.0666...	0.00010001...	1/16 + 1/256 . . .
1/16	0.0625 or 0.0624999...	0.0001 or 0.0000111...	1/32 + 1/64 + 1/128 . . .

Binary Arithmetic

Arithmetic in binary is much like arithmetic in other numeral systems. Addition, subtraction, multiplication, and division can be performed on binary numerals.

Addition:

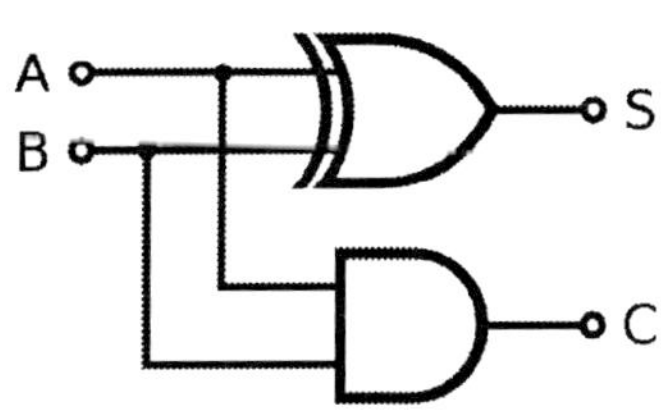

Figure: *The circuit diagram for a binary half adder, which adds two bits together, producing sum and carry bits.*

The simplest arithmetic operation in binary is addition. Adding two single-digit binary numbers is relatively simple, using a form of carrying:

$$0 + 0 \rightarrow 0$$

$$0 + 1 \rightarrow 1$$

$$1 + 0 \rightarrow 1$$

$$1 + 1 \rightarrow 0, \text{ carry } 1 \text{ (since } 1 + 1 = 0 + 1 \times \text{binary } 10)$$

Adding two "1" digits produces a digit "0", while 1 will have to be added to the next column. This is similar to what happens in decimal when certain single-digit numbers are added together; if the result equals or exceeds the value of the radix (10), the digit to the left is incremented:

$$5 + 5 \rightarrow 0, \text{ carry } 1 \text{ (since } 5 + 5 = 10 \text{ carry } 1)$$

$$7 + 9 \rightarrow 6, \text{ carry } 1 \text{ (since } 7 + 9 = 16 \text{ carry } 1)$$

This is known as *carrying*. When the result of an addition exceeds the value of a digit, the procedure is to "carry" the excess amount divided by the radix (that is, 10/10) to the left, adding it to the next positional value. This is correct since the next position has a weight that is higher by a factor equal to the radix. Carrying works the same way in binary:

```
 1 1 1 1 1    (carried digits)
   0 1 1 0 1
+  1 0 1 1 1
 ___________
= 1 0 0 1 0 0 = 36
```

In this example, two numerals are being added together: 01101_2 (13_{10}) and 10111_2 (23_{10}). The top row shows the carry bits used. Starting in the rightmost column, $1 + 1 = 10_2$. The 1 is carried to the left, and the 0 is written at the bottom of the rightmost column. The second column from the right is added: $1 + 0 + 1 = 10_2$ again; the 1 is carried, and 0 is written at the bottom. The third column: $1 + 1 + 1 = 11_2$. This time, a 1 is carried, and a 1 is written in the bottom row. Proceeding like this gives the final answer 100100_2 (36 decimal).

When computers must add two numbers, the rule that: x xor y = (x + y) mod 2 for any two bits x and y allows for very fast calculation, as well.

Long Carry Method

A simplification for many binary addition problems is the Long Carry Method or Brookhouse Method of Binary Addition. This method is generally useful in any binary addition where one of the numbers contains a long "string" of ones. It is based on the simple premise that under the binary system, when given a "string" of digits composed entirely of n ones (*where: n* is any integer length), adding 1 will result in the number 1 followed by a string of n zeros. That concept follows, logically, just as in the decimal system, where adding 1 to a string of n 9's will result in the number 1 followed by a string of n 0's:

Binary		Decimal
1 1 1 1 1	likewise	9 9 9 9 9
+ 1		+ 1
1 0 0 0 0 0		1 0 0 0 0 0

Such long strings are quite common in the binary system. From that one finds that large binary numbers can be added using two simple steps, without excessive carry operations. In the following example, two numerals are being added together: 1 1 1 0 1 1 1 1 1 0_2 (958_{10}) and 1 0 1 0 1 1 0 0 1 1_2 (691_{10}), using the traditional carry method on the left, and the long carry method on the right:

Traditional Carry Method	vs.	Long Carry Method
1 1 1 1 1 1 1 1	(carried digits)	1 ← 1 ←
1 1 1 0 1 1 1 1 1 0		~~1 1 1~~ 0 ~~1 1 1 1 1~~ 0
+ 1 0 1 0 1 1 0 0 1 1		+ 1 0 ~~1~~ 0 1 1 0 0 ~~1~~ 1
= 1 1 0 0 1 1 1 0 0 0 1		1 1 0 0 1 1 1 0 0 0 1

carry the 1 until it is one digit past the "string" below cross out the "string", and cross out the digit that was added to it

The top row shows the carry bits used. Instead of the standard carry from one column to the next, the lowest-ordered "1" with a "1" in the corresponding place value beneath it may be added and a "1" may be carried to one digit past the end of the series. The "used" numbers must be crossed off, since they are already added. Other long strings may likewise be cancelled using the same technique. Then, simply add together any remaining digits normally. Proceeding in this manner gives the final answer of 1 1 0 0 1 1 1 0 0 0 1_2 (1649_{10}). In our simple example using small numbers, the traditional carry method required eight carry operations, yet the long carry method required only two, representing a substantial reduction of effort.

Addition Table

	0	*1*
0	0	1
1	1	10

The binary addition table is similar, but not the same, as the truth table of the logical disjunction operation *V*. The difference is that $1 \vee 1 = 1$, while $1+1=10$.

Subtraction

Subtraction works in much the same way:

$0 - 0 \rightarrow 0$

$0 - 1 \rightarrow 1$, borrow 1

$1 - 0 \rightarrow 1$

$1 - 1 \rightarrow 0$

Subtracting a "1" digit from a "0" digit produces the digit "1", while 1 will have to be subtracted from the next column. This is known as *borrowing*. The principle is the same as for carrying. When the result of a subtraction is less than 0, the least possible value of a digit, the procedure is to "borrow" the deficit divided by the radix (that is, 10/10) from the left, subtracting it from the next positional value.

```
 *   * * *   (starred columns are borrowed from)
   1 1 0 1 1 1 0
 −     1 0 1 1 1
 ---------------
 = 1 0 1 0 1 1 1
```

Subtracting a positive number is equivalent to *adding* a negative number of equal absolute value; computers typically use two's complement notation to represent negative values. This notation eliminates the need for a separate "subtract" operation. Using two's complement notation subtraction can be summarized by the following formula:

A – B = A + not B + 1

Multiplication

Multiplication in binary is similar to its decimal counterpart. Two numbers A and B can be multiplied by partial products: for each digit in B, the product of that digit in A is calculated and written on a new line, shifted leftward so that its rightmost digit lines up with the digit in B that was used. The sum of all these partial products gives the final result.

Since there are only two digits in binary, there are only two possible outcomes of each partial multiplication:

- If the digit in B is 0, the partial product is also 0
- If the digit in B is 1, the partial product is equal to A

For example, the binary numbers 1011 and 1010 are multiplied as follows:

```
        1 0 1 1   (A)
      × 1 0 1 0   (B)
      ---------
        0 0 0 0   ← Corresponds to the rightmost 'zero' in B
  +   1 0 1 1     ← Corresponds to the next 'one' in B
  + 0 0 0 0
  + 1 0 1 1
  -------------
  = 1 1 0 1 1 1 0
```

Binary numbers can also be multiplied with bits after a binary point:

```
              1 0 1 . 1 0 1    A (5.625 in decimal)
            × 1 1 0 . 0 1      B (6.25  in decimal)
            -------------
                  1 . 0 1 1 0 1   ← Corresponds to a 'one' in B
  +              0 0 . 0 0 0 0    ← Corresponds to a 'zero' in B
  +             0 0 0 . 0 0 0
  +            1 0 1 1 . 0 1
  +           1 0 1 1 0 . 1
  ---------------------------
  =       1 0 0 0 1 1 . 0 0 1 0 1   (35.15625 in decimal)
```

Multiplication Table

	0	1
0	0	0
1	0	1

The binary multiplication table is the same as the Truth table of the Logical conjunction operation $\wedge$.

Division

Binary Division is Again Similar to its Decimal Counterpart: Here, the divisor is 101_2, or 5 decimal, while the dividend is 11011_2, or 27 decimal. The procedure is the same as that of decimal long division; here, the divisor 101_2 goes into the first three digits 110_2 of the dividend one time, so a "1" is written on the top line. This result is multiplied by the divisor, and subtracted from the first three digits of the dividend; the next digit (a "1") is included to obtain a new three-digit sequence:

```
              1
          __________
 1 0 1   ) 1 1 0 1 1
         - 1 0 1
           -------
             0 1 1
```

The procedure is then repeated with the new sequence, continuing until the digits in the dividend have been exhausted:

```
            1 0 1
          _________
 1 0 1   ) 1 1 0 1 1
         - 1 0 1
          -------
            0 1 1
          - 0 0 0
            -----
              1 1 1
            - 1 0 1
              -----
                1 0
```

Thus, the quotient of 11011_2 divided by 101_2 is 101_2, as shown on the top line, while the remainder, shown on the bottom line, is 10_2. In decimal, 27 divided by 5 is 5, with a remainder of 2.

Square Root

Binary square root is similar to its decimal counterpart too. But, it's simpler than that in decimal.

$$(10x+y)^{10}-100x^{10}=y^{10}+100xy=\begin{cases}0, & y=0\\ 100x+1, & y=1\end{cases}$$

for example

```
          1 0 0 1
        ---------
       √  1010001
            1
        ---------
101         01
             0
        ---------
1001        100
              0
        ---------
10001     10001
          10001
          -------
              0
```

Bitwise Operations

Though not directly related to the numerical interpretation of binary symbols, sequences of bits may be manipulated using Boolean logical operators. When a string of binary symbols is manipulated in this way, it is called a bitwise operation; the logical operators AND, OR, and XOR may be performed on corresponding bits in two binary numerals provided as input. The logical NOT operation may be performed on individual bits in a single binary numeral provided as input. Sometimes, such operations may be used as arithmetic short-cuts, and may have other computational benefits as well. For example, an arithmetic shift left of a binary number is the equivalent of multiplication by a (positive, integral) power of 2.

Conversion to and From Other Numeral Systems

Decimal: To convert from a base-10 integer numeral to its base-2 (binary) equivalent, the number is divided by two, and the remainder is the least-significant bit. The (integer) result is again divided by two, its remainder is the next least significant bit. This process repeats until the quotient becomes zero.

Conversion from base-2 to base-10 proceeds by applying the preceding algorithm, so to speak, in reverse. The bits of the binary number are used one by one, starting with the most significant (leftmost) bit. Beginning with the value 0, repeatedly double the prior value and add the next bit to produce the next value. This can be organized in a multi-column table. For example to convert 10010101101_2 to decimal:

Prior value	× 2 +	Next bit	Next value
0	× 2 +	**1**	= 1
1	× 2 +	**0**	= 2
2	× 2 +	**0**	= 4
4	× 2 +	**1**	= 9
9	× 2 +	**0**	= 18
18	× 2 +	**1**	= 37
37	× 2 +	**0**	= 74
74	× 2 +	**1**	= 149
149	× 2 +	**1**	= 299
299	× 2 +	**0**	= 598
598	× 2 +	**1**	= **1197**

The result is 1197_{10}. Note that the first Prior Value of 0 is simply an initial decimal value. This method is an application of the Horner scheme.

Binary	1	0	0	1	0	1	0	1	1	0	1	
Decimal	1×2^{10} +	0×2^9 +	0×2^8 +	1×2^7 +	0×2^6 +	1×2^5 +	0×2^4 +	1×2^3 +	1×2^2 +	0×2^1 +	1×2^0 =	1197

The fractional parts of a number are converted with similar methods. They are again based on the equivalence of shifting with doubling or halving.

In a fractional binary number such as 0.11010110101_2, the first digit is $\frac{1}{2}$, the second $(\frac{1}{2})^2 = \frac{1}{4}$, etc. So if there is a 1 in the first place after the decimal, then the number is at least , and vice versa. Double that number is at least 1. This suggests the algorithm: Repeatedly double the number to be converted, record if the result is at least $\frac{1}{2}$, and then throw away the integer part.

For example, $(\frac{1}{3})_{10}$, in binary, is:

Converting	Result
$\frac{1}{3}$	0.
$\frac{1}{3} \times 2 = \frac{2}{3} < 1$	0.0
$\frac{2}{3} \times 2 = 1\frac{1}{3} \geq 1$	0.01
$\frac{1}{3} \times 2 = \frac{2}{3} < 1$	0.010
$\frac{2}{3} \times 2 = 1\frac{1}{3} \geq 1$	0.0101

Thus the repeating decimal fraction 0.3... is equivalent to the repeating binary fraction 0.01... .

Or for example, 0.1_{10}, in binary, is:

Converting	Result
0.1	0.
$0.1 \times 2 = \mathbf{0.2} < 1$	0.0
$0.2 \times 2 = \mathbf{0.4} < 1$	0.00
$0.4 \times 2 = \mathbf{0.8} < 1$	0.000
$0.8 \times 2 = \mathbf{1.6} \geq 1$	0.0001
$0.6 \times 2 = \mathbf{1.2} \geq 1$	0.00011
$0.2 \times 2 = \mathbf{0.4} < 1$	0.000110
$0.4 \times 2 = \mathbf{0.8} < 1$	0.0001100
$0.8 \times 2 = \mathbf{1.6} \geq 1$	0.00011001
$0.6 \times 2 = \mathbf{1.2} \geq 1$	0.000110011
$0.2 \times 2 = \mathbf{0.4} < 1$	0.0001100110

This is also a repeating binary fraction 0.00011... . It may come as a surprise that terminating decimal fractions can have repeating expansions in binary.

It is for this reason that many are surprised to discover that 0.1 + ... + 0.1, (10 additions) differs from 1 in floating point arithmetic. In fact, the only binary fractions with terminating expansions are of the form of an integer divided by a power of 2, which 1/10 is not.

The final conversion is from binary to decimal fractions. The only difficulty arises with repeating fractions, but otherwise the method is to shift the fraction to an integer, convert it, and then divide by the appropriate power of two in the decimal base. For example:

$$\begin{aligned}
x &= 1100 && .1\overline{01110}\ldots \\
x\times 2^6 &= 1100101110 && .\overline{01110}\ldots \\
x\times 2 &= 11001 && .\overline{01110}\ldots \\
x\times(2^6-2) &= 1100010101 \\
x &= 1100010101/111110 \\
x &= (789/62)_{10}
\end{aligned}$$

Another way of converting from binary to decimal, often quicker for a person familiar with hexadecimal, is to do so indirectly—first converting (x in binary) into (x in hexadecimal) and then converting (x in hexadecimal) into (x in decimal).

For very large numbers, these simple methods are inefficient because they perform a large number of multiplications or divisions where one operand is very large.

A simple divide-and-conquer algorithm is more effective asymptotically: given a binary number, it is divided by 10^k, where k is chosen so that the quotient roughly equals the remainder; then each of these pieces is converted to decimal and the two are concatenated.

Given a decimal number, it can be split into two pieces of about the same size, each of which is converted to binary, whereupon the first converted piece is multiplied by 10^k and added to the second converted piece, where k is the number of decimal digits in the second, least-significant piece before conversion.

Hexadecimal

Binary may be converted to and from hexadecimal somewhat more easily. This is because the radix of the hexadecimal system (16) is a power of the radix of the binary system (2). More specifically, 16 = 2^4, so it takes four digits of binary to represent one digit of hexadecimal.

To convert a hexadecimal number into its binary equivalent, simply substitute the corresponding binary digits:

$$3A_{16} = 0011\ 1010_2$$

$$E7_{16} = 1110\ 0111_2$$

To convert a binary number into its hexadecimal equivalent, divide it into groups of four bits. If the number of bits isn't a multiple of four, simply insert extra 0 bits at the left (called padding). For example:

$1010010_2 = 0101\ 0010$ grouped with padding $= 52_{16}$

$11011101_2 = 1101\ 1101$ grouped $= DD_{16}$

To convert a hexadecimal number into its decimal equivalent, multiply the decimal equivalent of each hexadecimal digit by the corresponding power of 16 and add the resulting values:

$C0E7_{16} = (12 \times 16^3) + (0 \times 16^2) + (14 \times 16^1) + (7 \times 16^0) - (12 \times 4096) + (0 \times 256) + (14 \times 16) + (7 \times 1) = 49{,}383_{10}$

Octal

Binary is also easily converted to the octal numeral system, since octal uses a radix of 8, which is a power of two (namely, 2^3, so it takes exactly three binary digits to represent an octal digit). The correspondence between octal and binary numerals is the same as for the first eight digits of hexadecimal. Binary 000 is equivalent to the octal digit 0, binary 111 is equivalent to octal 7, and so forth.

Octal	*Binary*
0	000
1	001
2	010
3	011
4	100
5	101
6	110
7	111

Converting from octal to binary proceeds in the same fashion as it does for hexadecimal:

$65_8 = 110\ 101_2$

$17_8 = 001\ 111_2$

And from binary to octal:

$101100_2 = 101\ 100_2$ grouped $= 54_8$

$10011_2 = 010\ 011_2$ grouped with padding $= 23_8$

And from octal to decimal:

$65_8 = (6 \times 8^1) + (5 \times 8^0) = (6 \times 8) + (5 \times 1) = 53_{10}$

$127_8 = (1 \times 8^2) + (2 \times 8^1) + (7 \times 8^0) = (1 \times 64) + (2 \times 8) + (7 \times 1) = 87_{10}$

Representing Real Numbers

Non-integers can be represented by using negative powers, which are set off from the other digits by means of a radix point (called a decimal point in the decimal system). For example, the binary number 11.01_2 thus means:

$1 \times 2^1 (1 \times 2 = 2)$ plus

$1 \times 2^0 (1 \times 1 = 1)$ plus

0×2^{-1} $(0 \times ½ = 0)$ plus

1×2^{-2} $(1 \times ¼ = 0.25)$

For a total of 3.25 decimal.

All dyadic rational numbers $\frac{p}{2^a}$ have a *terminating* binary numeral—the binary representation has a finite number of terms after the radix point. Other rational numbers have binary representation, but instead of terminating, they *recur*, with a finite sequence of digits repeating indefinitely. For instance

$$\frac{1_{10}}{3_{10}} = \frac{1_2}{11_2} = 0.0101010101..._2$$

$$\frac{12_{10}}{17_{10}} = \frac{1100_2}{10001_2} = 0.10110100\ 10110100\ 10110100..._2$$

The phenomenon that the binary representation of any rational is either terminating or recurring also occurs in other radix-based numeral systems. Another similarity is the existence of alternative representations for any terminating representation, relying on the fact that 0.111111... is the sum of the geometric series $2^{-1} + 2^{-2} + 2^{-3} + ...$ which is 1.

Binary numerals which neither terminate nor recur represent irrational numbers. For instance,

- 0.10100100010000100000100... does have a pattern, but it is not a fixed-length recurring pattern, so the number is irrational
- 1.0110101000001001111001100110011111110... is the binary representation of $\sqrt{2}$, the square root of 2, another irrational. It has no discernible pattern.

Basic Logic Gates

A logic gate is an idealized or physical device implementing a Boolean function, that is, it performs a logical operation on one or more logical inputs, and produces a single logical output. Depending on the context, the term may refer to an ideal logic gate, one that has for instance zero rise time and unlimited fan-out, or it may refer to a non-ideal physical device.

Logic gates are primarily implemented using diodes or transistors acting as electronic switches, but can also be constructed using electromagnetic relays (relay logic), fluidic logic, pneumatic logic, optics, molecules, or even mechanical elements. With amplification, logic gates can be cascaded in the same way that Boolean functions can be composed, allowing the construction of a physical model of all of Boolean logic, and therefore, all of the algorithms and mathematics that can be described with Boolean logic.

Logic circuits include such devices as multiplexers, registers, arithmetic logic units (ALUs), and computer memory, all the way up through complete microprocessors, which may contain more than 100 million gates. In practice, the gates are made from field-effect transistors (FETs), particularly MOSFETs (metal–oxide–semiconductor field-effect transistors).

Compound logic gates AND-OR-Invert (AOI) and OR-AND-Invert (OAI) are often employed in circuit design because their construction using MOSFETs is simpler and more efficient than the sum of the individual gates.

In reversible logic, Toffoli gates are used.

Electronic Gates

To build a functionally complete logic system, relays, valves (vacuum tubes), or transistors can be used. The simplest family of logic gates using bipolar transistors is called resistor-transistor logic (RTL). Unlike diode logic gates, RTL gates can be cascaded indefinitely to produce more complex logic functions. These gates were used in early integrated circuits. For higher speed, the resistors used in RTL were replaced by diodes, leading to diode-transistor logic (DTL). Transistor-transistor logic (TTL) then supplanted DTL with the observation that one transistor could do the job of two diodes even more quickly, using only half the space. In virtually every type of contemporary chip implementation of digital systems, the bipolar transistors have been replaced by complementary field-effect transistors (MOSFETs) to reduce size and power consumption still further, thereby resulting in complementary metal–oxide–semiconductor (CMOS) logic.

For small-scale logic, designers now use prefabricated logic gates from families of devices such as the TTL 7400 series by Texas Instruments and the CMOS 4000 series by RCA, and their more recent descendants. Increasingly, these fixed-function logic gates are being replaced by programmable logic devices, which allow designers to pack a large number of mixed logic gates into a single integrated circuit. The field-programmable nature of programmable logic devices such as FPGAs has removed the 'hard' property of hardware; it is now possible to change the logic design of a hardware system by reprogramming some of its components, thus allowing the features or function of a hardware implementation of a logic system to be changed.

Electronic logic gates differ significantly from their relay-and-switch equivalents. They are much faster, consume much less power, and are much smaller (all by a factor of a million or more in most cases). Also, there is a fundamental structural difference. The switch circuit creates a continuous metallic path for current to flow (in either direction) between its input & its output. The semiconductor logic gate, on the other hand, acts as a high-gain voltage amplifier, which sinks a tiny current at its input and produces a low-impedance voltage at its output. It is not possible for current to flow between the output and the input of a semiconductor logic gate.

Another important advantage of standardized integrated circuit logic families, such as the 7400 and 4000 families, is that they can be cascaded. This means that the output of one gate can be wired to the inputs of one or several other gates, and so on. Systems with varying degrees of complexity can be built without great concern of the designer for the internal workings of the gates, provided the limitations of each integrated circuit are considered.

The output of one gate can only drive a finite number of inputs to other gates, a number called the 'fanout limit'. Also, there is always a delay, called the 'propagation delay', from a change in input of a gate to the corresponding change in its output. When gates are cascaded, the total propagation delay is approximately the sum of the individual delays, an effect which can become a problem in high-speed circuits. Additional delay can be caused when a large number of inputs are connected to an output, due to the distributed capacitance of all the inputs and wiring and the finite amount of current that each output can provide.

Universal Logic Gates

Figure: *The 7400 chip, containing four NANDs. The two additional pins supply power (+5 V) and connect the ground.*

Charles Sanders Peirce (winter of 1880–81) showed that NOR gates alone (or alternatively NAND gates alone) can be used to reproduce the functions of all the other logic gates, but his work on it was unpublished until 1933.

The first published proof was by Henry M. Sheffer in 1913, so the NAND logical operation is sometimes called Sheffer stroke; the logical NOR is sometimes called *Peirce's arrow*. Consequently, these gates are sometimes called *universal logic gates*.

De Morgan Equivalent Symbols

By use of De Morgan's theorem, an *AND* function is identical to an *OR* function with negated inputs and outputs. Likewise, an *OR* function is identical to an *AND* function with negated inputs and outputs. A NAND gate is equivalent to a OR gate with negated inputs, and a NOR gate is equivalent to a AND gate with negated inputs.

This leads to an alternative set of symbols for basic gates that use the opposite core symbol (*AND* or *OR*) but with the inputs and outputs negated . Use of these alternative symbols can make logic circuit diagrams much clearer and help to show accidental connection of an active high output to an active low input or vice-versa. Any connection that has logic negations at both ends can be replaced by a negationless connection and a suitable change of gate or vice-versa. Any connection that has a negation at one end and no negation at the other can be made easier to interpret by instead using the De Morgan equivalent symbol at either of the two ends. When negation or polarity indicators on both ends of a connection match, there is no logic negation in that path (effectively, bubbles "cancel"), making it easier to follow logic states from one symbol to the next. This is commonly seen in real logic diagrams - thus the reader must not get into the habit of associating the shapes exclusively as OR or AND shapes, but also take into account the bubbles at both inputs and outputs in order to determine the "true" logic function indicated.

De Morgan's theorem is most commonly used to implement logic gates as combinations of only NAND gates, or as combinations of only NOR gates, for economic reasons.

Data Storage

Logic gates can also be used to store data. A storage element can be constructed by connecting several gates in a "latch" circuit. More complicated designs that use clock signals and that change only on a rising or falling edge of the clock are called edge-triggered "flip-flops". The combination of multiple flip-flops in parallel, to store a multiple-bit value, is known as a register. When using any of these gate setups the overall system has memory; it is then called a sequential logic system since its output can be influenced by its previous state(s).

These logic circuits are known as computer memory. They vary in performance, based on factors of speed, complexity, and reliability of storage, and many different types of designs are used based on the application.

Three-state Logic Gates

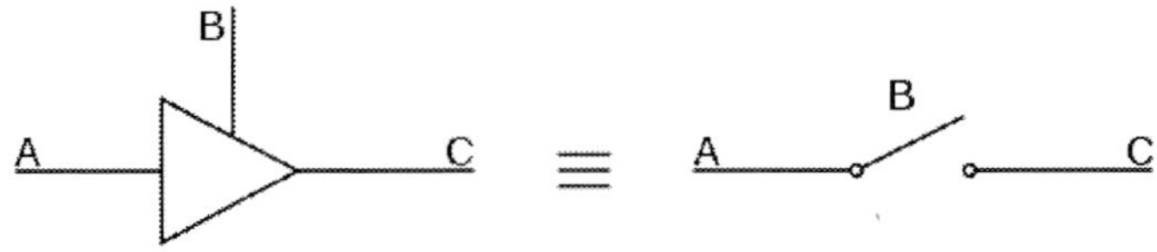

Figure: *A tristate buffer can be thought of as a switch. If* B *is on, the switch is closed. If B is off, the switch is open.*

A three-state logic gate is a type of logic gate that can have three different outputs: high (H), low (L) and high-impedance (Z). The high-impedance state plays no role in the logic, which is strictly binary. These devices are used on buses of the CPU to allow multiple chips to send data. A group of three-states driving a line with a suitable control circuit is basically equivalent to a multiplexer, which may be physically distributed over separate devices or plug-in cards.

In electronics, a high output would mean the output is sourcing current from the positive power terminal (positive voltage). A low output would mean the output is sinking current to the negative power terminal (zero voltage). High impedance would mean that the output is effectively disconnected from the circuit.

History and Development

In a 1886 letter, Charles Sanders Peirce described how logical operations could be carried out by electrical switching circuits. Eventually, vacuum tubes replaced relays for logic operations. Lee De Forest's modification, in 1907, of the Fleming valve can be used as AND logic gate. Ludwig Wittgenstein introduced a version of the 16-row truth table as proposition 5.101 of *Tractatus Logico-Philosophicus* (1921). Claude E. Shannon introduced the use of Boolean algebra in the analysis and design of switching circuits in 1937. Walther Bothe, inventor of the coincidence circuit, got part of the 1954 Nobel Prize in physics, for the first modern electronic AND gate in 1924. Active research is taking place in molecular logic gates.

Implementations

Since the 1990s, most logic gates are made in CMOS technology (i.e. NMOS and PMOS transistors are used). Often millions of logic gates are packaged in a single integrated circuit.

There are several logic families with different characteristics (power consumption, speed, cost, size) such as: RDL (resistor-diode logic), RTL (resistor-transistor logic), DTL (diode-transistor logic),

TTL (transistor-transistor logic) and CMOS (complementary metal oxide semiconductor).

There are also sub-variants, e.g. standard CMOS logic vs. advanced types using still CMOS technology, but with some optimizations for avoiding loss of speed due to slower PMOS transistors.

Non-electronic implementations are varied, though few of them are used in practical applications.

Many early electromechanical digital computers, such as the Harvard Mark I, were built from relay logic gates, using electro-mechanical relays. Logic gates can be made using pneumatic devices, such as the Sorteberg relay or mechanical logic gates, including on a molecular scale.

Logic gates have been made out of DNA and used to create a computer called MAYA. Logic gates can be made from quantum mechanical effects (though quantum computing usually diverges from boolean design). Photonic logic gates use non-linear optical effects.

Truth Tables

A truth table is a mathematical table used in logic—specifically in connection with Boolean algebra, boolean functions, and propositional calculus—to compute the functional values of logical expressions on each of their functional arguments, that is, on each combination of values taken by their logical variables (Enderton, 2001). In particular, truth tables can be used to tell whether a propositional expression is true for all legitimate input values, that is, logically valid.

Practically, a truth table is composed of one column for each input variable (for example, A and B), and one final column for all of the possible results of the logical operation that the table is meant to represent (for example, A XOR B).

Each row of the truth table therefore contains one possible configuration of the input variables (for instance, A=true B=false), and the result of the operation for those values. Ludwig Wittgenstein is often credited with their invention in the Tractatus Logico-Philosophicus.

Unary Operations

Logical Identity: Logical identity is an operation on one logical value, typically the value of a proposition, that produces a value of *true* if its operand is true and a value of *false* if its operand is false.

The truth table for the logical identity operator is as follows:

Logical	*Identity*
p	p
Operand	Value
T	T
F	F

Logical Negation

Logical negation is an operation on one logical value, typically the value of a proposition, that produces a value of *true* if its operand is false and a value of *false* if its operand is true.

The truth table for NOT p (also written as ¬p, Np, Fpq, or ~p) is as follows:

Logical	*Negation*
p	¬p
T	F
F	T

Binary Operations

Truth table for all binary logical operators:

Here is a truth table giving definitions of all 16 of the possible truth functions of 2 binary variables (P,Q are thus boolean variables):

P	*Q*		F 0	N OR 1	X q2	¬ p3	? 4	¬ q5	XO R6	NA ND7		A ND 8	XN OR9	q 10	if/th en11	p 12	the n/if1 3	O R1 4	T 15
T	T		F	F	F	F	F	F	F	F		T	T	T	T	T	T	T	T
T	F		F	F	F	F	T	T	T	T		F	F	F	F	T	T	T	T
F	T		F	F	T	T	F	F	T	T		F	F	T	T	F	F	T	T
F	F		F	T	F	T	F	T	F	T		F	T	F	T	F	T	F	T

where T = true and F = false.

Key:

		Operation	**name**	
0	O*pq*	xand	false	Contradiction
1	X*pq*	NOR	↓	Logical NOR
2	M*pq*	Xq		Converse nonimplication

Contd...

				Operation name
3	F*pq*	N*p*	¬p	Negation
4	L*pq*	Xp	↛	Material nonimplication
5	G*pq*	N*q*	¬q	Negation
6	J*pq*	XOR	⊕	Exclusive disjunction
7	D*pq*	NAND	↑	Logical NAND
8	K*pq*	AND	∧	Logical conjunction
9	E*pq*	XNOR	If and only if	Logical biconditional
10	H*pq*	q		Projection function
11	C*pq*	XNp	if/then	Material implication
12	I*pq*	p		Projection function
13	B*pq*	XNq	then/if	Converse implication
14	A*pq*	OR	∨	Logical disjunction
15	V*pq*	xnand	true	Tautology

Logical operators can also be visualized using Venn diagrams.

Logical Conjunction

Logical conjunction is an operation on two logical values, typically the values of two propositions, that produces a value of *true* if both of its operands are true. The truth table for p AND q (also written as $p \wedge q$, Kpq, p & q, or p .q) is as follows:

Logical Conjunction		
p	*q*	$p \wedge q$
T	T	T
T	F	F
F	T	F
F	F	F

In ordinary language terms, if both *p* and *q* are true, then the conjunction $p \wedge q$ is true. For all other assignments of logical values to *p* and to *q* the conjunction $p \wedge q$ is false.

It can also be said that if *p*, then $p \wedge q$ is *q*, otherwise $p \wedge q$ is *p*. There is maintance work here!

Logical Disjunction

Logical disjunction is an operation on two logical values, typically the values of two propositions, that produces a value of *true* if at least one of its operands is true. The truth table for p OR q (also written as $p \vee q$, Apq, p || q, or p + q) is as follows:

Logical Disjunction		
p	*q*	$p \vee q$
T	T	T
T	F	T
F	T	T
F	F	F

Stated in English, if *p*, then $p \vee q$ is *p*, otherwise $p \vee q$ is *q*.

Logical Implication

Logical implication and the material conditional are both associated with an operation on two logical values, typically the values of two propositions, that produces a value of *false* just in the singular case the first operand is true and the second operand is false.

The truth table associated with the material conditional if p then q (symbolized as $p \rightarrow q$) and the logical implication p implies q (symbolized as $p \Rightarrow q$, or Cpq) is as follows:

Logical Implication		
p	*q*	$p \rightarrow q$
T	T	T
T	F	F
F	T	T
F	F	T

It may also be useful to note that $p \rightarrow q$ is equivalent to $\neg p \vee q$.

Logical Equality

Logical equality (also known as biconditional) is an operation on two logical values, typically the values of two propositions, that produces a value of *true* if both operands are false or both operands are true.

The truth table for p XNOR q (also written as $p \leftrightarrow q$, Epq, p = q, or $p \equiv q$) is as follows:

Logical Equality		
p	q	$p \equiv q$
T	T	T
T	F	F
F	T	F
F	F	T

So p EQ q is true if p and q have the same truth value (both true or both false), and false if they have different truth values.

Exclusive Disjunction

Exclusive disjunction is an operation on two logical values, typically the values of two propositions, that produces a value of *true* if one but not both of its operands is true. The truth table for p XOR q (also written as $p \oplus q$, Jpq, or $p \neq q$) is as follows:

Exclusive Disjunction		
p	q	$p \oplus q$
T	T	F
T	F	T
F	T	T
F	F	F

For two propositions, XOR can also be written as $(p = 1 \wedge q = 0)$ $(p = 0 \wedge q = 1)$.

Logical NAND

The logical NAND is an operation on two logical values, typically the values of two propositions, that produces a value of *false* if both of its operands are true. In other words, it produces a value of *true* if at least one of its operands is false. The truth table for p NAND q (also written as $p \uparrow q$, Dpq, or $p | q$) is as follows:

Logical NAND		
p	q	$p \uparrow q$
T	T	F
T	F	T
F	T	T
F	F	T

It is frequently useful to express a logical operation as a compound operation, that is, as an operation that is built up or composed from other operations. Many such compositions are possible, depending on the operations that are taken as basic or "primitive" and the operations that are taken as composite or "derivative".

In the case of logical NAND, it is clearly expressible as a compound of NOT and AND. The negation of a conjunction: $\neg(p \wedge q)$, and the disjunction of negations: $(\neg p) \vee (\neg q)$ can be tabulated as follows:

p	q	$p \wedge q$	$\neg(p \wedge q)$	$\neg p$	$\neg q$	$(\neg p) \vee (\neg q)$
T	T	T	F	F	F	F
T	F	F	T	F	T	T
F	T	F	T	T	F	T
F	F	F	T	T	T	T

Logical NOR

The logical NOR is an operation on two logical values, typically the values of two propositions, that produces a value of *true* if both of its operands are false. In other words, it produces a value of *false* if at least one of its operands is true. "↓ is also known as the Peirce arrow after its inventor, Charles Sanders Peirce, and is a Sole sufficient operator. The truth table for p NOR q (also written as $p \downarrow q$, Xpq, or $p \perp q$) is as follows:

Logical NOR		
p	q	$p \downarrow q$
T	T	F
T	F	F
F	T	F
F	F	T

The negation of a disjunction $\neg(p \vee q)$, and the conjunction of negations $(\neg p) \wedge (\neg q)$ can be tabulated as follows:

p	q	$p \vee q$	$\neg(p \vee q)$	$\neg p$	$\neg q$	$(\neg p) \wedge (\neg q)$
T	T	T	F	F	F	F
T	F	T	F	F	T	F
F	T	T	F	T	F	F
F	F	F	T	T	T	T

Inspection of the tabular derivations for NAND and NOR, under each assignment of logical values to the functional arguments p and q, produces the identical patterns of functional values for $\neg(p \wedge q)$ as for $(\neg p)\vee(\neg q)$, and for $\neg(p \vee q)$ as for $(\neg p)\wedge(\neg q)$. Thus the first and second expressions in each pair are logically equivalent, and may be substituted for each other in all contexts that pertain solely to their logical values.

This equivalence is one of De Morgan's laws.

Applications

Truth tables can be used to prove many other logical equivalences. For example, consider the following truth table:

Logical Equivalence : $(p \rightarrow q) = (\neg p \vee q)$				
p	q	$\neg p$	$\neg p \vee q$	$p \rightarrow q$
T	T	F	T	T
T	F	F	F	F
F	T	T	T	T
F	F	T	T	T

This demonstrates the fact that $p \rightarrow q$ is logically equivalent to $\neg p \vee q$.

Truth Table for Most Commonly Used Logical Operators

Here is a truth table giving definitions of the most commonly used 6 of the 16 possible truth functions of 2 binary variables (P,Q are thus boolean variables):

P	Q	$P \wedge Q$	$P \vee Q$	$P \underline{\vee} Q$	$P \underline{\wedge} Q$	$P \Rightarrow Q$	$P \Leftarrow Q$	$P \iff Q$
T	T	T	T	F	T	T	T	T
T	F	F	T	T	F	F	T	F
F	T	F	T	T	F	T	F	F
F	F	F	F	F	T	T	T	T

Key:

T = true, F = false

$\wedge$ = AND (logical conjunction)

$\vee$ = OR (logical disjunction)

$\underline{\vee}$ = XOR (exclusive or)

$\underline{\wedge}$ = XNOR (exclusive nor)

$\rightarrow$ = conditional "if-then"

$\leftarrow$ = conditional "(then)-if"

$\Leftrightarrow$ biconditional or "if-and-only-if" is logically equivalent to : XNOR (exclusive nor).

Logical operators can also be visualized using Venn diagrams.

Condensed Truth Tables for Binary Operators

For binary operators, a condensed form of truth table is also used, where the row headings and the column headings specify the operands and the table cells specify the result. For example Boolean logic uses this condensed truth table notation:

∧	F	T
F	F	F
T	F	T

∨	F	T
F	F	T
T	T	T

This notation is useful especially if the operations are commutative, although one can additionally specify that the rows are the first operand and the columns are the second operand. This condensed notation is particularly useful in discussing multi-valued extensions of logic, as it significantly cuts down on combinatoric explosion of the number of rows otherwise needed. It also provides for quickly recognizable characteristic "shape" of the distribution of the values in the table which can assist the reader in grasping the rules more quickly.

Truth Tables in Digital Logic

Truth tables are also used to specify the functionality of hardware look-up tables (LUTs) in digital logic circuitry. For an n-input LUT, the truth table will have 2^n values (or rows in the above tabular format), completely specifying a boolean function for the LUT. By representing each boolean value as a bit in a binary number, truth table values can be efficiently encoded as integer values in electronic design automation (EDA) software. For example, a 32-bit integer can encode the truth table for a LUT with up to 5 inputs.

When using an integer representation of a truth table, the output value of the LUT can be obtained by calculating a bit index k based on the input values of the LUT, in which case the LUT's output value is the kth bit of the integer. For example, to evaluate the output value of a LUT given an array of n boolean input values, the bit index of the truth table's output value can be computed as follows: if the ith input is true, let V*i* = 1, else let V*i* = 0. Then the *k*th bit of the binary representation of the truth table is the LUT's output value, where *k* = V0*2^0 + V1*2^1 + V2*2^2 + ... + V*n**2^*n*.

Truth tables are a simple and straightforward way to encode boolean functions, however given the exponential growth in size as the number of inputs increase, they are not suitable for functions with a large number of inputs. Other representations which are more memory efficient are text equations and binary decision diagrams.

Applications of Truth Tables in Digital Electronics

In digital electronics and computer science (fields of applied logic engineering and mathematics), truth tables can be used to reduce basic boolean operations to simple correlations of inputs to outputs, without the use of logic gates or code. For example, a binary addition can be represented with the truth table:

```
A B | C R

1 1 | 1 0

1 0 | 0 1

0 1 | 0 1

0 0 | 0 0


where


A = First Operand

B = Second Operand

C = Carry

R = Result
```

This truth table is read left to right:

- Value pair (A,B) equals value pair (C,R).
- Or for this example, A plus B equal result R, with the Carry C.

Note that this table does not describe the logic operations necessary to implement this operation, rather it simply specifies the function of inputs to output values.

With respect to the result, this example may be arithmetically viewed as modulo 2 binary addition, and as logically equivalent to the exclusive-or (exclusive disjunction) binary logic operation.

In this case it can be used for only very simple inputs and outputs, such as 1's and 0's, however if the number of types of values one can have on the inputs increases, the size of the truth table will increase.

For instance, in an addition operation, one needs two operands, A and B. Each can have one of two values, zero or one. The number of combinations of these two values is 2x2, or four. So the result is four possible outputs of C and R. If one were to use base 3, the size would increase to 3x3, or nine possible outputs.

The first "addition" example above is called a half-adder. A full-adder is when the carry from the previous operation is provided as input to the next adder. Thus, a truth table of eight rows would be needed to describe a full adder's logic:

```
A B C*  | C R

0 0 0   | 0 0

0 1 0   | 0 1

1 0 0   | 0 1

1 1 0   | 1 0

0 0 1   | 0 1

0 1 1   | 1 0

1 0 1   | 1 0

1 1 1   | 1 1


Same as previous, but..

C* = Carry from previous adder
```

Diagrammatic Representations

Venn Diagrams: A Venn diagram is a representation of a Boolean operation using shaded overlapping regions. There is one region for each variable, all circular in the examples here.

The interior and exterior of region x corresponds respectively to the values 1 (true) and 0 (false) for variable x.

The shading indicates the value of the operation for each combination of regions, with dark denoting 1 and light 0 (some authors use the opposite convention).

The three Venn diagrams in the figure below represent respectively conjunction $x \wedge y$, disjunction $x \vee y$, and complement $\neg x$.

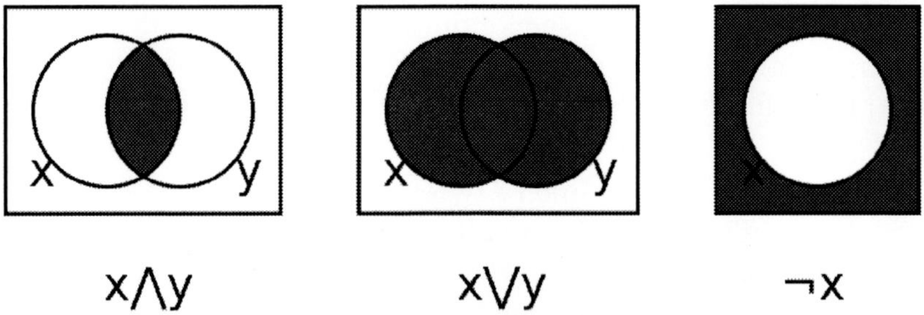

Figure: *Venn diagrams for conjunction, disjunction, and complement*

For conjunction, the region inside both circles is shaded to indicate that $x \wedge y$ is 1 when both variables are 1. The other regions are left unshaded to indicate that $x \wedge y$ is 0 for the other three combinations.

The second diagram represents disjunction $x \vee y$ by shading those regions that lie inside either or both circles. The third diagram represents complement $\neg x$ by shading the region *not* inside the circle.

While we have not shown the Venn diagrams for the constants 0 and 1, they are trivial, being respectively a white box and a dark box, neither one containing a circle. However we could put a circle for x in those boxes, in which case each would denote a function of one argument, x, which returns the same value independently of x, called a constant function. As far as their outputs are concerned, constants and constant functions are indistinguishable; the difference is that a constant takes no arguments, called a *zeroary* or *nullary* operation, while a constant function takes one argument, which it ignores, and is a *unary* operation.

Venn diagrams are helpful in visualizing laws. The commutativity laws for $\wedge$ and $\vee$ can be seen from the symmetry of the diagrams: a binary operation that was not commutative would not have a symmetric diagram because interchanging x and y would have the effect of reflecting the diagram horizontally and any failure of commutativity would then appear as a failure of symmetry.

Idempotence of $\wedge$ and $\vee$ can be visualized by sliding the two circles together and noting that the shaded area then becomes the whole circle, for both $\wedge$ and $\vee$.

To see the first absorption law, $x \wedge (x \vee y) = x$, start with the diagram in the middle for $x \vee y$ and note that the portion of the shaded area in common with the x circle is the whole of the x circle. For the second absorption law, $x \vee (x \wedge y) = x$, start with the left diagram for $x \wedge y$ and note that shading the whole of the x circle results in just the x circle being shaded, since the previous shading was inside the x circle.

The double negation law can be seen by complementing the shading in the third diagram for $\neg x$, which shades the x circle.

To visualize the first De Morgan's law, $(\neg x) \wedge (\neg y) = \neg(x \vee y)$, start with the middle diagram for $x \vee y$ and complement its shading so that only the region outside both circles is shaded, which is what the right hand side of the law describes. The result is the same as if we shaded that region which is both outside the x circle *and* outside the y circle, i.e. the conjunction of their exteriors, which is what the left hand side of the law describes.

The second De Morgan's law, $(\neg x) \vee (\neg y) = \neg(x \wedge y)$, works the same way with the two diagrams interchanged.

The first complement law, $x \wedge \neg x = 0$, says that the interior and exterior of the x circle have no overlap. The second complement law, $x \vee \neg x = 1$, says that everything is either inside or outside the x circle.

Digital Logic Gates

Digital logic is the application of the Boolean algebra of 0 and 1 to electronic hardware consisting of logic gates connected to form a circuit diagram. Each gate implements a Boolean operation, and is depicted schematically by a shape indicating the operation. The shapes associated with the gates for conjunction (AND-gates), disjunction (OR-gates), and complement (inverters) are as follows.

Figure: *Logic gates*

The lines on the left of each gate represent input wires or *ports*. The value of the input is represented by a voltage on the lead. For so-called "active-high" logic 0 is represented by a voltage close to zero or "ground" while 1 is represented by a voltage close to the supply voltage; active-low reverses this. The line on the right of each gate represents the output port, which normally follows the same voltage conventions as the input ports.

Complement is implemented with an inverter gate. The triangle denotes the operation that simply copies the input to the output; the small circle on the output denotes the actual inversion complementing the input. The convention of putting such a circle on any port means that the signal passing through this port is complemented on the way through, whether it is an input or output port.

There being eight ways of labelling the three ports of an AND-gate or OR-gate with inverters, this convention gives a wide range of possible Boolean operations realized as such gates so decorated. Not all combinations are distinct however: any labelling of AND-gate ports with inverters realizes the same Boolean operation as the opposite labelling of OR-gate ports (a given port of the AND-gate is labelled with an inverter if and only if the corresponding port of the OR-gate is not so labelled). This follows from De Morgan's laws.

If we complement all ports on every gate, and interchange AND-gates and OR-gates, as in Figure below, we end up with the same operations as we started with, illustrating both De Morgan's laws and the Duality Principle. Note that we did not need to change the triangle part of the inverter, illustrating self-duality for complement.

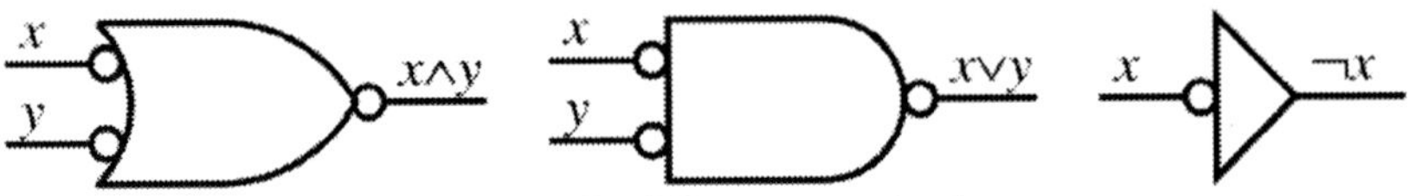

***Figure:** De Morgan equivalents*

Because of the pairwise identification of gates via the Duality Principle, even though 16 schematic symbols can be manufactured from the two basic binary gates AND and OR by furnishing their ports with inverters (circles), they only represent eight Boolean operations, namely those operations with an odd number of ones in their truth table. Altogether there are 16 binary Boolean operations, the other eight being those with an even number of ones in their truth table, namely the following. The constant 0, viewed as a binary operation that ignores both its inputs, has no ones, the six operations x, y, $\neg x$, $\neg y$ (as binary operations that ignore one input), *x*□□*y*, and *x*a”*y* have two ones, and the constant 1 has four ones.

Boolean Algebras

The term "algebra" denotes both a subject, namely the subject of algebra, and an object, namely an algebraic structure. Whereas the foregoing has addressed the subject of Boolean algebra, this section deals with mathematical objects called Boolean algebras, defined in full generality as any model of the Boolean laws. We begin with a special case of the notion definable without reference to the laws, namely concrete Boolean algebras, and then give the formal definition of the general notion.

Concrete Boolean Algebras

A concrete Boolean algebra or field of sets is any nonempty set of subsets of a given set X closed under the set operations of union, intersection, and complement relative to X.

(As an aside, historically X itself was required to be nonempty as well to exclude the degenerate or one-element Boolean algebra, which is the one exception to the rule that all Boolean algebras satisfy the same equations since the degenerate algebra satisfies every equation. However this exclusion conflicts with the preferred purely equational definition of "Boolean algebra," there being no way to rule out the one-element algebra using only equations— $0 \neq 1$ does not count, being a negated equation. Hence modern authors allow the degenerate Boolean algebra and let X be empty.)

Example 1. The power set 2 of X, consisting of all subsets of X. Here X may be any set: empty, finite, infinite, or even uncountable.

Example 2. The empty set and X. This two-element algebra shows that a concrete Boolean algebra can be finite even when it consists of subsets of an infinite set. It can be seen that every field of subsets of X must contain the empty set and X. Hence no smaller example is possible, other than the degenerate algebra obtained by taking X to be empty so as to make the empty set and X coincide.

Example 3. The set of finite and cofinite sets of integers, where a cofinite set is one omitting only finitely many integers. This is clearly closed under complement, and is closed under union because the union of a cofinite set with any set is cofinite, while the union of two finite sets is finite. Intersection behaves like union with "finite" and "cofinite" interchanged.

Example 4. For a less trivial example of the point made by Example 2, consider a Venn diagram formed by n closed curves partitioning the diagram into 2^n regions, and let X be the (infinite) set of all points in the plane not on any curve but somewhere within the diagram. The interior of each region is thus an infinite subset of X, and every point in X is in exactly one region. Then the set of all 2^{2n} possible unions of regions (including the empty set obtained as the union of the empty set of regions and X obtained as the union of all 2^n regions) is closed under union, intersection, and complement relative to X and therefore forms a concrete Boolean algebra. Again we have finitely many subsets of an infinite set forming a concrete Boolean algebra, with Example 2 arising as the case $n = 0$ of no curves.

Subsets as bit Vectors

A subset Y of X can be identified with an indexed family of bits with index set X, with the bit indexed by $x \in X$ being 1 or 0 according to whether or not $x \in X$. (This is the so-called characteristic function notion of a subset.) For example a 32-bit computer word consists of 32 bits indexed by the set {0,1,2,...,31}, with 0 and 31 indexing the low and high order bits respectively. For a smaller example, if $X = \{a,b,c\}$ where a, b, c are viewed as bit positions in that order from left to right, the eight subsets , {c}, {b}, {b,c}, {a}, {a,c}, {a,b}, and {a,b,c} of X can be identified with the respective bit vectors 000, 001, 010, 011, 100, 101, 110, and 111. Bit vectors indexed by the set of natural numbers are infinite sequences of bits, while those indexed by the reals in the unit interval [0,1] are packed too densely to be able to write conventionally but nonetheless form well-defined indexed families (imagine colouring every point of the interval [0,1] either black or white independently; the black points then form an arbitrary subset of [0,1]).

From this bit vector viewpoint, a concrete Boolean algebra can be defined equivalently as a nonempty set of bit vectors all of the same length (more generally, indexed by the same set) and closed under the bit vector operations of bitwise $\wedge, \vee$, and $\neg$, as in $1010 \wedge 0110 = 0010$, $1010 \vee 0110 = 1110$, and $\neg 1010 = 0101$, the bit vector realizations of intersection, union, and complement respectively.

The prototypical Boolean Algebra

The set {0,1} and its Boolean operations as treated above can be understood as the special case of bit vectors of length one, which by the identification of bit vectors with subsets can also be understood as the two subsets of a one-element set. We call this the prototypical Boolean algebra, justified by the following observation.

The laws satisfied by all nondegenerate concrete Boolean algebras coincide with those satisfied by the prototypical Boolean algebra.

This observation is easily proved as follows. Certainly any law satisfied by all concrete Boolean algebras is satisfied by the prototypical one since it is concrete. Conversely any law that fails for some concrete Boolean algebra must have failed at a particular bit position, in which case that position by itself furnishes a one-bit counterexample to that law. Nondegeneracy ensures the existence of at least one bit position because there is only one empty bit vector.

The final goal of the next section can be understood as eliminating "concrete" from the above observation. We shall however reach that goal via the surprisingly stronger observation that, up to isomorphism, all Boolean algebras are concrete.

Boolean Algebras: the Definition

The Boolean algebras we have seen so far have all been concrete, consisting of bit vectors or equivalently of subsets of some set. Such a Boolean algebra consists of a set and operations on that set which can be *shown* to satisfy the laws of Boolean algebra.

Instead of showing that the Boolean laws are satisfied, we can instead postulate a set X, two binary operations on X, and one unary operation, and *require* that those operations satisfy the laws of Boolean algebra. The elements of X need not be bit vectors or subsets but can be anything at all. This leads to the more general *abstract* definition.

A Boolean algebra is any set with binary operations "" and (" and a unary operation ¬ thereon satisfying the Boolean laws.

For the purposes of this definition it is irrelevant how the operations came to satisfy the laws, whether by fiat or proof. All concrete Boolean algebras satisfy the laws (by proof rather than fiat), whence every concrete Boolean algebra is a Boolean algebra according to our definitions. This axiomatic definition of a Boolean algebra as a set and certain operations satisfying certain laws or axioms *by fiat* is entirely analogous to the abstract definitions of group, ring, field etc. characteristic of modern or abstract algebra.

Given any complete axiomatization of Boolean algebra, such as the axioms for a complemented distributive lattice, a sufficient condition for an algebraic structure of this kind to satisfy all the Boolean laws is that it satisfy just those axioms. The following is therefore an equivalent definition.

A Boolean algebra is a complemented distributive lattice.

The section on axiomatization lists other axiomatizations, any of which can be made the basis of an equivalent definition.

Representable Boolean Algebras

Although every concrete Boolean algebra is a Boolean algebra, not every Boolean algebra need be concrete. Let n be a square-free positive integer, one not divisible by the square of an integer, for example 30 but not 12. The operations of greatest common divisor, least common multiple, and division into n (that is, $\neg x = n/x$), can be shown to satisfy all the Boolean laws when their arguments range over the positive divisors of n. Hence those divisors form a Boolean algebra. These divisors are not subsets of a set, making the divisors of n a Boolean algebra that is not concrete according to our definitions.

However if we *represent* each divisor of n by the set of its prime factors, we find that this nonconcrete Boolean algebra is isomorphic to the concrete Boolean algebra consisting of all sets of prime factors of n, with union corresponding to least common multiple, intersection to greatest common divisor, and complement to division into n. So this example while not technically concrete is at least "morally" concrete via this representation, called an isomorphism. This example is an instance of the following notion.

A Boolean algebra is called representable when it is isomorphic to a concrete Boolean algebra.

The obvious next question is answered positively as follows.

Every Boolean algebra is representable.

That is, up to isomorphism, abstract and concrete Boolean algebras are the same thing. This quite nontrivial result depends on the Boolean prime ideal theorem, a choice principle slightly weaker than the axiom of choice, and is treated in more detail in the article Stone's representation theorem for Boolean algebras. This strong relationship implies a weaker result strengthening the observation in the previous subsection to the following easy consequence of representability.

The laws satisfied by all Boolean algebras coincide with those satisfied by the prototypical Boolean algebra.

It is weaker in the sense that it does not of itself imply representability. Boolean algebras are special here, for example a relation algebra is a Boolean algebra with additional structure but it is not the case that every relation algebra is representable in the sense appropriate to relation algebras.

Axiomatizing Boolean Algebra

The above definition of an abstract Boolean algebra as a set and operations satisfying "the" Boolean laws raises the question, what are those laws? A simple-minded answer is "all Boolean laws," which can be defined as all equations that hold for the Boolean algebra of 0 and 1. Since there are infinitely many such laws this is not a terribly satisfactory answer in practice, leading to the next question: does it suffice to require only finitely many laws to hold?

In the case of Boolean algebras the answer is yes. In particular the finitely many equations we have listed above suffice. We say that Boolean algebra is finitely axiomatizable or finitely based.

Can this list be made shorter yet? Again the answer is yes. To begin with, some of the above laws are implied by some of the others. A sufficient subset of the above laws consists of the pairs of associativity, commutativity, and absorption laws, distributivity of $\wedge$ over $\vee$ (or the other distributivity law—one suffices), and the two complement laws. In fact this is the traditional axiomatization of Boolean algebra as a complemented distributive lattice.

By introducing additional laws not listed above it becomes possible to shorten the list yet further. In 1933 Edward Huntington showed that if the basic operations are taken to be $x \vee y$ and $\neg x$, with $x \wedge y$ considered a derived operation (e.g. via De Morgan's law in the form $x \wedge y = \neg(\neg x \vee \neg y)$, then the equation $\neg(\neg x \vee \neg y) \vee \neg(\neg x \vee y) = x$ along with the two equations expressing associativity and commutativity of $\vee$ completely axiomatized Boolean algebra. When the only basic operation is the binary NAND operation $\neg(x \wedge y)$, Stephen Wolfram has proposed in his book *A New Kind of Science* the single axiom $(((xy)z)(x((xz)x))) = z$ as a one-equation axiomatization of Boolean algebra, where for convenience here xy denotes the NAND rather than the AND of x and y.

Propositional Logic

Propositional logic is a logical system that is intimately connected to Boolean algebra. Many syntactic concepts of Boolean algebra carry over to propositional logic with only minor changes in notation and terminology, while the semantics of propositional logic are defined via Boolean algebras in a way that the tautologies (theorems) of propositional logic correspond to equational theorems of Boolean algebra.

Syntactically, every Boolean term corresponds to a propositional formula of propositional logic. In this translation between Boolean algebra and propositional logic, Boolean variables x,y... become propositional variables (or atoms) P,Q,..., Boolean terms such as $x \vee y$ become propositional formulas $P \vee Q$, 0 becomes *false* or $\perp$, and 1 becomes *true* or T. It is convenient when referring to generic propositions to use Greek letters Φ, Ψ ,... as metavariables (variables outside the language of propositional calculus, used when talking *about* propositional calculus) to denote propositions.

The semantics of propositional logic rely on truth assignments. The essential idea of a truth assignment is that the propositional variables are mapped to elements of a fixed Boolean algebra, and then the truth value of a propositional formula using these letters is the element of the Boolean algebra that is obtained by computing the value of the Boolean term corresponding to the formula. In classical semantics, only the two-element Boolean algebra is used, while in Boolean-valued semantics arbitrary Boolean algebras are considered. A tautology is a propositional formula that is assigned truth value *1* by every truth assignment of its propositional variables to an arbitrary Boolean algebra (or, equivalently, every truth assignment to the two element Boolean algebra).

These semantics permit a translation between tautologies of propositional logic and equational theorems of Boolean algebra. Every tautology Φ of propositional logic can be expressed as the Boolean equation $\Phi = 1$, which will be a theorem of Boolean algebra. Conversely every theorem $\Phi = \Psi$ of Boolean algebra corresponds to the tautologies $(\Phi \vee \neg\Psi) \wedge (\neg\Phi \vee \Psi)$ and $(\Phi \wedge \Psi) \vee (\neg\Phi \wedge \neg\Psi)$. If $\rightarrow$ is in the language these last tautologies can also be written as $(\Phi \rightarrow \Psi) \wedge (\Psi \rightarrow \Phi)$, or as two separate theorems $\Phi \rightarrow \Psi$ and $\Psi \rightarrow \Phi$; if $\equiv$ is available then the single tautology $\Phi \equiv \Psi$ can be used.

Applications

One motivating application of propositional calculus is the analysis of propositions and deductive arguments in natural language. Whereas the proposition "if $x = 3$ then $x+1 = 4$" depends on the meanings of such symbols as + and 1, the proposition "if $x = 3$ then $x = 3$" does not; it is true merely by virtue of its structure, and remains true whether "$x = 3$" is replaced by "$x = 4$" or "the moon is made of green cheese." The generic or abstract form of this tautology is "if *P* then *P*", or in the language of Boolean algebra, "$P \rightarrow P$".

Replacing *P* by $x = 3$ or any other proposition is called instantiation of *P* by that proposition. The result of instantiating *P* in an abstract proposition is called an instance of the proposition. Thus "$x = 3 \rightarrow x = 3$" is a tautology by virtue of being an instance of the abstract tautology "$P \rightarrow P$". All occurrences of the instantiated variable must be instantiated with the same proposition, to avoid such nonsense as $P \rightarrow x = 3$ or $x = 3 \rightarrow x = 4$.

Propositional calculus restricts attention to abstract propositions, those built up from propositional variables using Boolean operations. Instantiation is still possible within propositional calculus, but only by instantiating propositional variables by abstract propositions, such as instantiating Q by $Q \rightarrow P$ in $P \rightarrow (Q \rightarrow P)$ to yield the instance $P \rightarrow ((Q \rightarrow P) \rightarrow P)$.

(The availability of instantiation as part of the machinery of propositional calculus avoids the need for metavariables within the language of propositional calculus, since ordinary propositional variables can be considered within the language to denote arbitrary propositions. The metavariables themselves are outside the reach of instantiation, not being part of the language of propositional calculus but rather part of the same language for talking about it that this sentence is written in, where we need to be able to distinguish propositional variables and their instantiations as being distinct syntactic entities.)

Deductive Systems for Propositional Logic

An axiomatization of propositional calculus is a set of tautologies called axioms and one or more inference rules for producing new tautologies from old. A *proof* in an axiom system A is a finite nonempty sequence of propositions each of which is either an instance of an axiom of A or follows by some rule of A from propositions appearing earlier in the proof (thereby disallowing circular reasoning). The last proposition is the theorem proved by the proof. Every nonempty initial segment of a proof is itself a proof, whence every proposition in a proof is itself a theorem. An axiomatization is sound when every theorem is a tautology, and complete when every tautology is a theorem.

Sequent Calculus

Propositional calculus is commonly organized as a Hilbert system, whose operations are just those of Boolean algebra and whose theorems are Boolean tautologies, those Boolean terms equal to the Boolean constant 1. Another form is sequent calculus, which has two sorts, propositions as in ordinary propositional calculus, and pairs of lists of propositions called sequents, such as $A \vee B, A \wedge C, \ldots \vdash A, B \rightarrow C, \ldots$. The two halves of a sequent are called the antecedent and the succedent respectively. The customary metavariable denoting an antecedent or part thereof is Γ, and for a succedent Δ; thus $\Gamma, A \vdash \Delta$ would denote a sequent whose succedent is a list Δ and whose antecedent is a list Γ with an additional proposition A appended after it. The antecedent is interpreted as the conjunction of its propositions, the succedent as the disjunction of its propositions, and the sequent itself as the entailment of the succedent by the antecedent.

Entailment differs from implication in that whereas the latter is a binary *operation* that returns a value in a Boolean algebra, the former is a binary *relation* which either holds or does not hold. In this sense entailment is an *external* form of implication, meaning external to the Boolean algebra, thinking of the reader of the sequent as also being external and interpreting and comparing antecedents and succedents in some Boolean algebra.

The natural interpretation of is as $\leq$ in the partial order of the Boolean algebra defined by $x \leq y$ just when $x \vee y = y$. This ability to mix external implication $\vdash$ and internal implication $\rightarrow$ in the one logic is among the essential differences between sequent calculus and propositional calculus.

Applications

Two-valued Logic: Boolean algebra as the calculus of two values is fundamental to digital logic, computer programming, and mathematical logic, and is also used in other areas of mathematics such as set theory and statistics.

Digital logic codes its symbols in various ways: as voltages on wires in high-speed circuits and capacitive storage devices, as orientations of a magnetic domain in ferromagnetic storage devices, as holes in punched cards or paper tape, and so on. Now it is possible to code more than two symbols in any given medium. For example one might use respectively 0, 1, 2, and 3 volts to code a four-symbol alphabet on a wire, or holes of different sizes in a punched card. In practice however the tight constraints of high speed, small size, and low power combine to make noise a major factor. This makes it hard to distinguish between symbols when there are many of them at a single site. Rather than attempting to distinguish between four voltages on one wire, digital designers have settled on two voltages per wire, high and low. To obtain four symbols one uses two wires, and so on.

Programmers programming in machine code, assembly language, and other programming languages that expose the low-level digital structure of the data registers operate on whatever symbols were chosen for the hardware, invariably bit vectors in modern computers for the above reasons. Such languages support both the numeric operations of addition, multiplication, etc. performed on words interpreted as integers, as well as the logical operations of disjunction, conjunction, etc. performed bit-wise on words interpreted as bit vectors. Programmers therefore have the option of working in and applying the laws of either numeric algebra or Boolean algebra as needed. A core differentiating feature is carry propagation with the former but not the latter.

Other areas where two values is a good choice are the law and mathematics. In everyday relaxed conversation, nuanced or complex answers such as "maybe" or "only on the weekend" are acceptable. In more focused situations such as a court of law or theorem-based mathematics however it is deemed advantageous to frame questions so as to admit a simple yes-or-no answer—is the defendant guilty or not guilty, is the proposition true or false—and to disallow any other answer. However much of a straitjacket this might prove in practice for the respondent, the principle of the simple yes-no question has become a central feature of both judicial and mathematical logic, making two-valued logic deserving of organization and study in its own right.

A central concept of set theory is membership. Now an organization may permit multiple degrees of membership, such as novice, associate, and full. With sets however an element is either in or out. The candidates for membership in a set work just like the wires in a digital computer: each candidate is either a member or a nonmember, just as each wire is either high or low.

Algebra being a fundamental tool in any area amenable to mathematical treatment, these considerations combine to make the algebra of two values of fundamental importance to computer hardware, mathematical logic, and set theory.

Two-valued logic can be extended to multi-valued logic, notably by replacing the Boolean domain {0, 1} with the unit interval [0,1], in which case rather than only taking values 0 or 1, any value between and including 0 and 1 can be assumed. Algebraically, negation (NOT) is replaced with $1 - x$, conjunction (AND) is replaced with multiplication (xy), and disjunction (OR) is defined via De Morgan's law. Interpreting these values as logical truth values yields a multi-valued logic, which forms the basis for fuzzy logic and probabilistic logic. In these interpretations, a value is interpreted as the "degree" of truth – to what extent a proposition is true, or the probability that the proposition is true.

Boolean Operations

The original application for Boolean operations was mathematical logic, where it combines the truth values, true or false, of individual formulas. Natural languages such as English have words for several Boolean operations, in particular conjunction (*and*), disjunction (*or*), negation (*not*), and implication (*implies*). *But not* is synonymous with *and not*. When used to combine situational assertions such as "the block is on the table" and "cats drink milk," which naively are either true or false, the meanings of these logical connectives often have the meaning of their logical counterparts. However with descriptions of behaviour such as "Jim walked through the door", one starts to notice differences such as failure of commutativity, for example the conjunction of "Jim opened the door" with "Jim walked through the door" in that order is not equivalent to their conjunction in the other order, since *and* usually means *and then* in such cases. Questions can be similar: the order "Is the sky blue, and why is the sky blue?" makes more sense than the reverse order. Conjunctive commands about behaviour are like behavioural assertions, as in *get dressed and go to school.* Disjunctive commands such *love me or leave me* or *fish or cut bait* tend to be asymmetric via the implication that one alternative is less preferable.

Conjoined nouns such as *tea and milk* generally describe aggregation as with set union while *tea or milk* is a choice. However context can reverse these senses, as in *your choices are coffee and tea* which usually means the same as *your choices are coffee or tea* (alternatives). Double negation as in "I don't not like milk" rarely means literally "I do like milk" but rather conveys some sort of hedging, as though to imply that there is a third possibility. "Not not P" can be loosely interpreted as "surely P", and although *P* necessarily implies "not not *P*" the converse is suspect in English, much as with intuitionistic logic. In view of the highly idiosyncratic usage of conjunctions in natural languages, Boolean algebra cannot be considered a reliable framework for interpreting them.

Boolean operations are used in digital logic to combine the bits carried on individual wires, thereby interpreting them over {0,1}. When a vector of n identical binary gates are used to combine two bit vectors each of n bits, the individual bit operations can be understood collectively as a single operation on values from a Boolean algebra with 2^n elements.

Naive set theory interprets Boolean operations as acting on subsets of a given set X. As we saw earlier this behaviour exactly parallels the coordinate-wise combinations of bit vectors, with the union of two sets corresponding to the disjunction of two bit vectors and so on.

The 256-element free Boolean algebra on three generators is deployed in computer displays based on raster graphics, which use bit blit to manipulate whole regions consisting of pixels, relying on Boolean operations to specify how the source region should be combined with the destination, typically with the help of a third region called the mask. Modern video cards offer all 2^{23} = 256 ternary operations for this purpose, with the choice of operation being a one-byte (8-bit) parameter. The constants SRC = 0xaa or 10101010, DST = 0xcc or 11001100, and MSK = 0xf0 or 11110000 allow Boolean operations such as (SRC^DST)&MSK (meaning XOR the source and destination and then AND the result with the mask) to be written directly as a constant denoting a byte calculated at compile time, 0x60 in the (SRC^DST)&MSK example, 0x66 if just SRC^DST, etc. At run time the video card interprets the byte as the raster operation indicated by the original expression in a uniform way that requires remarkably little hardware and which takes time completely independent of the complexity of the expression.

Solid modelling systems for computer aided design offer a variety of methods for building objects from other objects, combination by Boolean operations being one of them. In this method the space in which objects exist is understood as a set S of voxels (the three-dimensional analogue of pixels in two-dimensional graphics) and shapes are defined as subsets of S, allowing objects to be combined as sets via union, intersection, etc. One obvious use is in building a complex shape from simple shapes simply as the union of the latter. Another use is in sculpting understood as removal of material: any grinding, milling, routing, or drilling operation that can be performed with physical machinery on physical materials can be simulated on the computer with the Boolean operation $x \wedge \neg y \text{or} x - y$, which in set theory is set difference, remove the elements of y from those of x. Thus given two shapes one to be machined and the other the material to be removed, the result of machining the former to remove the latter is described simply as their set difference.

Boolean Searches

Search engine queries also employ Boolean logic. For this application, each web page on the Internet may be considered to be an "element" of a "set". The following examples use a syntax supported by Google.

- Doublequotes are used to combine whitespace-separated words into a single search term.
- Whitespace is used to specify logical AND, as it is the default operator for joining search terms:

"Search term 1" "Search term 2"

- The OR keyword is used for logical OR:

"Search term 1" OR "Search term 2"

- The minus sign is used for logical NOT (AND NOT):

"Search term 1" – "Search term 2"

De Morgan s Theorems

A mathematician named DeMorgan developed a pair of important rules regarding group complementation in Boolean algebra. By *group* complementation, I'm referring to the complement of a group of terms, represented by a long bar over more than one variable.

You should recall from the chapter on logic gates that inverting all inputs to a gate reverses that gate's essential function from AND to OR, or vice versa, and also inverts the output. So, an OR gate with all inputs inverted (a Negative-OR gate) behaves the same as a NAND gate, and an AND gate with all inputs inverted (a Negative-AND gate) behaves the same as a NOR gate. DeMorgan's theorems state the same equivalence in "backward" form: that inverting the output of any gate results in the same function as the opposite type of gate (AND vs. OR) with inverted inputs:

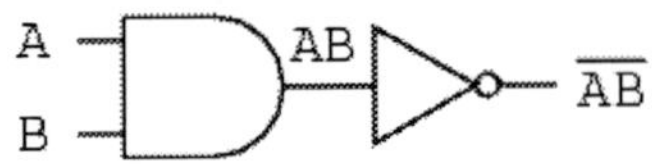

. . . *is equivalent to* . . .

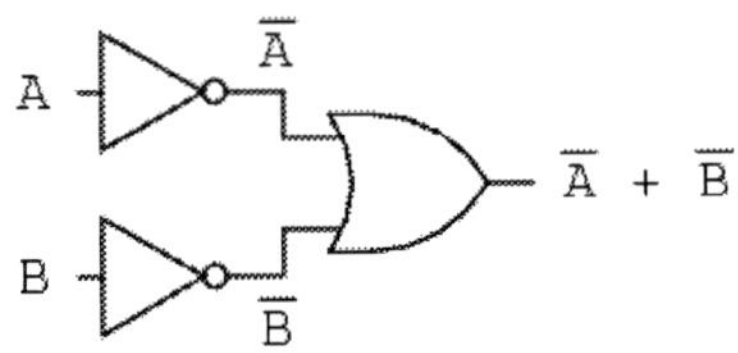

$$\overline{AB} = \overline{A} + \overline{B}$$

A long bar extending over the term AB acts as a grouping symbol, and as such is entirely different from the product of A and B independently inverted. In other words, (AB)' is not equal to A'B'. Because the "prime" symbol (') cannot be stretched over two variables like a bar can, we are forced to use parentheses to make it apply to the whole term AB in the previous sentence. A bar, however, acts as its own grouping symbol when stretched over more than one variable. This has profound impact on how Boolean expressions are evaluated and reduced, as we shall see.

DeMorgan's theorem may be thought of in terms of *breaking* a long bar symbol. When a long bar is broken, the operation directly underneath the break changes from addition to multiplication, or vice versa, and the broken bar pieces remain over the individual variables. To illustrate:

DeMorgan's Theorems

NAND to *Negative-OR* *NOR* to *Negative-AND*

When multiple "layers" of bars exist in an expression, you may only break *one bar at a time*, and it is generally easier to begin simplification by breaking the longest (uppermost) bar first. To illustrate, let's take the expression (A + (BC)')' and reduce it using DeMorgan's Theorems:

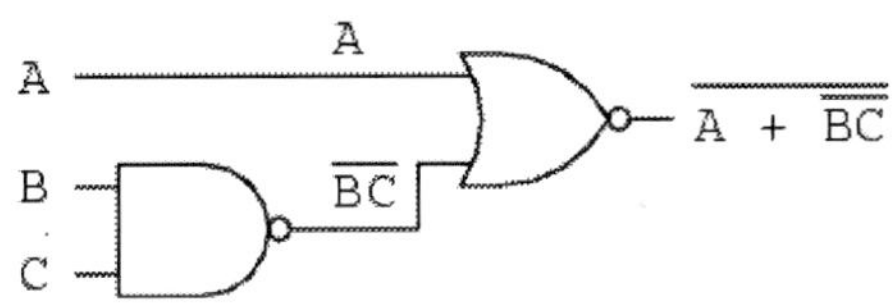

Following the advice of breaking the longest (uppermost) bar first, I'll begin by breaking the bar covering the entire expression as a first step:

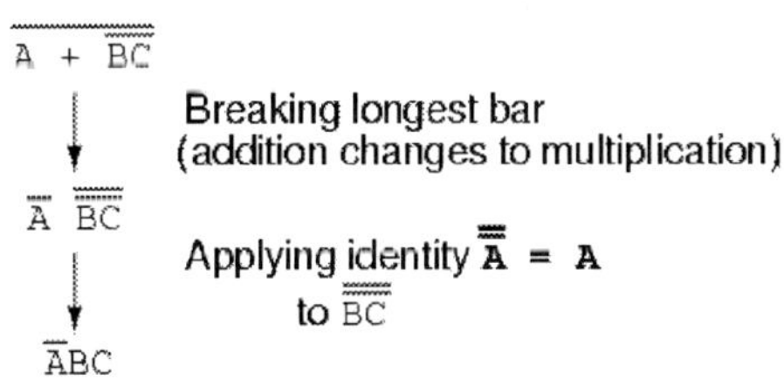

As a result, the original circuit is reduced to a three-input AND gate with the A input inverted:

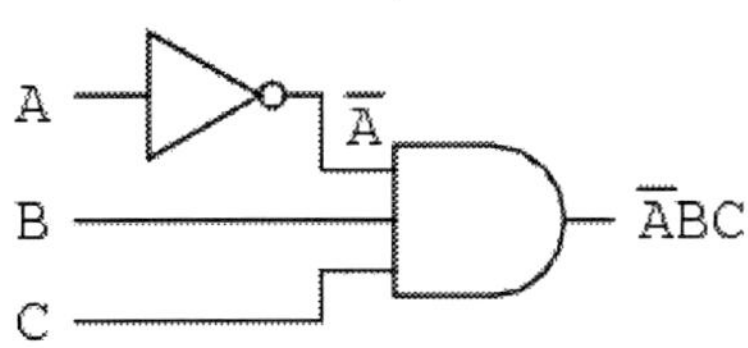

You should *never* break more than one bar in a single step, as illustrated here:

$$\overline{A + \overline{BC}}$$

Incorrect step! ↓ Breaking long bar between A and B; Breaking both bars between B and C

$$\overline{A}\ \overline{\overline{B}} + \overline{\overline{C}}$$

↓ Applying identity $\overline{\overline{A}} = A$ to $\overline{\overline{B}}$ and $\overline{\overline{C}}$

Incorrect answer: $\overline{A}B + C$

As tempting as it may be to conserve steps and break more than one bar at a time, it often leads to an incorrect result, so don't do it!

It is possible to properly reduce this expression by breaking the short bar first, rather than the long bar first:

$$\overline{A + \overline{BC}}$$

↓ Breaking shortest bar (multiplication changes to addition)

$$\overline{A + (\overline{B} + \overline{C})}$$

↓ Applying associative property to remove parentheses

$$\overline{A + \overline{B} + \overline{C}}$$

↓ Breaking long bar in two places, between 1st and 2nd terms; between 2nd and 3rd terms

$$\overline{A}\ \overline{\overline{B}}\ \overline{\overline{C}}$$

↓ Applying identity $\overline{\overline{A}} = A$ to $\overline{\overline{B}}$ and $\overline{\overline{C}}$

$$\overline{A}BC$$

The end result is the same, but more steps are required compared to using the first method, where the longest bar was broken first. Note how in the third step we broke the long bar in two places. This is a legitimate mathematical operation, and not the same as breaking two bars in one step! The prohibition against breaking more than one bar in one step is *not* a prohibition against breaking a bar in more than one place. Breaking in more than one *place* in a single step is okay; breaking more than one *bar* in a single step is not.

You might be wondering why parentheses were placed around the sub-expression B' + C', considering the fact that I just removed them in the next step. I did this to emphasize an important but easily neglected aspect of DeMorgan's theorem. Since a long bar functions as a grouping symbol, the variables formerly grouped by a broken bar must remain grouped lest proper precedence (order of operation) be lost. In this example, it really wouldn't matter if I forgot to put parentheses in after breaking the short bar, but in other cases it might. Consider this example, starting with a different expression:

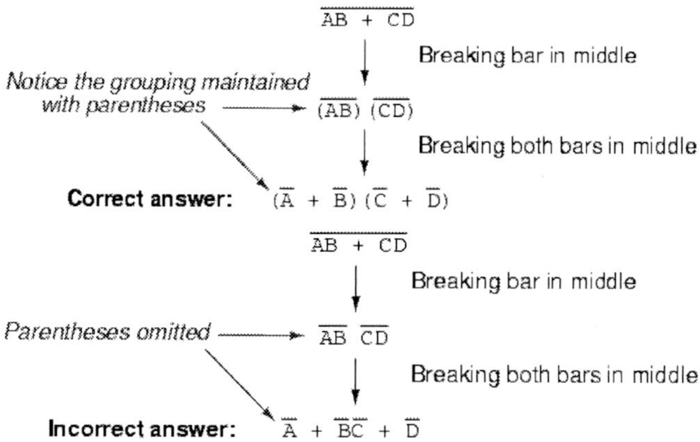

As you can see, maintaining the grouping implied by the complementation bars for this expression is crucial to obtaining the correct answer.

Let's apply the principles of DeMorgan's theorems to the simplification of a gate circuit:

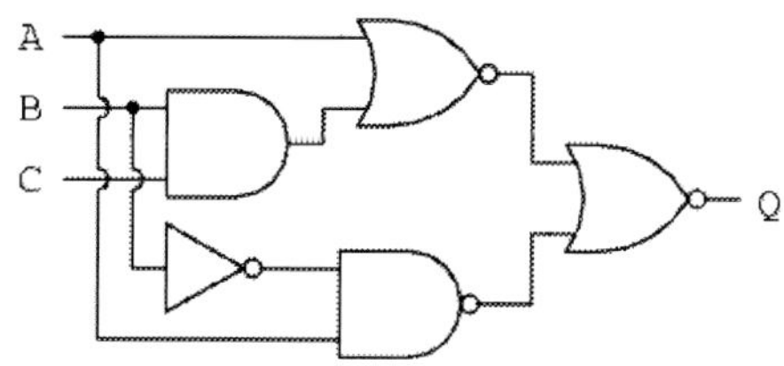

As always, our first step in simplifying this circuit must be to generate an equivalent Boolean expression. We can do this by placing a sub-expression label at the output of each gate, as the inputs become known. Here's the first step in this process:

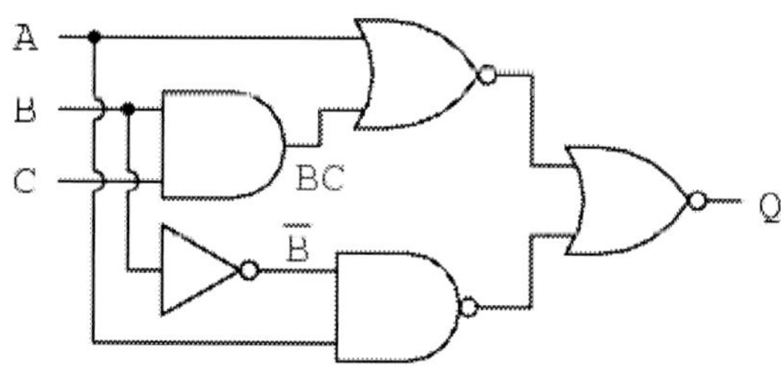

Next, we can label the outputs of the first NOR gate and the NAND gate. When dealing with inverted-output gates, I find it easier to write an expression for the gate's output *without* the final inversion, with an arrow pointing to just before the inversion bubble. Then, at the wire leading out of the gate (after the bubble), I write the full, complemented expression. This helps ensure I don't forget a complementing bar in the sub-expression, by forcing myself to split the expression-writing task into two steps:

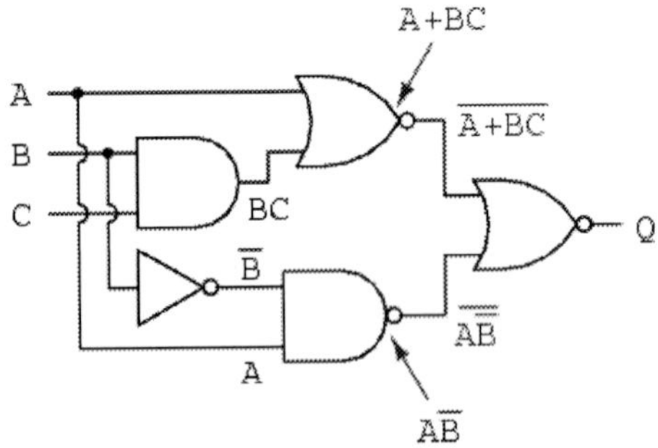

Finally, we write an expression (or pair of expressions) for the last NOR gate:

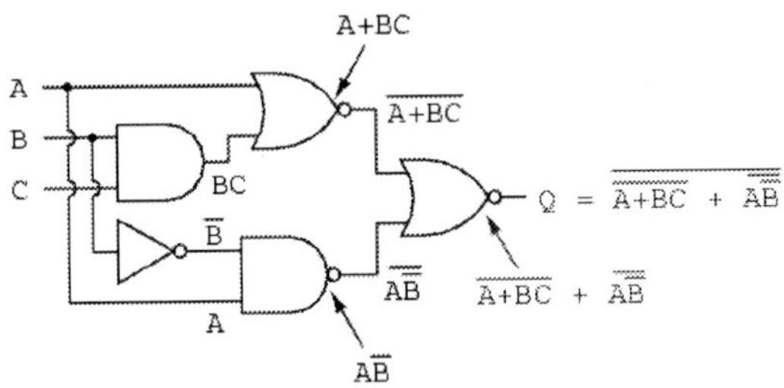

Now, we reduce this expression using the identities, properties, rules, and theorems (DeMorgan's) of Boolean algebra:

$$\overline{\overline{A + BC} + \overline{A\overline{B}}}$$

↓ Breaking longest bar

$$(\overline{\overline{A + BC}})\ (\overline{\overline{A\overline{B}}})$$

↓ Applying identity $\overline{\overline{A}} = A$ wherever double bars of equal length are found

$$(A + BC)\ (A\overline{B})$$

↓ Distributive property

$$AA\overline{B} + BCA\overline{B}$$

↓ Applying identity **AA = A** to left term; applying identity $A\overline{A} = 0$ to B and $\overline{B}$ in right term

$$A\overline{B} + 0$$

↓ Applying identity **A + 0 = A**

$$A\overline{B}$$

The equivalent gate circuit for this much-simplified expression is as follows:

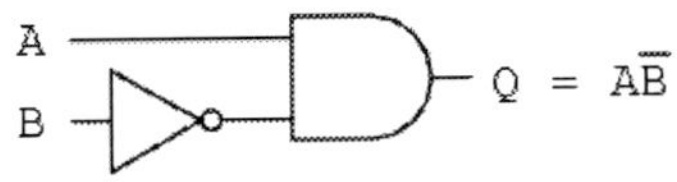

Flip-Flops

In electronics, a flip-flop or latch is a circuit that has two stable states and can be used to store state information. A flip-flop is a bistable multivibrator. The circuit can be made to change state by signals applied to one or more control inputs and will have one or two outputs. It is the basic storage element in sequential logic. Flip-flops and latches are a fundamental building block of digital electronics systems used in computers, communications, and many other types of systems.

Flip-flops and latches are used as data storage elements. Such data storage can be used for storage of *state*, and such a circuit is described as sequential logic. When used in a finite-state machine, the output and next state depend not only on its current input, but also on its current state (and hence, previous inputs). It can also be used for counting of pulses, and for synchronizing variably-timed input signals to some reference timing signal.

Flip-flops can be either simple (transparent or opaque) or clocked (synchronous or edge-triggered); the simple ones are commonly called latches. The word *latch* is mainly used for storage elements, while clocked devices are described as *flip-flops*. A latch is level-sensitive, whereas a flip-flop is edge-sensitive. That is, when a latch is enabled it becomes transparent, while a flip flop's output only changes on a single type (positive going or negative going) of clock edge.

History

The first electronic flip-flop was invented in 1918 by William Eccles and F. W. Jordan. It was initially called the *Eccles–Jordan trigger circuit* and consisted of two active elements (vacuum tubes). Such circuits and their transistorized versions were common in computers even after the introduction of integrated circuits, though flip-flops made from logic gates are also common now. Early flip-flops were known variously as trigger circuits or multivibrators.

According to P. L. Lindley, a JPL engineer, the flip-flop types discussed below (RS, D, T, JK) were first discussed in a 1954 UCLA course on computer design by Montgomery Phister, and then appeared in his book *Logical Design of Digital Computers.* Lindley was at the time working at Hughes Aircraft under Dr. Eldred Nelson, who had coined the term JK for a flip-flop which changed states when both inputs were on. The other names were coined by Phister. They differ slightly from some of the definitions given below. Lindley explains that he heard the story of the JK flip-flop from Dr. Eldred Nelson, who is responsible for coining the term while working at Hughes Aircraft. Flip-flops in use at Hughes at the time were all of the type that came to be known as J-K. In designing a logical system, Dr. Nelson assigned letters to flip-flop inputs as follows: #1: A & B, #2: C & D, #3: E & F, #4: G & H, #5: J & K. Nelson used the notations "*j*-input" and "*k*-input" in a patent application filed in 1953.

Implementation

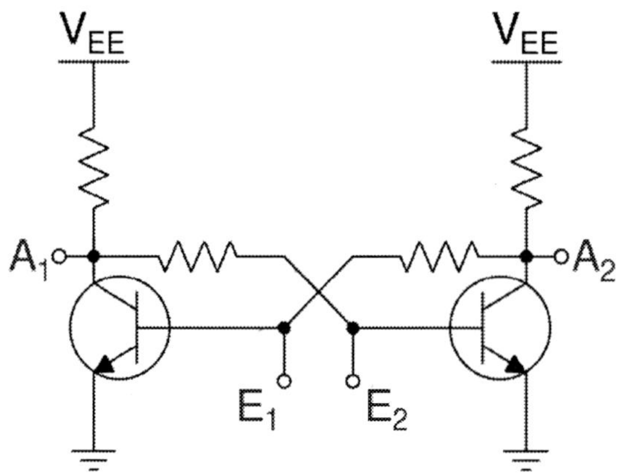

Figure: *A traditional flip-flop circuit based on bipolar junction transistors*

Flip-flops can be either simple (transparent or asynchronous) or clocked (synchronous); the transparent ones are commonly called latches. The word *latch* is mainly used for storage elements, while clocked devices are described as *flip-flops.*

Simple flip-flops can be built around a pair of cross-coupled inverting elements: vacuum tubes, bipolar transistors, field effect transistors, inverters, and inverting logic gates have all been used in practical circuits. Clocked devices are specially designed for synchronous systems; such devices ignore their inputs except at the transition of a dedicated clock signal (known as clocking, pulsing, or strobing). Clocking causes the flip-flop to either change or retain its output signal based upon the values of the input signals at the transition. Some flip-flops change output on the rising edge of the clock, others on the falling edge.

Since the elementary amplifying stages are inverting, two stages can be connected in succession (as a cascade) to form the needed non-inverting amplifier. In this configuration, each amplifier may be considered as an active inverting feedback network for the other inverting amplifier. Thus the two stages are connected in a non-inverting loop although the circuit diagram is usually drawn as a symmetric cross-coupled pair (both the drawings are initially introduced in the Eccles–Jordan patent).

Flip-flop types

Flip-flops can be divided into common types: the SR ("set-reset"), D ("data" or "delay"), T ("toggle"), and JK types are the common ones. The behaviour of a particular type can be described by what is termed the characteristic equation, which derives the "next" (i.e., after the next clock pulse) output, Q_{next} in terms of the input signal(s) and/or the current output, Q.

Simple set-reset latches

SR NOR latch: When using static gates as building blocks, the most fundamental latch is the simple *SR latch*, where S and R stand for *set* and *reset*. It can be constructed from a pair of cross-coupled NOR logic gates. The stored bit is present on the output marked Q.

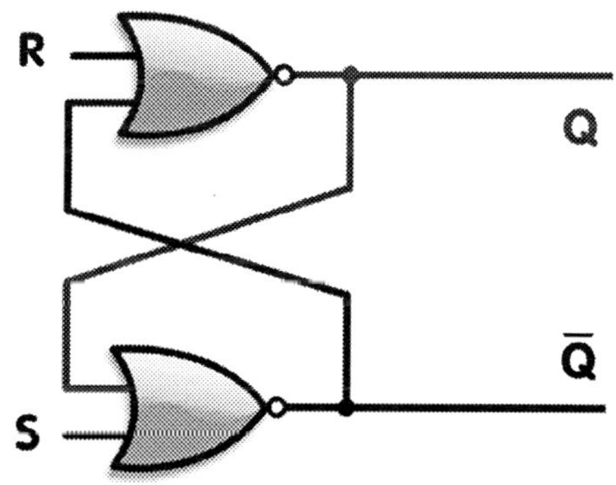

Figure: *An SR latch, constructed from a pair of cross-coupled NOR gates (an animated picture). Red and black mean logical '1' and '0', respectively.*

While the S and R inputs are both low, feedback maintains the Q and Q outputs in a constant state, with Q the complement of Q. If S (*Set*) is pulsed high while R (*Reset*) is held low, then the Q output is forced high, and stays high when S returns to low; similarly, if R is pulsed high while S is held low, then the Q output is forced low, and stays low when R returns to low.

SR latch operation[12]								
Characteristic table				Excitation table				
S	R	Q_{next}	Action	Q	Q_{next}	S	R	
0	0	Q	hold state	0	0	0	X	
0	1	0	reset	0	1	1	0	
1	0	1	set	1	0	0	1	
1	1	X	not allowed	1	1	X	0	

The R = S = 1 combination is called a restricted combination or a forbidden state because, as both NOR gates then output zeros, it breaks the logical equation Q = not Q. The combination is also inappropriate in circuits where *both* inputs may go low *simultaneously* (i.e. a transition from *restricted* to *keep*). The output would lock at either 1 or 0 depending on the propagation time relations between the gates (a race condition).

To overcome the restricted combination, one can add gates to the inputs that would convert (S,R) = (1,1) to one of the non-restricted combinations. That can be:

- Q = 1 (1,0) – referred to as an *S (dominated)-latch*
- Q = 0 (0,1) – referred to as an *R (dominated)-latch*

This is done in nearly every Programmable logic controller.

- Keep state (0,0) – referred to as an *E-latch*

Alternatively, the restricted combination can be made to *toggle* the output. The result is the JK latch.

Characteristic: Q+ = R'Q + R'S or Q+ = R'Q + S.

SR NAND Latch

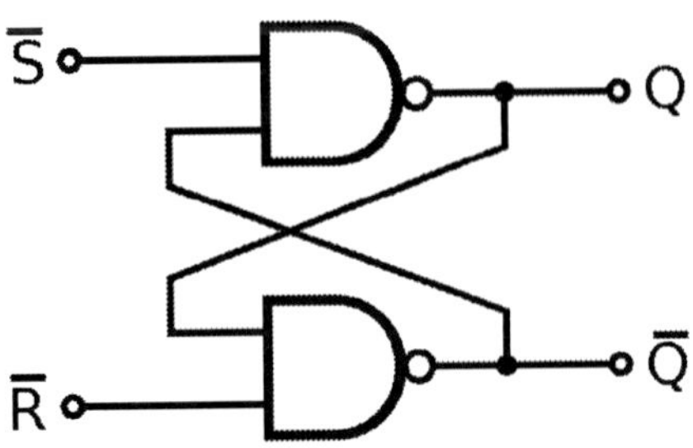

Figure: *An SR latch*

This is an alternate model of the simple SR latch which is built with NAND logic gates. *Set* and *reset* now become active low signals, denoted S and R respectively. Otherwise, operation is identical to that of the SR latch. Historically, SR-latches have been predominant despite the notational inconvenience of active-low inputs.

SR latch operationSRAction00Restricted combination01Q = 110Q = 011No Change

SR latch operation						
S	**R**	**Action**				
0	0	Restricted combination				
0	1	Q = 1				
1	0	Q = 0				
1	1	No Change				

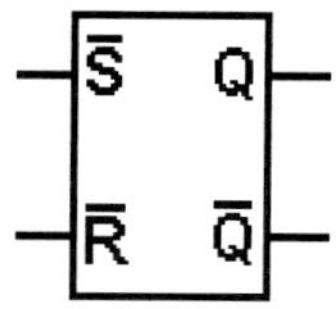

Symbol for an SR NAND latch

JK Latch

The JK latch is much less frequently used than the JK flip-flop. The JK latch follows the following state table:

JK latch truth table				
J	**K**	$\mathbf{Q_{next}}$	**Comment**	
0	0	Q	No change	
0	1	0	Reset	
1	0	1	Set	
1	1	Q	Toggle	

Hence, the JK latch is an SR latch that is made to *toggle* its output when passed the restricted combination of 11. Unlike the JK flip-flop, the 11 input combination for the SR latch is not useful because there is no clock that directs toggling.

Gated Latches and Conditional Transparency

Latches are designed to be *transparent*. That is, input signal changes cause immediate changes in output; when several *transparent* latches follow each other, using the same clock signal, signals can propagate through all of them at once. Alternatively, additional logic can be added to a simple transparent latch to make it *non-transparent* or *opaque* when another input (an "enable" input) is not asserted. By following a *transparent-high* latch with a *transparent-low* (or *opaque-high*) latch, a master–slave flip-flop is implemented.

Gated SR Latch

A *synchronous SR latch* (sometimes *clocked SR flip-flop*) can be made by adding a second level of NAND gates to the inverted SR latch (or a second level of AND gates to the direct SR latch). The extra gates further invert the inputs so the simple SR latch becomes a gated SR latch (and a simple SR latch would transform into a gated SR latch with inverted enable).

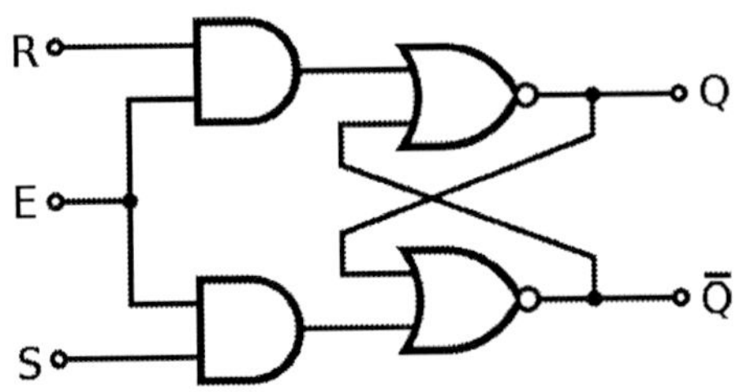

Figure: *A gated SR latch circuit diagram constructed from NOR gates.*

With E high (*enable* true), the signals can pass through the input gates to the encapsulated latch; all signal combinations except for (0,0) = *hold* then immediately reproduce on the (Q,Q) output, i.e. the latch is *transparent*.

With E low (*enable* false) the latch is *closed (opaque)* and remains in the state it was left the last time E was high.

Gated SR latch operationE/CAction0No action (keep state)1The same as non-clocked SR latch

The *enable* input is sometimes a clock signal, but more often a read or write strobe.

Gated SR latch operation	
E/C	**Action**
0	No action (keep state)
1	The same as non-clocked SR latch

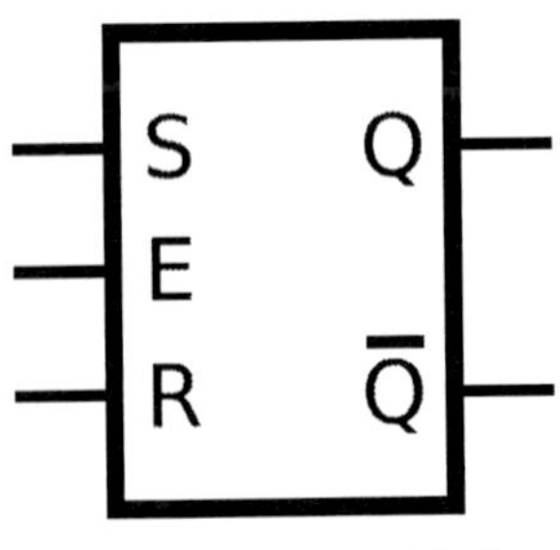

Symbol for a gated SR latch

Gated D Latch

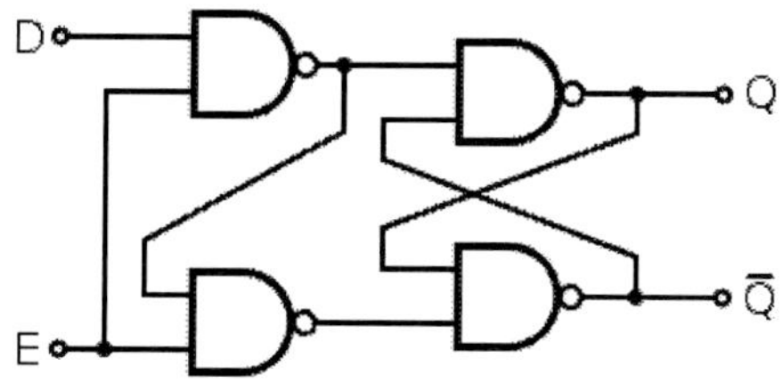

Figure: *A D-type transparent latch based on an SR NAND latch*

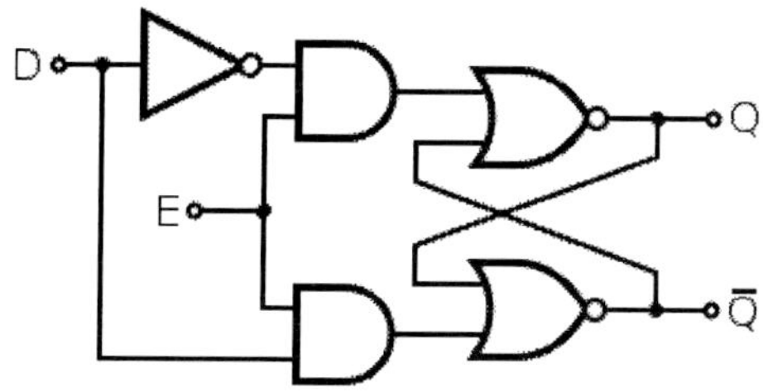

Figure: *A gated D latch based on an SR NOR latch*

This latch exploits the fact that, in the two active input combinations (01 and 10) of a gated SR latch, R is the complement of S. The input NAND stage converts the two D input states (0 and 1) to these two input combinations for the next SR latch by inverting the data input signal. The low state of the *enable* signal produces the inactive "11" combination.

Thus a gated D-latch may be considered as a *one-input synchronous SR latch*. This configuration prevents application of the restricted input combination. It is also known as *transparent latch*, *data latch*, or simply *gated latch*. It has a *data* input and an *enable* signal (sometimes named *clock*, or *control*). The word *transparent* comes from the fact that, when the enable input is on, the signal propagates directly through the circuit, from the input D to the output Q.

Transparent latches are typically used as I/O ports or in asynchronous systems, or in synchronous two-phase systems (synchronous systems that use a two-phase clock), where two latches operating on different clock phases prevent data transparency as in a master–slave flip-flop.

Latches are available as integrated circuits, usually with multiple latches per chip. For example, 74HC75 is a quadruple transparent latch in the 7400 series.

Gated D latch truth tableE/CDQQComment0X$Q_{prev}$$Q_{prev}$No change1001Reset1110Set

Gated D latch truth table					
E/C	**D**		**Q**	**Q**	**Comment**
0	X		Q_{prev}	Q_{prev}	No change
1	0		0	1	Reset
1	1		1	0	Set

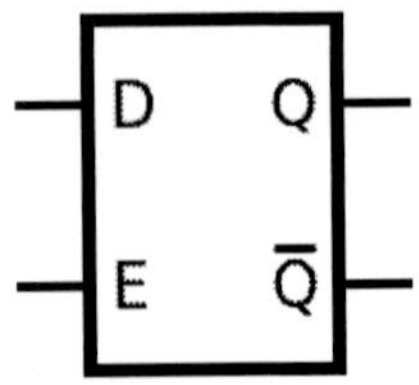

Symbol for a gated D latch

The truth table shows that when the *e*nable/*c*lock input is 0, the D input has no effect on the output. When E/C is high, the output equals D.

Earle Latch

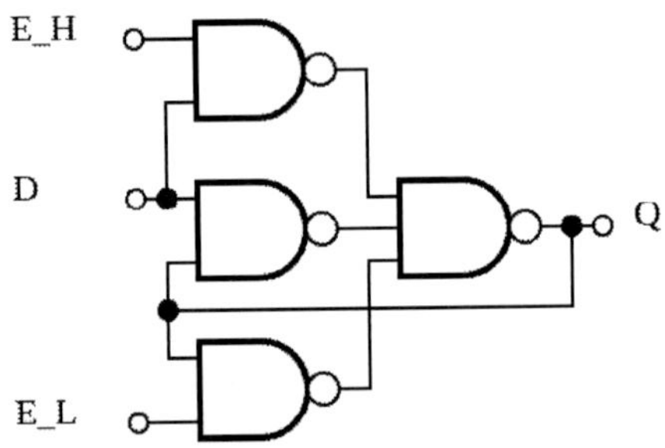

Figure: *Earle latch uses complementary enable inputs: enable active low (E_L) and enable active high (E_H)*

The classic gated latch designs have some undesirable characteristics. They require double-rail logic or an inverter. The input-to-output propagation may take up to three gate delays. The input-to-output propagation is not constant – some outputs take two gate delays while others take three.

Designers looked for alternatives. A successful alternative is the Earle latch. It requires only a single data input, and its output takes a constant two gate delays. In addition, the two gate levels of the Earle latch can be merged with the last two gate levels of the circuits driving the latch. Merging the latch function can implement the latch with no additional gate delays.

The Earle latch is hazard free. If the middle NAND gate is omitted, then one gets the polarity hold latch, which is commonly used because it demands less logic.

However, it is susceptible to logic hazard. Intentionally skewing the clock signal can avoid the hazard.

D flip-flop

The D flip-flop is widely used. It is also known as a *data* or *delay* flip-flop.

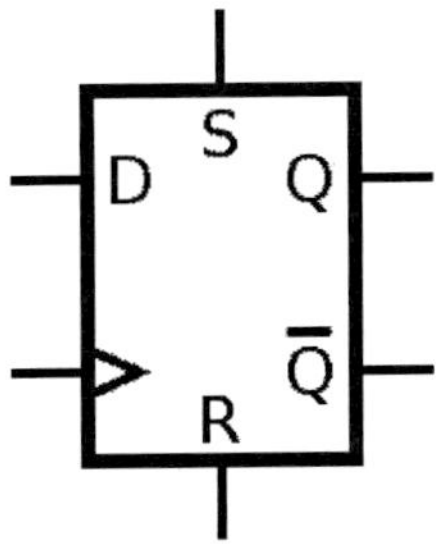

Figure: *D flip-flop symbol*

The D flip-flop captures the value of the D-input at a definite portion of the clock cycle (such as the rising edge of the clock). That captured value becomes the Q output. At other times, the output Q does not change. The D flip-flop can be viewed as a memory cell, a zero-order hold, or a delay line.

Clock	D	Q_{next}
Rising edge	0	0
Rising edge	1	1
Non-Rising	X	Q

('X' denotes a *Don't care* condition, meaning the signal is irrelevant)

Most D-type flip-flops in ICs have the capability to be forced to the set or reset state (which ignores the D and clock inputs), much like an SR flip-flop. Usually, the illegal S = R = 1 condition is resolved in D-type flip-flops. By setting S = R = 0, the flip-flop can be used as described above. Here is the truth table for the others S and R possible configurations:

Inputs				Outputs	
S	**R**	**D**	**>**	**Q**	**Q'**
0	1	X	X	0	1
1	0	X	X	1	0
1	1	X	X	1	1

These flip-flops are very useful, as they form the basis for shift registers, which are an essential part of many electronic devices. The advantage of the D flip-flop over the D-type "transparent latch" is that the signal on the D input pin is captured the moment the flip-flop is clocked, and subsequent changes on the D input will be ignored until the next clock event. An exception is that some flip-flops have a "reset" signal input, which will reset Q (to zero), and may be either asynchronous or synchronous with the clock.

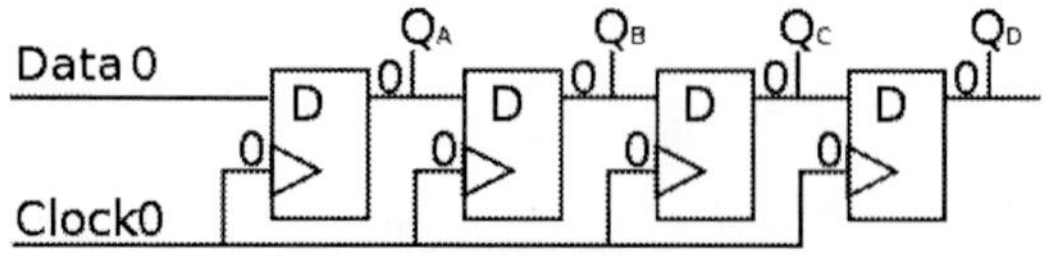

Figure: *4-bit serial-in, parallel-out (SIPO) shift register*

The above circuit shifts the contents of the register to the right, one bit position on each active transition of the clock. The input X is shifted into the leftmost bit position.

Classical Positive-edge-triggered D flip-flop

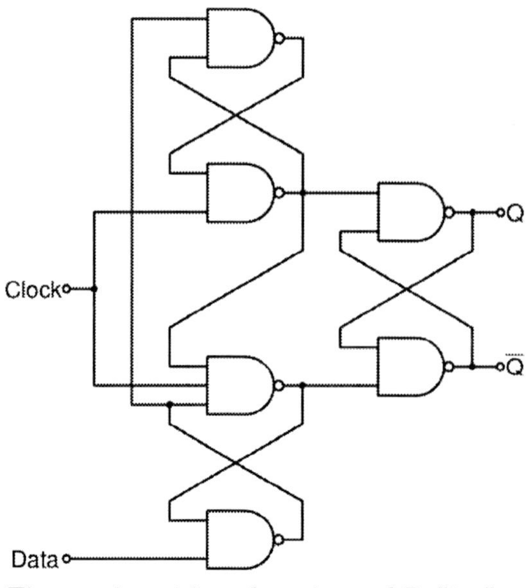

Figure: *A positive-edge-triggered D flip-flop*

This circuit consists of two stages implemented by SR NAND latches. The input stage (the two latches on the left) processes the clock and data signals to ensure correct input signals for the output stage (the single latch on the right). If the clock is low, both the output signals of the input stage are high regardless of the data input; the output latch is unaffected and it stores the previous state. When the clock signal changes from low to high, only one of the output voltages (depending on the data signal) goes low and sets/resets the output latch: if D = 0, the lower output becomes low; if D = 1, the upper output becomes low. If the clock signal continues staying high, the outputs keep their states regardless of the data input and force the output latch to stay in the corresponding state as the input logical zero remains active while the clock is high. Hence the role of the output latch is to store the data only while the clock is low.

The circuit is closely related to the gated D latch as both the circuits convert the two D input states (0 and 1) to two input combinations (01 and 10) for the output SR latch by inverting the data input signal (both the circuits split the single D signal in two complementary S and R signals). The difference is that in the gated D latch simple NAND logical gates are used while in the positive-edge-triggered D flip-flop SR NAND latches are used for this purpose. The role of these latches is to "lock" the active output producing low voltage (a logical zero); thus the positive-edge-triggered D flip-flop can be thought of as a gated D latch with latched input gates.

Master–slave edge-triggered D flip-flop

A master–slave D flip-flop is created by connecting two gated D latches in series, and inverting the *enable* input to one of them. It is called master–slave because the second latch in the series only changes in response to a change in the first (master) latch.

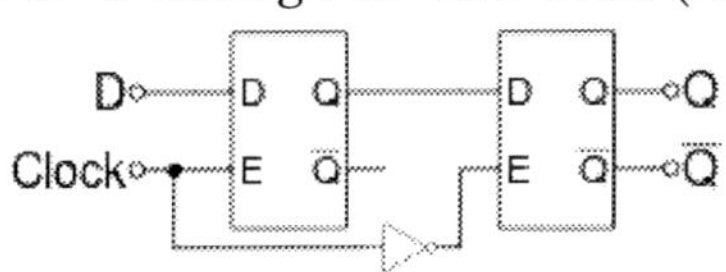

Figure: *A master–slave D flip-flop. It responds on the falling edge of the* enable *input (usually a clock)*

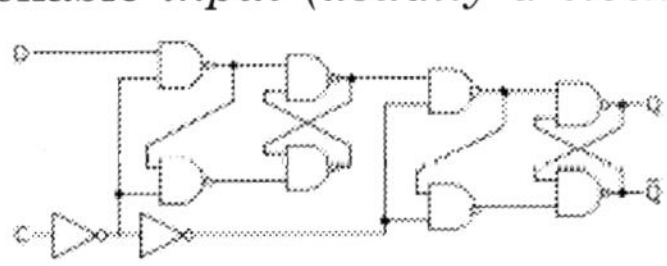

Figure: *An implementation of a master–slave D flip-flop that is triggered on the rising edge of the clock*

For a positive-edge triggered master–slave D flip-flop, when the clock signal is low (logical 0) the "enable" seen by the first or "master" D latch (the inverted clock signal) is high (logical 1). This allows the "master" latch to store the input value when the clock signal transitions from low to high. As the clock signal goes high (0 to 1) the inverted "enable" of the first latch goes low (1 to 0) and the value seen at the input to the master latch is "locked". Nearly simultaneously, the twice inverted "enable" of the second or "slave" D latch transitions from low to high (0 to 1) with the clock signal. This allows the signal captured at the rising edge of the clock by the now "locked" master latch to pass through the "slave" latch. When the clock signal returns to low (1 to 0), the output of the "slave" latch is "locked", and the value seen at the last rising edge of the clock is held while the "master" latch begins to accept new values in preparation for the next rising clock edge.

By removing the leftmost inverter in the circuit at side, a D-type flip-flop that strobes on the *falling edge* of a clock signal can be obtained. This has a truth table like this:

D	Q	>	Q_{next}
0	X	Falling	0
1	X	Falling	1

Edge-triggered Dynamic D storage Element

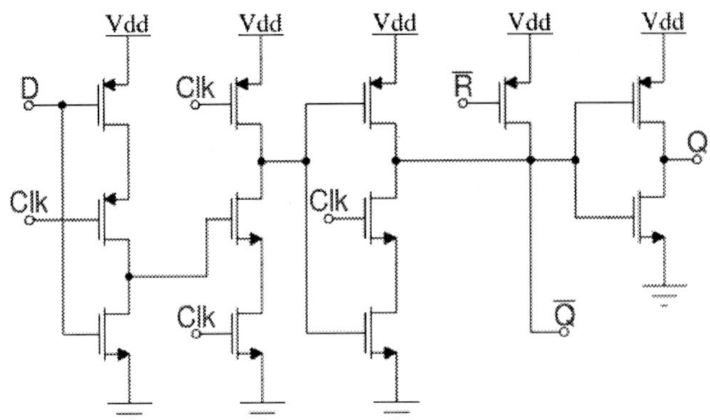

Figure: A CMOS IC implementation of a "true single-phase edge-triggered flip-flop with reset"

An efficient functional alternative to a D flip-flop can be made with dynamic circuits as long as it is clocked often enough; while not a true flip-flop, it is still called a flip-flop for its functional role. While the master–slave D element is triggered on the edge of a clock, its components are each triggered by clock levels. The "edge-triggered D flip-flop", as it is called even though it is not a true flip-flop, does not have the master–slave properties.

Edge-triggered D flip-flops are often implemented in integrated high-speed operations using dynamic logic. This means that the digital output is stored on parasitic device capacitance while the device is not transitioning. This design of dynamic flip flops also enables simple resetting since the reset operation can be performed by simply discharging one or more internal nodes. A common dynamic flip-flop variety is the true single-phase clock (TSPC) type which performs the flip-flop operation with little power and at high speeds. However, dynamic flip-flops will typically not work at static or low clock speeds: given enough time, leakage paths may discharge the parasitic capacitance enough to cause the flip-flop to enter invalid states.

T flip-flop

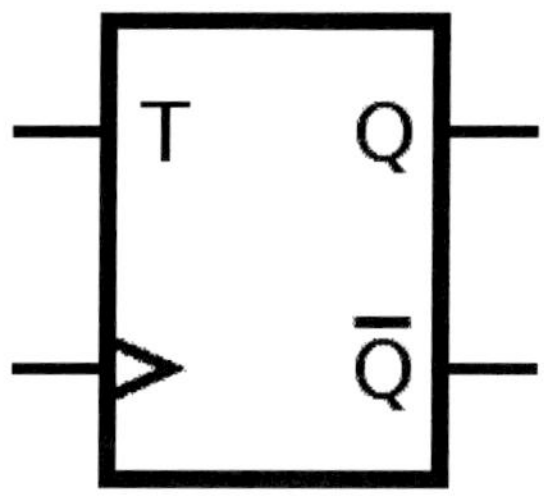

Figure: *A circuit symbol for a T-type flip-flop*

If the T input is high, the T flip-flop changes state ("toggles") whenever the clock input is strobed. If the T input is low, the flip-flop holds the previous value. This behaviour is described by the characteristic equation:

$Q_{next} = T \oplus Q = T\overline{Q} + \overline{T}Q$ (expanding the XOR operator)

and can be described in a truth table:

T flip-flop operation								
Characteristic table				**Excitation table**				
T	Q	Q_{next}	**Comment**	Q	Q_{next}	T	**Comment**	
0	0	0	hold state (no clk)	0	0	0	No change	
0	1	1	hold state (no clk)	1	1	0	No change	
1	0	1	toggle	0	1	1	Complement	
1	1	0	toggle	1	0	1	Complement	

When T is held high, the toggle flip-flop divides the clock frequency by two; that is, if clock frequency is 4 MHz, the output frequency obtained from the flip-flop will be 2 MHz. This "divide by" feature has application in various types of digital counters. A T flip-flop can also be built using a JK flip-flop (J & K pins are connected together and act as T) or D flip-flop (T input and $Q_{previous}$ is connected to the D input through an XOR gate).

JK flip-flop

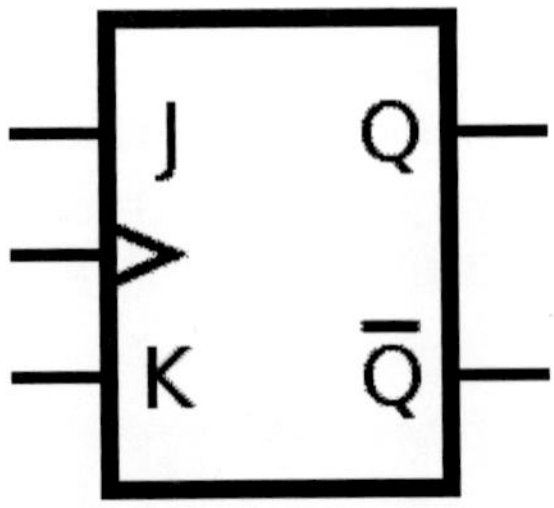

Figure: *A circuit symbol for a positive-edge-triggered JK flip-flop*

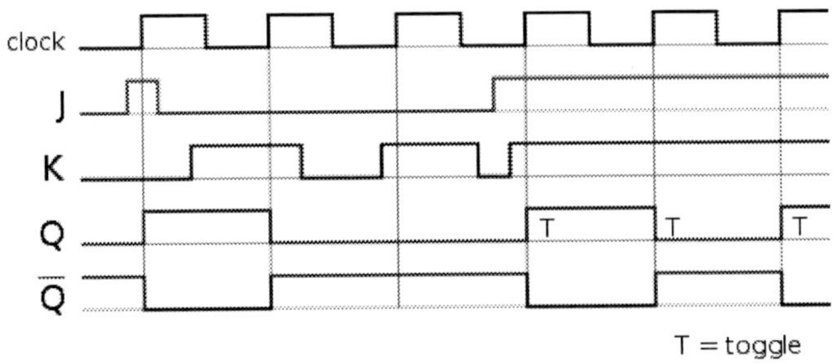

Figure: *JK flip-flop timing diagram*

The JK flip-flop augments the behaviour of the SR flip-flop (J=Set, K=Reset) by interpreting the S = R = 1 condition as a “flip” or toggle command.

Specifically, the combination J = 1, K = 0 is a command to set the flip-flop; the combination J = 0, K = 1 is a command to reset the flip-flop; and the combination J = K = 1 is a command to toggle the flip-flop, i.e., change its output to the logical complement of its current value. Setting J = K = 0 does NOT result in a D flip-flop, but rather, will hold the current state.

To synthesize a D flip-flop, simply set K equal to the complement of J. Similarly, to synthesize a T flip-flop, set K equal to J. The JK flip-flop is therefore a universal flip-flop, because it can be configured to work as an SR flip-flop, a D flip-flop, or a T flip-flop.

The characteristic equation of the JK flip-flop is:

$$Q_{next} = J\overline{Q} + \overline{K}Q$$

and the corresponding truth table is:

JK flip-flop operation								
Characteristic table				Excitation table				
J	K	Comment	Q_{next}	Q	J	K	Comment	Q_{next}
0	0	hold state	Q	0	0	X	No Change	0
0	1	reset	0	0	1	X	Set	1
1	0	set	1	1	X	1	Reset	0
1	1	toggle	Q	1	X	0	No Change	1

Metastability

Flip-flops are subject to a problem called metastability, which can happen when two inputs, such as data and clock or clock and reset, are changing at about the same time. When the order is not clear, within appropriate timing constraints, the result is that the output may behave unpredictably, taking many times longer than normal to settle to one state or the other, or even oscillating several times before settling. Theoretically, the time to settle down is not bounded. In a computor system, this metastability can cause corruption of data or a programme crash, if the state is not stable before another circuit uses its value; in particular, if two different logical paths use the output of a flip-flop, one path can interpret it as a 0 and the other as a 1 when it has not resolved to stable state, putting the machine into an inconsistent state.

Timing Considerations

Setup, hold, recovery, removal times:

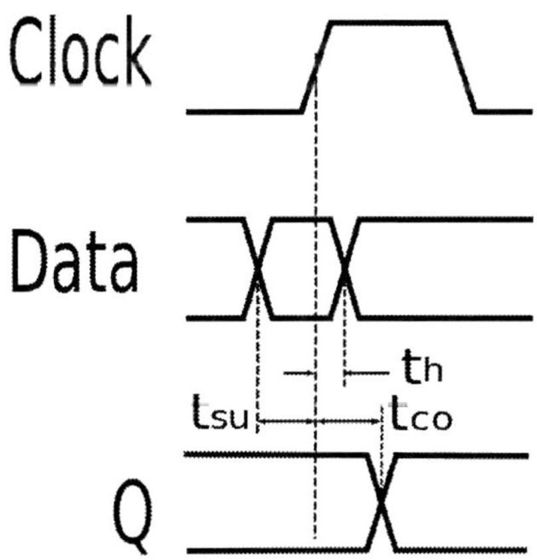

***Figure:** Flip-flop setup, hold and clock-to-output timing parameters*

Setup time is the minimum amount of time the data signal should be held steady before the clock event so that the data are reliably sampled by the clock. This applies to synchronous input signals to the flip-flop.

Hold time is the minimum amount of time the data signal should be held steady after the clock event so that the data are reliably sampled. This applies to synchronous input signals to the flip-flop.

Synchronous signals (like Data) should be held steady from the set-up time to the hold time, where both times are relative to the clock signal.

Recovery time is like setup time for asynchronous ports (set, reset). It is the time available between the asynchronous signals going inactive and the active clock edge.

Removal time is like hold time for asynchronous ports (set, reset). It is the time between active clock edge and asynchronous signal going inactive.

Short impulses applied to asynchronous inputs (set, reset) should not be applied completely within the recovery-removal period, or else it becomes entirely indeterminable whether the flip-flop will transition to the appropriate state. In another case, where an asynchronous signal simply makes one transition that happens to fall between the recovery/removal time, eventually the asynchronous signal will be applied, but in that case it is also possible that a very short glitch may appear on the output, dependent on the synchronous input signal. This second situation may or may not have significance to a circuit design.

Set and Reset (and other) signals may be either synchronous or asynchronous and therefore may be characterized with either Setup/Hold or Recovery/Removal times, and synchronicity is very dependent on the TTL design of the flip-flop.

Differentiation between Setup/Hold and Recovery/Removal times is often necessary when verifying the timing of larger circuits because asynchronous signals may be found to be less critical than synchronous signals. The differentiation offers circuit designers the ability to define the verification conditions for these types of signals independently.

The metastability in flip-flops can be avoided by ensuring that the data and control inputs are held valid and constant for specified periods before and after the clock pulse, called the setup time (t_{su}) and the hold time (t_h) respectively. These times are specified in the data sheet for the device, and are typically between a few nanoseconds and a few hundred picoseconds for modern devices.

Unfortunately, it is not always possible to meet the setup and hold criteria, because the flip-flop may be connected to a real-time signal that could change at any time, outside the control of the designer. In this case, the best the designer can do is to reduce the probability of error to a certain level, depending on the required reliability of the circuit.

One technique for suppressing metastability is to connect two or more flip-flops in a chain, so that the output of each one feeds the data input of the next, and all devices share a common clock. With this method, the probability of a metastable event can be reduced to a negligible value, but never to zero. The probability of metastability gets closer and closer to zero as the number of flip-flops connected in series is increased.

So-called metastable-hardened flip-flops are available, which work by reducing the setup and hold times as much as possible, but even these cannot eliminate the problem entirely. This is because metastability is more than simply a matter of circuit design. When the transitions in the clock and the data are close together in time, the flip-flop is forced to decide which event happened first. However fast we make the device, there is always the possibility that the input events will be so close together that it cannot detect which one happened first. It is therefore logically impossible to build a perfectly metastable-proof flip-flop.

Propagation Delay

Another important timing value for a flip-flop is the clock-to-output delay (common symbol in data sheets: t_{CO}) or propagation delay (t_P), which is the time a flip-flop takes to change its output after the clock edge. The time for a high-to-low transition (t_{PHL}) is sometimes different from the time for a low-to-high transition (t_{PLH}).

When cascading flip-flops which share the same clock (as in a shift register), it is important to ensure that the t_{CO} of a preceding flip-flop is longer than the hold time (t_h) of the following flip-flop, so data present at the input of the succeeding flip-flop is properly "shifted in" following the active edge of the clock. This relationship between t_{CO} and t_h is normally guaranteed if the flip-flops are physically identical. Furthermore, for correct operation, it is easy to verify that the clock period has to be greater than the sum $t_{su} + t_h$.

Generalizations

Flip-flops can be generalized in at least two ways: by making them 1-of-N instead of 1-of-2, and by adapting them to logic with more than two states. In the special cases of 1-of-3 encoding, or multi-valued ternary logic, these elements may be referred to as *flip-flap-flops*.

In a conventional flip-flop, exactly one of the two complementary outputs is high. This can be generalized to a memory element with N outputs, exactly one of which is high (alternatively, where exactly one of N is low). The output is therefore always a one-hot (respectively *one-cold*) representation. The construction is similar to a conventional cross-coupled flip-flop; each output, when high, inhibits all the other outputs. Alternatively, more or less conventional flip-flops can be used, one per output, with additional circuitry to make sure only one at a time can be true.

Another generalization of the conventional flip-flop is a memory element for multi-valued logic. In this case the memory element retains exactly one of the logic states until the control inputs induce a change. In addition, a multiple-valued clock can also be used, leading to new possible clock transitions.

Counters

In digital logic and computing, a counter is a device which stores (and sometimes displays) the number of times a particular event or process has occurred, often in relationship to a clock signal.

Electronic Counters

In electronics, counters can be implemented quite easily using register-type circuits such as the flip-flop, and a wide variety of classifications exist:

- Asynchronous (ripple) counter – changing state bits are used as clocks to subsequent state flip-flops
- Synchronous counter – all state bits change under control of a single clock
- Decade counter – counts through ten states per stage
- Up/down counter – counts both up and down, under command of a control input
- Ring counter – formed by a shift register with feedback connection in a ring
- Johnson counter – a *twisted* ring counter

- Cascaded counter
- modulas counter.

Each is useful for different applications. Usually, counter circuits are digital in nature, and count in natural binary. Many types of counter circuits are available as digital building blocks, for example a number of chips in the 4000 series implement different counters.

Occasionally there are advantages to using a counting sequence other than the natural binary sequence—such as the binary coded decimal counter, a linear feedback shift register counter, or a Gray-code counter.

Counters are useful for digital clocks and timers, and in oven timers, VCR clocks, etc.

Asynchronous (Ripple) Counter

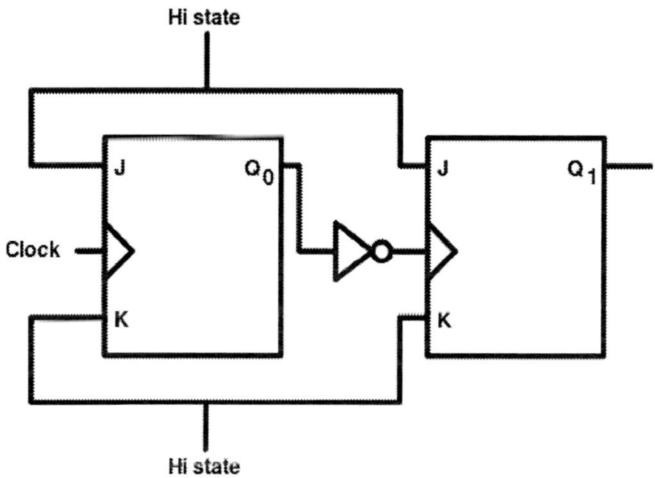

Figure: *Asynchronous counter created from two JK flip-flops*

An asynchronous (ripple) counter is a single JK-type flip-flop, with its J (data) input fed from its own inverted output. This circuit can store one bit, and hence can count from zero to one before it overflows (starts over from 0). This counter will increment once for every clock cycle and takes two clock cycles to overflow, so every cycle it will alternate between a transition from 0 to 1 and a transition from 1 to 0. Notice that this creates a new clock with a 50% duty cycle at exactly half the frequency of the input clock. If this output is then used as the clock signal for a similarly arranged D flip-flop (remembering to invert the output to the input), one will get another 1 bit counter that counts half as fast. Putting them together yields a two-bit counter:

Cycle	Q1	Q0	(Q1:Q0)dec
0	0	0	0
1	0	1	1
2	1	0	2
3	1	1	3
4	0	0	0

You can continue to add additional flip-flops, always inverting the output to its own input, and using the output from the previous flip-flop as the clock signal. The result is called a ripple counter, which can count to $2^n - 1$ where n is the number of bits (flip-flop stages) in the counter. Ripple counters suffer from unstable outputs as the overflows "Ripple" from stage to stage, but they do find frequent application as dividers for clock signals, where the instantaneous count is unimportant, but the division ratio overall is (to clarify this, a 1-bit counter is exactly equivalent to a divide by two circuit; the output frequency is exactly half that of the input when fed with a regular train of clock pulses).

The use of flip-flop outputs as clocks leads to timing skew between the count data bits, making this ripple technique incompatible with normal synchronous circuit design styles.

Synchronous Counter

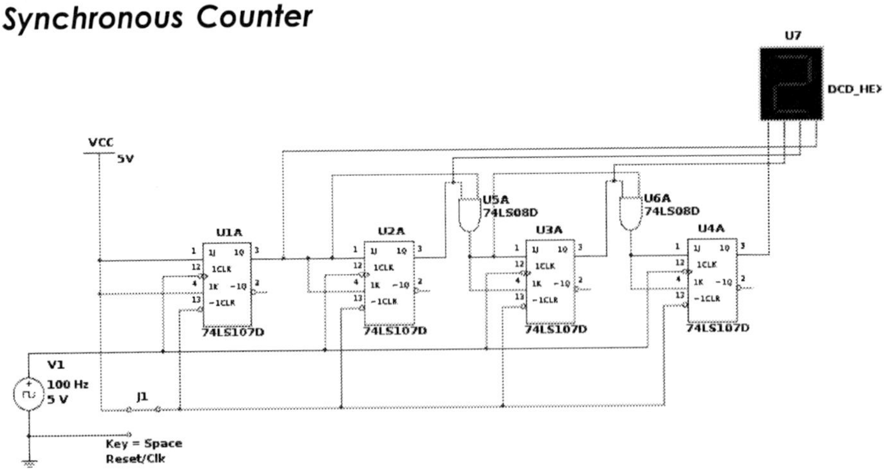

Figure: *A 4-bit synchronous counter using JK flip-flops*

Synchronous Counters

In synchronous counters, the clock inputs of all the flip-flops are connected together and are triggered by the input pulses. Thus, all the flip-flops change state simultaneously (in parallel). The circuit below is a 3-bit synchronous counter. The J and K inputs of FF0 are connected to HIGH. FF1 has its J and K inputs connected to the output of FF0, and the J and K inputs of FF2 are connected to the output of an AND gate that is fed by the outputs of FF0 and FF1.

A simple way of implementing the logic for each bit of an ascending counter (which is what is depicted in the image to the right) is for each bit to toggle when all of the less significant bits are at a logic high state. For example, bit 1 toggles when bit 0 is logic high; bit 2 toggles when both bit 1 and bit 0 are logic high; bit 3 toggles when bit 2, bit 1 and bit 0 are all high; and so on.

Synchronous counters can also be implemented with hardware finite state machines, which are more complex but allow for smoother, more stable transitions.

Hardware-based counters are of this type.

Decade Counter

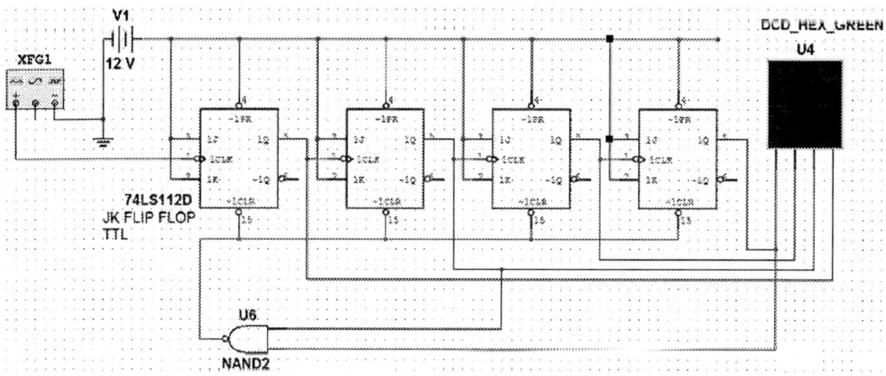

***Figure:** A circuit diagram of decade counter using JK FlipFlops (74LS112D)*

A decade counter is one that counts in decimal digits, rather than binary. A decade counter may have each digit binary encoded (that is, it may count in binary-coded decimal, as the 7490 integrated circuit did) or other binary encodings (such as the bi-quinary encoding of the 7490 integrated circuit).

Alternatively, it may have a "fully decoded" or one-hot output code in which each output goes high in turn (the 4017 is such a circuit). The latter type of circuit finds applications in multiplexers and demultiplexers, or wherever a scanning type of behaviour is useful. Similar counters with different numbers of outputs are also common.

The decade counter is also known as a mod-counter when it counts to ten (0, 1, 2, 3, 4, 5, 6, 7, 8, 9). A Mod Counter that counts to 64 stops at 63 because 0 counts as a valid digit.

Up/down Counter

A counter that can change state in either direction, under the control of an up or down selector input, is known as an up/down counter. When the selector is in the up state, the counter increments its value. When the selector is in the down state, the counter decrements the count.

Ring Counter

A ring counter is a circular shift register which is initiated such that only one of its flip-flops is the state one while others are in their zero states. A ring counter is a Shift Register (a cascade connection of flip-flops) with the output of the last one connected to the input of the first, that is, in a ring. Typically, a pattern consisting of a single bit is circulated so the state repeats every n clock cycles if n flip-flops are used. It can be used as a cycle counter of n states.

Johnson Counter

A Johnson counter (or switchtail ring counter, twisted-ring counter, walking-ring counter, or Moebius counter) is a modified ring counter, where the output from the last stage is inverted and fed back as input to the first stage. The register cycles through a sequence of bit-patterns, whose length is equal to twice the length of the shift register, continuing indefinitely. These counters find specialist applications, including those similar to the decade counter, digital-to-analog conversion, etc. They can be implemented easily using D- or JK-type flip-flops.

Computer Science Counters

In computability theory, a counter is considered a type of memory. A counter stores a single natural number (initially zero) and can be arbitrarily long. A counter is usually considered in conjunction with a finite-state machine (FSM), which can perform the following operations on the counter:

- Check whether the counter is zero
- Increment the counter by one.
- Decrement the counter by one (if it's already zero, this leaves it unchanged).

The following machines are listed in order of power, with each one being strictly more powerful than the one below it:

1. Deterministic or non-deterministic FSM plus two counters
2. Non-deterministic FSM plus one stack
3. Non-deterministic FSM plus one counter
4. Deterministic FSM plus one counter
5. Deterministic or non-deterministic FSM.

For the first and last, it doesn't matter whether the FSM is a deterministic finite automaton or a nondeterministic finite automaton. They have power. The first two and the last one are levels of the Chomsky hierarchy.

The first machine, an FSM plus two counters, is equivalent in power to a Turing machine.

Mechanical Counters

Long before electronics became common, mechanical devices were used to count events. These are known as tally counters. They typically consist of a series of disks mounted on an axle, with the digits 0 through 9 marked on their edge. The right most disk moves one increment with each event. Each disk except the left-most has a protrusion that, after the completion of one revolution, moves the next disk to the left one increment. Such counters were originally used to control manufacturing processes, but were later used as odometers for bicycles and cars and in tape recorders and fuel dispensers. One of the largest manufacturers was the Veeder-Root company, and their name was often used for this type of counter.

Shift Registers

In digital circuits, a shift register is a cascade of flip flops, sharing the same clock, in which the output of each flip-flop is connected to the "data" input of the next flip-flop in the chain, resulting in a circuit that shifts by one position the "bit array" stored in it, *shifting in* the data present at its input and *shifting out* the last bit in the array, at each transition of the clock input. More generally, a shift register may be multidimensional, such that its "data in" and stage outputs are themselves bit arrays: this is implemented simply by running several shift registers of the same bit-length in parallel.

Shift registers can have both parallel and serial inputs and outputs. These are often configured as serial-in, parallel-out (SIPO) or as parallel-in, serial-out (PISO). There are also types that have both serial and parallel input and types with serial and parallel output. There are also bi-directional shift registers which allow shifting in both directions: L→R or R→L. The serial input and last output of a shift register can also be connected to create a circular shift register.

Serial-in, Serial-out (SISO)

Destructive readout: These are the simplest kind of shift registers. The data string is presented at 'Data In', and is shifted right one stage each time 'Data Advance' is brought high. At each advance, the bit on the far left (i.e. 'Data In') is shifted into the first flip-flop's output. The bit on the far right (i.e. 'Data Out') is shifted out and lost.

The data are stored after each flip-flop on the 'Q' output, so there are four storage 'slots' available in this arrangement, hence it is a 4-Bit Register. To give an idea of the shifting pattern, imagine that the register holds 0000 (so all storage slots are empty). As 'Data In' presents 1,0,1,1,0,0,0,0 (in that order, with a pulse at 'Data Advance' each time—this is called clocking or strobing) to the register, this is the result. The left hand column corresponds to the left-most flip-flop's output pin, and so on.

So the serial output of the entire register is 10110000. It can be seen that if data were to be continued to input, it would get exactly what was put in, but offset by four 'Data Advance' cycles. This arrangement is the hardware equivalent of a queue. Also, at any time, the whole register can be set to zero by bringing the reset (R) pins high. This arrangement performs *destructive readout* - each datum is lost once it has been shifted out of the right-most bit.

Serial-in, Parallel-out (SIPO)

This configuration allows conversion from serial to parallel format. Data is input serially. Once the data has been input, it may be either read off at each output simultaneously, or it can be shifted out and replaced.

In cases where the parallel outputs should not change during the serial loading process, it is desirable to use a latched output. In a latched shift register (such as the 74595) the serial data is first loaded into an internal shift register, then upon receipt of a load signal the state of the shift register is copied into a set of output registers.

Bibliography

Amy, Friedlander: *Power and Light: Electricity in the U.S. Energy Infrastructure, 1870-1940*, Corporation for National Research Initiatives, Reston, VA, 1996.

Arnold Zankl: *Milestones in Automation: From the Transistor to the Digital Factory*, Wiley-VCH, NJ, 2006.

Bamberger, Robert L.: *The National Energy Strategy: Summary and Analysis*, Congressional Research Service, Library of Congress, Washington, DC, 1991.

Bard, A.J.; Faulkner, L.R.: *Electrochemical Methods: Fundamentals and Applications*, John Wiley & Sons, New York, 2005.

Belanger, P.R.: *Control Engineering*, Saunders, UK, 1995.

Bird, John: *Electrical and Electronic Principles and Technology*, Newnes, UK, 2007.

Christopher Kilian: *Modern Control Technology*, Thompson Delmar Learning, UK, 2005.

Christopher Kilian: *Modern Control Technology*, Thompson Delmar Learning, UK, 2005.

Desoer, C. A.; M. Vidyasagar: *Feedback Systems: Input-Output Properties*, SIAM, 2009.

Dhawan, B C : *Textbook of Electrophoresis*, Pearl Books, Delhi, 2008.

Duncan, Wallace L.: *Federal Restructuring of The Electric Power Industry*, Weinberg, Miller & Pembroke, P.C., Duncan, 1996.

Edwin, J. Houston and Arthur Kennelly: *Recent Types of Dynamo-Electric Machinery*, P.F. Collier and Sons New York, 1902.

Eveleigh, E.W.: *Introduction to Control Systems Design*, McGraw-Hill, NY, 1972.

Franklin, G.F., J.D. Powell, and A. Emami-Naeini: *Feedback Control of Dynamic Systems*, Addison-Wesley, London, 1994.

Glazer, A.M.: *Space Groups for Scientists and Engineers,* Academic Press, Boston, 1990.

Herman, Stephen L.: *Electric Motor Control*, Cengage Learning, Delmar, 2009.

Issa Batarseh: *Power Electronic Circuits,* John Wiley, NY, 2003.

John, N. Chiasson: *Modeling and High Performance Control of Electric Machines*, Wiley-IEEE Press, New York, 2005.

Kuo, B.C., *Automatic Control Systems*, Prentice-Hall, NJ, 1995.

Liptak, Bela G.: *Instrument Engineers' Handbook: Process Control and Optimization*, CRC Press, 2005.

Mazda, F. F.: *Thyristor Control*, Halsted Press Div. of John Wiley & Sons, New York, 1973.

Nivola, Pietro S.: *The Politics of Energy Conservation,* Brooking Institution, Washington, DC, 1986.

Ouellette, Robert: *Automation Impacts on Industry*, Ann Arbor Science Publishers, MI, USA, 1983.

Patterson, Walter C.: *Transforming Electricity: The Coming Generation of Change*, Earthscan, 1999.

Phillips, C.L. and Harbor, R.D., *Feedback Control Systems*, Prentice Hall, NJ, 2000.

Rajput, R.K.: *A Course in Electrical Engineering Materials*, Laxmi Publications, Delhi, 2009.

Reddy, K. Krishna: *Electrical Measurements*, Scitech Pub, Delhi, 2009.

Shanefield, D. J.: *Industrial Electronics for Engineers, Chemists, and Technicians,* William Andrew Publishing, Norwich, NY, 2001.

Smith, Bruce A.: *Technical Innovation in Electric Power Generation, 1950-1970*, Div. Of Research, Michigan State University, East Lansing, 1977.

Thaler, G.J.: *Automatic Control Systems*, West, US, 1989.

Trevathan, Vernon L.: *A Guide to the Automation Body of Knowledge,* International Society of Automation, NC, USA, 2006.

Ueha, S.; Tomikawa, Y.; Kurosawa, M.; Nakamura, N.: *Ultrasonic Motors: Theory and Applications*, Clarendon Press, UK, 1993.

Ulrich, Dr. Nicolai, Dr. Tobias Reimann, Prof. Jürgen Petzoldt, Josef Lutz: *Application Manual IGBT- and MOSFET-power modules*, ISLE Verlag, 1998.

H

I

J

L

M

N

O

P

Q

R

S

T

V

W

Z

❑❑❑

Index

A

B

C

D

E

F

G